MOLYBDENUM DISULPHIDE LUBRICATION

Vol. 6	Friction and Wear of Polymers (Bartenev and Lavrentev)
Vol. 10	Microstructure and Wear of Materials (Zum Gahr)
Vol. 11	Fluid Film Lubrication – Osborne Reynolds Centenary (Dowson et al., Editors)
Vol. 12	Interface Dynamics (Dowson et al., Editors)
Vol. 13	Tribology of Miniature Systems (Rymuza)
Vol. 14	Tribological Design of Machine Elements (Dowson et al., Editors)
Vol. 15	Encyclopedia of Tribology (Kajdas et al.)
Vol. 16	Tribology of Plastic Materials (Yamaguchi)
Vol. 17	Mechanics of Coatings (Dowson et al., Editors)
Vol. 18	Vehicle Tribology (Dowson et al., Editors)
Vol. 19	Rheology and Elastohydrodynamic Lubrication (Jacobson)
Vol. 20	Materials for Tribology (Glaeser)
Vol. 21	Wear Particles: From the Cradle to the Grave (Dowson et al., Editors)
Vol. 22	Hydrostatic Lubrication (Bassani and Piccigallo)
Vol. 23	Lubricants and Special Fluids (Stepina and Vesely)
Vol. 24	Engineering Tribology (Stachowiak and Batchelor)
Vol. 25	Thin Films in Tribology (Dowson et al., Editors)
Vol. 26	Engine Tribology (Taylor, Editor)
Vol. 27	Dissipative Processes in Tribology (Dowson et al., Editors)
Vol. 28	Coatings Tribology – Properties, Techniques and Applications in Surface Engineering (Holmberg and Matthews)
Vol. 29	Friction Surface Phenomena (Shpenkov)
Vol. 30	Lubricants and Lubrication (Dowson et al., Editors)
Vol. 31	The Third Body Concept: Interpretation of Tribological Phenomena (Dowson et al., Editors)
Vol. 32	Elastohydrodynamics - '96: Fundamentals and Applications in Lubrication and Traction (Dowson et al., Editors)
Vol. 33	Hydrodynamic Lubrication - Bearings and Thrust Bearings (Frêne et al.)
Vol. 34	Tribology For Energy Conservation (Dowson et al., Editors)

TRIBOLOGY SERIES, 35
EDITOR: D. DOWSON

MOLYBDENUM DISULPHIDE LUBRICATION

A.R. Lansdown
Swansea, UK

1999

ELSEVIER
Amsterdam • Lausanne • New York • Oxford • Shannon • Singapore • Tokyo

ELSEVIER SCIENCE B.V.
Sara Burgerhartstraat 25
P.O. Box 211, 1000 AE Amsterdam, The Netherlands

First edition 1999

Library of Congress Cataloging in Publication Data
A catalog record from the Library of Congress has been applied for.

ISBN: 0-444-50032-4

∞The paper used in this publication meets the requirements of ANSI/NISO Z39.48-1992 (Permanence of Paper).

Transferred to digital printing 2006

To Elaine,
with thanks

" Among the Minerals not yet examined, that which chiefly deserves our consideration is, the Molybdaena . . . . ."

J.A.Cramer, 1764

PREFACE

This book follows the series of six reports on molybdenum disulphide lubrication which I wrote between 1970 and 1984 for the European Space Research Organisation (ESRO) and its successor the European Space Agency (ESA).

At that time there was an enormous volume of research and development effort in the subject, much of it supported by national governments for the benefit of defence, aviation or space activities. There were already some well-established practical guidelines for deciding when and how to use molybdenum disulphide, but there was still a considerable lack of universally-accepted theoretical understanding of some of the important and fundamental aspects of molybdenum disulphide technology and the state of knowledge was growing rapidly. In some respects, theories could change quite quickly and dramatically in response to new information and inferences.

It was therefore a very productive period for writing reviews and reports, but a much less satisfactory time for attempting any sort of definitive textbook. The reports which I wrote then represented little more than progress reports on a rapidly-developing technology. Nevertheless, they have proved to be a useful starting-point for writing this more comprehensive and up-dated publication, and to that extent the contribution of the two space organisations is gratefully acknowledged.

In the past fifteen years the situation with regard to the technology of molybdenum disulphide lubrication has stabilised in many respects, and a measure of consensus has been reached about some of the mechanisms involved. The use of molybdenum disulphide has become routine in some industries, and there are many well-established and reputable commercial products available. Except in the high-technology field of physical deposition techniques, especially sputtering, the output of new research publications has fallen from perhaps two hundred a year in the nineteen-seventies to fewer than ten a year in the nineteen-nineties.

In spite of this maturing of the subject, it is clear that there are still many aspects in which disagreements persist about the mechanisms involved, and which as a result are unclear or misunderstood among users, and perhaps even more importantly, among potential users. These aspects range from the mechanism of action of the important additive antimony trioxide to the behaviour of molybdenum disulphide in the presence of liquids, and the critical importance of consolidation of molybdenum disulphide films.

In 1976 when I was Director of the Swansea Tribology Centre, we were asked to advise a major French company on the possible use of dry lubrication for the third stage rocket motor of the Ariane launcher system. For the particular operating conditions, our recommendation was to use a specific commercial inorganic-bonded molybdenum disulphide film. Shortly afterwards I arrived at the company's headquarters for technical discussions, and was greeted with some suspicion, if not hostility, by their engineers. They had coated steel test plates with our recommended bonded film, and showed me that the film was so soft that it could easily be scraped off with a finger-nail. Fortunately I was able to show them that the film was easily consolidated by proper running-in, or even by drawing the same finger-nail backwards across it under pressure. Once consolidated, it could hardly be scraped off with a knife, and only with considerable effort with a file.

There had obviously been a serious communication gap with respect to the consolidation of bonded films. The phenomenon of burnishing of powder had been described almost twenty years earlier, and the effects of running-in for bonded films had been known to scientists for several years. In spite of this, important users had not been aware of the necessity, either from trade articles, or the product manufacturer's literature, or, I was ashamed to realise, through my own first two ESRO reports.

During the following twenty years, reading and re-reading of most of the existing literature on molybdenum disulphide lubrication has confirmed that very few authors have ever made clear the importance of proper running-in, or burnishing, of films, or its effects on friction and film life. Most publications on bonded films or on films produced from dispersions have simply reported the test conditions and the performance without any attempt to clarify the effects or extent of running-in or film consolidation.

One of the primary objectives of this book is therefore to analyse the various aspects of molybdenum disulphide lubrication technology concerning which there are still disagreements or controversy, and to attempt to come to firm conclusions about some of the mechanisms involved. In particular, it will place emphasis on the importance and effects of burnishing and film consolidation.

In addition this is, I believe, a suitable time for publishing a book on molybdenum disulphide in general. In most respects the state of knowledge of the subject is on a stable plateau, in which radical changes in the short term are unlikely. In the special case of sputtering, and other physical deposition techniques, the highly active state of research may lead to radical developments at any time. Hopefully, this may be a topic which a greater specialist could effectively describe in some future book.

ACKNOWLEDGEMENTS

A great deal of literature on this subject has been published by the American Society of Lubrication Engineers, now the Society of Tribologists and Lubrication Engineers, and their permission to reproduce a number of items from their publications is gratefully acknowledged.

~~~~~~~~~~~~~~~~
~~~~~~~~~~~~~~~~

CONTENTS

~~~~~~~~~~~~~~~
~~~~~~~~~~~~~~~

FIGURES

~ ~ ~ ~ ~ ~ ~ ~ ~ ~ ~ ~ ~ ~ ~

TABLES

~~~~~~~~~~~~~~~~
~~~~~~~~~~~~~~~~

CHAPTER 1

HISTORY

1.1 EARLY BEGINNINGS

Lubrication is probably almost as old as intelligent man. Dowson[1] has described early evidence of lubricants in potters' wheels, on chariot axles, on sledges, and between stone blocks in the construction of buildings. All of these are from the Sumerian and Egyptian civilisations, between 5000 and 3500 years ago.

But these applications mainly relate to machines of various types, and machines themselves only date from the same periods. In other words, from the first times when man's way of life permitted the design and construction of machines, it was only a relatively few generations before we find firm evidence that he was using lubricants in those machines. Should we therefore assume that his use of lubricants followed his invention of machines, or did his first machines make use of the knowledge he already had of lubricants and lubrication?

In fact it seems very unlikely that some of man's early machines would have worked at all without lubricants. One of the fastest ways to produce wood dust ("saw-dust") is to operate a wooden bearing without a lubricant. A very effective way to soften and weld metals is to operate a metallic bearing without a lubricant. Without lubricants the early geniuses who invented potter's wheels or chariot axles would have generated such rapid wear or seizure that they would probably have given up and gone back to some useful activity like hunting.

A lubricant is any material introduced between surfaces to reduce friction, and the earliest reasoning men must have been aware, often painfully aware, of some of the materials which reduced friction. Water itself is a poor lubricant for bare feet, wood or stone, but such natural materials as wet clay, ice, loose sand and smooth pebbles would have made him slip and injure himself, and blood or animal fats would have made his cutting blades slip and cut him. Even pre-intelligent man and other animals would have learned from such harmful experiences.

Recognition and use of slippery situations is certainly not restricted to man. Otters will take advantage of a powdering of snow on ice to make slides up to six metres long, as well as having invented their well-known mud slides. To progress from here to the deliberate application of such lubricants is a very small step indeed, for example from slipping oneself to making one's enemy or prey slip into a pit. So there must be at least a reasonable probability that man's use of lubricants goes back to the Paleolithic Period, perhaps as much as 100,000 years ago.

Where does molybdenum disulphide fit into this picture? Certainly the use of solid lubricants must be ancient. Loose sand, ice and powdered snow have already been mentioned, but other slippery solids which were available and even locally abundant were graphite, mica, talc and molybdenum disulphide.

For centuries molybdenum disulphide was called "molybdena" or "plumbago", meaning lead-like, and both words occurred in the Greek and Roman civilisations of 2,000 years ago. Pliny (24-79 A.D.) refers[2] to molybdaena in the context of lead sulphide or the leaden dross in the smelting of silver. Dioscorides also writes[3] about molubdaina or plumbago, and various forms of "molubdos" (lead or its products). Molybdenum disulphide is also a possible substance to be present in the slag in copper smelting, which by Roman times had been in use for hundreds of years. Agricola (Georg Bauer) in his classic "De Re Metallica", published in 1556, also mentions[4] molybdena and plumbago in a similar context to Pliny and Dioscorides.

Unfortunately both words were used to describe several different things. "Plumbago" meant graphite, which was easily confused with molybdenum disulphide. "Molybdena" (or molybdaena) also meant graphite, as well as various ores or salts of lead. At certain times and places molybdenite would have been more readily available than graphite, and may well be the subject of some of the early references, but the lack of continuity in the written record makes it impossible to establish when or how the name became more closely associated with molybdenum disulphide.

The word "molybdenite" which is now clearly identified with natural molybdenum disulphide, and sometimes more generally used for any sample of it in the same crystalline form, is undoubtedly derived from "molybdena". There is therefore a firm connection from the Greek and Roman usage to the first clear identification of molybdenum disulphide, although its use prior to the seventeenth century AD can only be considered probable and not definitely proved.

1.2 RECORDED HISTORY

The earliest written account which can be definitely identified with the use of molybdenite as a lubricant is in "Elements of the Art of Assaying Metals" by John Andrew Cramer[5], published in 1764, and some of his references go back a further 150 years to the early seventeenth century. One passage is worth quoting verbatim, as it gives a delightful illustration of both the developing technology and the confusion which still existed.

"Among the Minerals not yet examined, that which chiefly deserves our Consideration is, the Molybdaena, or otherwise called Cerussa nigra, Plumbum marinum in English Wad or Black-Lead, in German Wasser-Bley: it must not be confounded with the Galaena or Steel-grained Lead Ore, which, though commonly called by the same Name, yet is altogether different from it. The Black-Lead is a Mineral of a Lead Colour, consisting of small shining Scales, soft, so as to be easily scraped with a Knife. ***It is much heavier than the glimmer-Stones, of which it has almost the whole Texture.*** It feels much like Soap, and its rubbing against solid Bodies, renders them slippery: whence, Workmen rub their Presses, and other Tools, with Black-Lead instead of Soap, partly to facilitate Motion, and partly to cover and keep off Rust, by such a Lay of a shining black Colour. It is likewise commonly used for Writing-Pencils. ***It hardly suffers any Alteration in the strongest open Fire; except that, being thus divided into very small Particles, it loses its Colour entirely, and becomes of a Consistence somewhat softer."***

The two sections I have put in italics must refer to molybdenum disulphide and not graphite. "Glimmer-Stones" are micas and the various micas all have specific gravities between 2.7 and 3.3, Molybdenite has a specific gravity of 4.6 to 4.75, while that of natural graphites varies between 2.05 and 2.25. Incidentally, the comparison between Molybdaena and mica is very acute, in view of their crystallographic similarity. Similarly the effect of "the strongest open fire" on molybdenite would be to oxidise it to the white or yellow molybdenum trioxide, whereas natural graphite would be oxidised to carbon dioxide, leaving a relatively small solid residue. The reference to Black-Lead keeping off rust is fascinating in view of the many reports in more recent times about both molybdenum disulphide and graphite actually causing corrosion.

It was in 1778 that molybdenite was finally, and clearly, distinguished from graphite, when Scheele found that on heating with nitric acid it gave a white residue, whereas graphite was unchanged[6]. It seems surprising that there was no immediate resulting increase in the use of molybdenite as a lubricant. Its lubricating properties were known, it had become a material of interest to scientists and engineers, and the other available lubricants all had serious defects. For example talc and graphite were both recommended as lubricants during this period, and the lubricating properties of both are inferior to those of molybdenum disulphide.

This was also a period of great interest in lubrication associated with the needs of the machinery of the Industrial Revolution. Dowson mentions several patents for lubricant compositions of considerable complexity which were granted between 1800 and 1850. Nevertheless the next references to the use of molybdenite as a lubricant are by gold miners in the Colorado gold rush of 1858-62, who are said[7] to have used it to lubricate the axles of their wagons. This can hardly be considered a new development, since it was probably only a repetition of use in primitive times.

Although petroleum products had been used earlier, their use only became important from the middle of the nineteenth century. They then slowly revolutionised lubrication because of their effectiveness, stability, availability and cheapness, and because of the wide range of viscosity grades which could be easily produced. Vegetable oils and animal fats continued to be used as alternatives, especially where there was a need for high load-carrying capacity or low friction, but otherwise little effort was made to find other types of lubricant for many years.

Molybdenum disulphide became readily available in reasonable purity after 1918, but interest in its use was still slow to develop. Johnson[7] has suggested that technical consideration of it mainly followed the establishment of its crystal structure by Pauling and Dickinson[8] in 1923, but if that is so then progress was still very sparse. Koehler[9] used molybdenum disulphide in a composition, patented in 1927, which also included talc, mica and in some cases graphite, but in retrospect that seems more of a witch's brew than a technical development. It seems much more reasonable to suggest that the real turning-point was in 1934, when a clear understanding of the potential value of the crystal structure of molybdenum disulphide for low friction became evident for the first time. This was in the work of Cooper and Damerell[10], who patented its use in oils and greases.

The first major expansion in interest took place in 1938-9, when several industrial organisations started technical investigations. These included Standard Oil Company (Indiana), Cleveland Graphite Bronze Company, International Silver Company, and especially Westinghouse Electric Company. The Cleveland Graphite Bronze patent[11] was for incorporation in a resinous binder, but at the time this was intended to be used as a solid composite, and its potential as a possible thin bonded film was not recognised until much later. Thus by 1939 most of the present forms of molybdenum disulphide lubricant had been devised, including free powder, dispersion in oils and greases, organic and inorganic composites, and a potential bonded film.

The outstanding work in this period was by Bell and Findlay at Westinghouse Electric. They were looking for a lubricant for the bearings in the high vacuum conditions of a rotating anode X-ray tube. The lubricating properties of molybdenite attracted their attention, in conjunction with its chemical stability and its low vapour pressure. Its successful operation was reported[12] in 1941, in a paper whose title "Molybdenite as a New Lubricant" may have been a little naive. It was certainly far from new.

That paper and their many associated patents show the depth of understanding which Bell and Findlay reached of the mechanism of action and of the importance of the crystal structure and bond energies. Westinghouse also reported on methods of producing pure material and of making satisfactory films, and showed for the first time[13] the very high load-carrying capacity obtainable, in experiments at contact pressures up to 600,000 psi.

This expansion of interest was very well supported by Climax Molybdenum Company. The company obviously had a vested interest in increasing utilisation of molybdenite or any molybdenum derivatives, but the methods which it used from the early years were a model of responsible technical encouragement. Samples suitable for lubrication studies were made available, and circulation of technical papers, reports and literature has continued until very recently. In more recent years the company has also contributed to the technical study and development of molybdenum disulphide lubrication, but for the greater part of the period from 1945 to 1955 their major contribution was to provide a communication link for other workers. One interesting early paper from Climax[14] gave the first specific suggestion of a bonded thin film, using a binder consisting of corn syrup.

The most important single development in the use of molybdenum disulphide as a lubricant was probably the initiation of studies by the US National Advisory Committee for Aeronautics (NACA) in 1946. Their first report[15] was published in 1948. This work by NACA and its successor the National Aeronautics and Space Administration (NASA) laid the foundations for the great expansion in use during the past forty years. The overall increase in activity in this period was so rapid that by 1952 Climax published a list of 154 different applications.

The first military uses began in 1950, and the first military specification, MIL-L-7866 for dry powder, was issued in 1952. In general these early uses were on relatively non-critical components such as hinges, clips, latches, etc. and tended to be concerned with anti-seizure or anti-galling rather than conventional lubrication. The range of military applications grew rapidly and by 1965 there were nine US and five British military specifications covering molybdenum disulphide-based materials, including powders, bonded films, greases and anti-seize compounds.

Applications in aircraft also increased very quickly. In 1959 Boeing reported[16] from 150 to 200 applications of solid-film lubricants in B-52, KC-135 and Boeing 707 aircraft without any unserviceability reports, and the applications included critical aircraft components. By 1966 over 1000 applications of solid-film lubricants were reported on the North American B-70, and many of these involved molybdenum disulphide. Applications on the General Dynamics F-111 included the heavily-loaded variable geometry wing pivot.

Van Wyk[17] reported an increase of 100% in sales of molybdenum disulphide as a solid lubricant between 1962 and 1972. However, this rapid expansion led to a number of adverse reports of its performance. BOAC reported accelerated corrosion of Boeing 707 undercarriage bogeys associated with its use, although other reports indicated that corrosion problems on the bogeys disappeared when conventional lubricants were replaced by molybdenum disulphide. This situation was fairly typical of the complaints during the late 1950's and early 1960's, in that different operators and investigators reported conflicting results, many of which could not be repeated in controlled laboratory experiments. The corrosion issue is discussed in more detail in Chapter 15, but it seems likely that at best some of the service complaints about corrosion occurred because designers and operators failed to appreciate that, unlike conventional lubricants, molybdenum disulphide gives no protection against corrosion. There may also have been instances where fretting was confused with ordinary corrosion.

During the same period use in road vehicles had become widespread. The first reported application was to the leaf-springs of Rolls-Royce cars[18] in 1955, but by 1962 applications were reported by many major car and commercial vehicle manufacturers. Most of these were concerned with such components as ball-joints, shackles, pins, and steering linkages. There was also an increasing use of molybdenum disulphide dispersions in engine oils, but this was generally initiated by the user rather than the vehicle manufacturer.

The one remaining important application technique devised so far was vacuum sputtering, which was first described[19] in 1967.

1.3 RANGE OF APPLICATIONS

In terms of volume, the most important area of application of molybdenum disulphide lubrication is now the automotive field. A major part of this volume consists of molybdenum disulphide greases, and these applications are discussed in more detail in Chapter 13. There is little doubt that their use has made a significant contribution to the extended chassis lubrication intervals in vehicles.

Utilisation of molybdenum disulphide generally has been increasing steadily, and it seems clear that in many areas its use has achieved technical respectibility after the exaggerated claims and complaints of the 1950's and early 1960's. The aviation industry has always been a leading user, but there is now a more widespread acceptance of molybdenum disulphide in various forms. Among the other industries which have accepted its use in a wide variety of applications are metalworking and railways.

Table 1.1 Some Spacecraft Applications of Molybdenum Disulphide

Format	Spacecraft	Applications
Sodium silicate-bonded MoS_2 and graphite	OGO 1 Pegasus1,2	Louvre shaft and springs Gears and bearings
Bonded MoS_2 film	Mariner3,4 Nimbus 1 Mercury Apollo	Solar panel actuator Panel hinge pins Heat shield mechanism Legs on Lunar Module
Drilube bonded MoS_2	Ranger Mariner Surveyor	Stepping motor,antenna Instrument gears Hinges, latch mechanism
MoS_2 in sintered bronze	OGO	Wabble drive gear
75%Silver,20%graphite, MoS_2	OSO 1-v	Slip ring brushes
85%silver,2.5%copper, 12.5%MoS_2	Nimbus 4	Slip ring brushes
Burnished MoS_2 powder	Snap 10A	Control drum bearings
80%MoS_2 with Mo and Ta	Surveyor	DC motor brushes on moon landing vehicles
85%silver,3%carbon, 12%MoS_2	TACSAT	Electrical brushes
33.3%silver,50%MoS_2, 16.7%nickel	EXOS-A	Electric motor brushes
Sputtered MoS_2	TRIAD SMS-1,2 GEOS-1	Orbit sensing mechanism Gimbal bearings

Table 1.2 Some Applications of Molybdenum Disulphide

Application	Form	Conditions
Baking oven chains	Dispersion	Air, moisture to 300°C
Furnace bogies	Paste	Air to 600°C
Railway centre plates	Oil dispersion, paste	Heavy load, water
Screw threads	Paste	Heavy load to 700°C
Pickle plant conveyors	Paste	Hot alkali, hot acid
Expansion joints	Powder and paste	High load, heat
Slideways	Powder	High load, low speed
Oxygen valves	Bonded film	Oxygen atmosphere
Reactor hinge pins	Paste	Carbon dioxide atmosphere
Vehicle ball joints	Grease	Suspensions, water, dirt
Piston rings	Composite	Compressors, hot air
Vehicle leaf springs	Paste, grease	Suspensions, water, dirt
Stopcocks	Grease, paste	Gases, solvents, low speed
Shaft packings	Composite	Solvents, acids
Vehicle tie rod ends	Grease	Suspensions, water, dirt
Splines	Dispersion, grease	High loads, fretting
Wire ropes	Grease	High loads, dirt
Camera shutters	Grease	Light loads
Bridge bearings	Paste	High loads, weather
Hot forging	Powder, paste	Heat, high loads
Hot or cold extrusion	Paste, bonded film	Heat, high loads
Vehicle engines	Oil dispersion	Heat, oxidation

A vast amount of information has been published on the testing of molybdenum disulphide materials for space use. It seems probable that most if not all American satellites and spacecraft have contained some application of molybdenum disulphide, and a number of space applications are listed in Table 1.1. A notable early example was its use on the extendible legs of the Apollo Lunar Module in 1969. Application of molybdenum disulphide in more conventional bearing systems is described in Chapters 9 to 13, but the wide variety of lubricant uses is shown in Table 1.2 by a list of applications not described in more detail elsewhere in the book.

Many reviews of solid lubrication have been published since 1970, and this may be an indication of the growing importance of the subject. Some useful general ones are those by Clauss[20], Ducas[21], Wunsch[22], Lipp[23] and Lancaster[24]. A review by Vetter[25]is particularly related to gear applications, and Campbell et al[26] produced a handbook on applications in space containing much information which is still useful. Three reviews on molybdenum disulphide specifically are those by Farr[27], Church[28] and Winer[29]. Overall, several thousand papers and reports have been published about molybdenum disulphide in the past fifty years.

~ ~ ~ ~ ~ ~ ~ ~ ~ ~ ~ ~ ~ ~ ~ ~ ~

CHAPTER 2

OCCURRENCE AND EXTRACTION

2.1 OCCURRENCE

Molybdenum disulphide occurs naturally in very large quantities as the mineral molybdenite. Because of this ready availability, there is little incentive to develop any alternative sources, but small amounts have been produced synthetically, and the synthetic processes will be described later. Molybdenite is the most common naturally-occurring molybdenum compound, and the most important source of molybdenum metal. It occurs in many parts of the world, including the United States, Australia, Peru, Germany, Rumania, Canada and China. In 1915 two-thirds of the world's production was from Australia[30], but increased demand during the first world war led to the development of the huge deposits at Climax, Colorado, and these are now the principal source.

Molybdenite occurs principally as thin veins in altered granite, in low concentration. The deposits at Climax contain between 0.3 and 0.6% of molybdenum disulphide. Occasionally massive pieces of relatively pure molybdenite have been found, up to several kilograms in weight, and the occurrence of such material in outcrops gives some support to the idea that it may have been recognised and used in ancient times. Several other molybdenum-containing minerals are listed in Table 2.1. Of these wulfenite, molybdite, powellite and ilsemannite have been worked commercially, but none of them is now of great importance. Apart from primary molybdenite, the only other significant source of molybdenum is as a by-product from the extraction of other metals, especially copper. Typically in recent years by-product production has represented 40-45% of total world production, and the bulk of this is also in the form of molybdenite.

Braithwaite[31] has reviewed the world supply and production of molybdenum. He referred to current estimates of the total availability of molybdenum as four to five million tons, which he considered an under-estimate. It is in fact considerably lower than the estimate of seven million tonnes quoted by Ulmanns Encyclopedia in 1987.

Table 2.1 Molybdenum-Containing Minerals

Mineral	Chemical Composition	
Molybdenite	Molybdenum disulphide	MoS_2
Wulfenite	Lead molybdate	$PbMoO_4$
Molybdite	Hydrated iron molybdate	$FeO_3MoO_3 + H_2O$
Powellite	Calcium tungsto-molybdate	$Ca(MoW)O_4$
Ilsemannite	Molybdenum oxides	$MoO_2.4MoO_3$(variable)
Chillagite	Lead tungsto-molybdate	$3PbWO_4.PbMoO_4$
Koechlinite	Bismuth molybdate	$(BiO)_2.MoO_4$
Lindgrenite	Copper hydroxy-molybdate	$Cu_3(MoO_4)_2(OH)_2$
Bilonesite	Magnesium molybdate	$MgMoO_4$
Pateraite	Cobalt molybdate	$CoMoO_4$

Braithwaite pointed out that China and the Confederation of Independent States were now introducing substantial quantities of by-product molybdenum. The estimated ore reserves in the CIS alone are about 1.6 x 10^9 tonnes, with a molybdenum content varying from 0.015% to 0.09%. This represents a total molybdenum content of about 800,000 tonnes not previously included in Western estimates. Inclusion of similar quantities from China and the remainder of the Far East would raise the estimated world total to about seven to nine million tonnes.

Prior to 1925 the production of molybdenum was very irregular. It was only about 100 tonnes in 1914, reached 8,000 tonnes in 1918, and virtually ceased from 1920 to 1925. Since then it has increased dramatically[32], to 2,200 tonnes in 1933, 9,000 tonnes in 1938, 31,000 tonnes in 1943, 58,000 tonnes in 1966, and more than 100,000 tonnes by 1989. During the nineteen seventies the demand for molybdenum in the Western world had outstripped the supply[33], as shown in Table 2.2, and several important new mines were brought into operation, at Henderson and Mount Emmons in Colorado and at Kitsault in British Columbia. During the nineteen nineties increased by-product production in the CIS has led to that area becoming a net exporter, and this represents an additional source[31]. In recent years recovery of molybdenum from spent petroleum catalyst has represented 2% of total production.

Table 2.2 Western World Molybdenum Demand and Supply 1973-1989 (Thousand Tonnes)

Year	Demand	Mine Production			Deficit/ (Surplus)
		Primary	By-Product	Total	
1973	82	37	35	72	10
1974	94	40	33	73	21
1975	76	40	34	74	2
1976	80	42	36	78	2
1977	83	45	38	83	0
1978	90	48	40	88	2
1979	91	48	41	89	2
1980	87			99	(12)
1981	84			99	(15)
1982	69			80	(11)
1983	66			48	18
1984	77			80	(3)
1985	78			84	(6)
1986	77			82	(5)
1987	79			77	2
1988	94			82	12
1989	96			105	(9)

2.2 EXTRACTION OF MOLYBDENUM DISULPHIDE

Molybdenite is separated by crushing and liquid flotation from the felspar and quartzite which constitute the bulk of the ore. The molybdenite is finely dispersed in the ore, and most of it is closely associated with quartz in very fine veins. It must therefore be ground very fine, and the final grinding is to about 200 mesh (75μm particle size). Molybdenite is very easily separated by flotation, and was originally floated with pure pine oil. More recently the flotation oil has consisted largely of hydrocarbon oil with a wetting agent to give effective wetting of the particles and assist frothing[6]. Minor constituents are added to inhibit flotation of copper-containing minerals and pyrite. Recovery of well over 90% can be achieved with a mill feed containing only 0.6%.

This first refining stage gives a product containing 85 - 90% molybdenum disulphide, the remainder being largely silicaceous. In the early days this grade of material was sometimes used as a lubricant, and produced disastrously high wear rates, probably giving rise to some of the adverse comments on molybdenum disulphide lubrication.

By-product processing is also largely by crushing and flotation, but the flotation processes are more specialised because of the variety of ores involved and the need to separate the small molybdenum content from the major proportion of copper or other primary metals. Because the product is mainly used in steelmaking, oxidation to molybdic oxide is acceptable, and an intermediate roasting may also be used.

For lubricant use the concentrate is now further ground, acid treated, milled and finally dried and graded. Residual traces of flotation oil are then removed either by solvent extraction or by heating, the latter leaving carbonaceous impurities. The other residual impurities at this stage are usually silica and iron and copper compounds, but other materials may be present, and the nature and quantity of impurities depend on the ore source and the refining process. Since 1950 purified powders have been available with less than 2% impurities, and half of this may be carbon from the flotation oils used in the purification process. The carbon does not seriously degrade the friction and wear performance, and the availability of these purer powders coincided with the great expansion in the use of molybdenum disulphide for lubrication. The main problem associated with impurities is abrasion, and specifications place restrictions on the amount of insoluble contaminants, the limit in the United States specification MIL-M-7866B being 1%.

In the British specification DEF-2304 the assumption was made that abrasivity was directly linked to silica and a limit of 0.02% was placed on the silica content. This low limit can usually only be achieved by the use of an additional hydrofluoric acid treatment, which increases the cost of the product. Direct measurement of the abrasiveness[34] suggested that in fact there was no significant difference between products meeting MIL-M-7866B and DEF-2304, and that the more stringent limit might therefore not be justified. Nevertheless, there can be enormous (>5000 fold) differences in the abrasiveness of even high quality molybdenum disulphide powders, and at present a direct abrasion test seems to be the only reliable way of distinguishing between them.

Table 2.3 Analysis of Commercial Lubricant-Grade Molybdenum Disulphide Powder (Ref.35)

	Technical	Technical Fine	Superfine
Particle size, μm	3 to 4	0.45 to 0.75	0.40 to 0.50
Acid number	0.05	0.05	0.10
Composition, wt.%			
MoS_2	98	98	97
Acid insolubles	0.50	0.50	0.75
Iron	0.30	0.30	0.40
Molybdic oxide	0.05	0.05	0.05
Carbon	1.90	1.80	2.70
Oil	0.05	0.40	0.70
Water	0.02	0.05	0.10

Typical analyses of three commercial powders are shown in Table 2.3[35]. They differ mainly in that the fine grinding of the Superfine powder increased the acidity and allowed it to pick up some oil and water contamination. The acid insolubles include siliceous material, and these may possibly be abrasive, but there is no direct correlation between silica content and abrasiveness. The molybdic oxide, formed by oxidation of molybdenum disulphide, is less abrasive than most technical grades of the disulphide itself, as shown in Table 2.4 and it is only the acidity associated with the sulphur oxides which is potentially damaging. Any increase in friction due to the slight reduction in MoS_2 content is hardly likely to be detectable.

Table 2.4 Relative Abrasiveness of Materials (Data from Ref.33)

Material	Wear rate ($\times 10^{-15} m^3/kg.m.$)
Impure MoS_2	720, 2800
Purified standard MoS_2 (MIL-M-7866A)	2.6, 6, 18
Purified standard MoS_2 (DEF-2304)	2.4, 12, 15
Purified micronated MoS_2 (MIL-M-7866A)	130
Purified micronated MoS_2 (DEF-2304)	4.4
Tungsten disulphide	4
Molybdenum diselenide	17, 30
Niobium diselenide	8.3
Technical mica	0.7
Natural graphite	55
Synthetic graphite	1.4 (with transfer)
Molybdenum trioxide	<0.5
Molybdenum dioxide	130

In recent years large quantities of crude molybdenum disulphide have been available as a byproduct of copper mining and processing. Ritsko, Laferty and Hubbell examined[36] a process for chemically upgrading this material by digesting with acid at high temperature, water washing, drying and sieving, and compared the lubricating properties with those of a commercial lubricant grade product. The thermogravimetric analyses and X-Ray diffraction patterns of the two materials showed no significant differences. The comparative chemical analyses are shown in Table 2.5, and the differences in lubricating properties were slight.

The purpose of the study was to assess this material as a potential source for lubricant grade molybdenum disulphide, because of the steady increase in its use in lubricants. However, the total consumption for lubricant use represents less than 4% of the primary molybdenite production. It seems unlikely, therefore, that the demand will justify even the low quoted cost of upgrading the by-product material, except where the availability of the indigenous source is important.

Table 2.5 Chemical Properties of Commercial and Upgraded Molybdenum Disulphide (Ref.36)

	Commercial MoS_2	Chemically Upgraded MoS_2
Carbon	1.30%	0.31%
Iron	0.17%	0.05%
Silica	0.16%	0.18%
Particle size	. 4.05μm	4.50μm

2.3 EXTRACTION OF MOLYBDENUM

The bulk of the concentrate separated from molybdenite ore by flotation is further processed to produce molybdenum. A typical extraction and purification procedure is outlined in Figure 2.1. The concentrate is roasted to convert the molybdenum disulphide to molybdic oxide. The product is called roasted concentrate, and about 30% is marketed as Technical Oxide, mainly for alloy manufacture. A typical range of compositions is shown in Table 2.6.

Between 40% and 50% of the roasted concentrate is converted to ferromolybdenum, either by means of an electric furnace or by a thermite process. The thermite process involves ignition of a mixture of the roasted concentrate with aluminium and an iron source (iron ore and ferrosilicon) together with a flux. The resulting ferromolybdenum contains between 55% and 70% of molybdenum, and is used in alloy steel and cast iron manufacture. Some of the roasted concentrate is converted to briquettes by pressing with a pitch binder. The briquettes, weighing about 5 kg., are also used in manufacture of alloy steels and cast irons.

Purified molybdic oxide is produced by volatilisation in a stream of air in sand-hearth furnaces at about 1200°C. Molybdic oxide melts at 795°C, but its vapour pressure is very high, and the volatilisation process is sometimes referred to as sublimation rather than distillation. The product is a fine powder containing from 99.5% to 99.97% of molybdenum trioxide.

Ammonium molybdate is manufactured by dissolving technical molybdic oxide in hot ammonia. It can be highly purified to over 99.9% purity, and

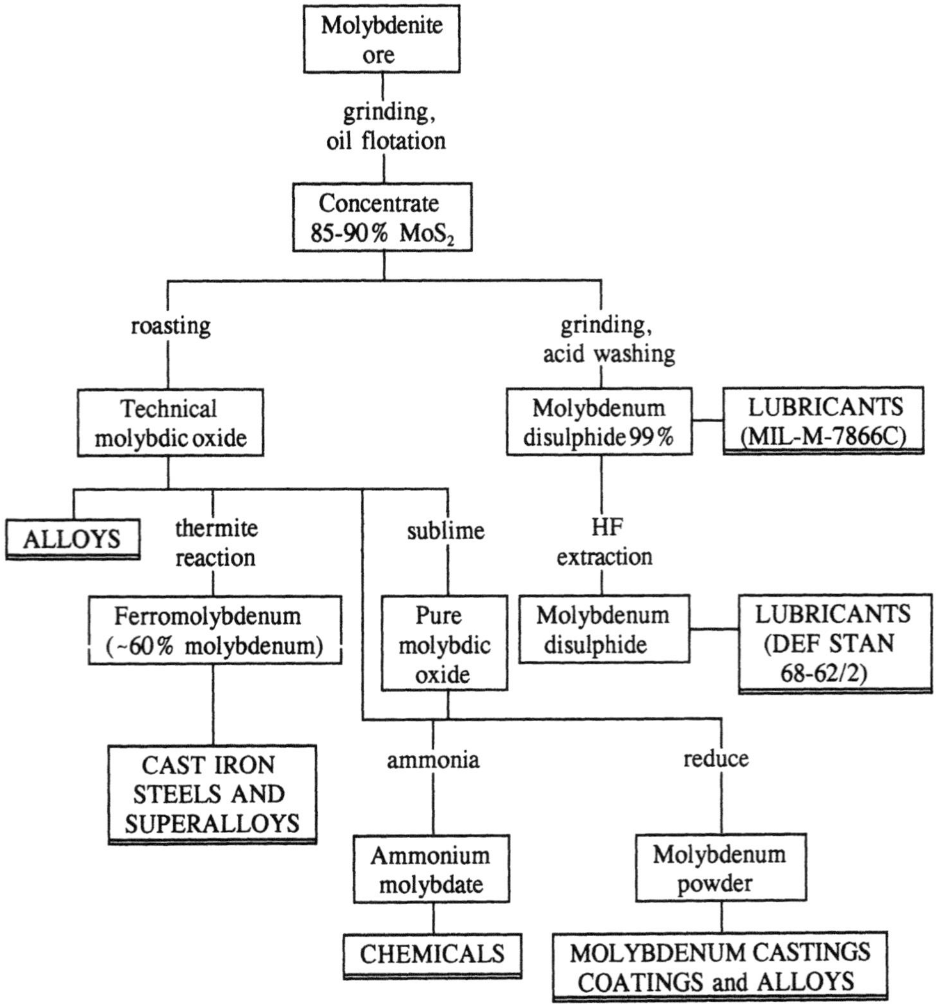

Figure 2.1 Typical Flow Chart for Molybdenite Processing

represents a source of high-purity molybdenum. Sodium molybdate is also manufactured in small quantity by a similar reaction with hot caustic soda. Both salts are used in the manufacture of other molybdenum compounds. Molybdenum metal is produced by high temperature reduction of purified molybdic oxide or ammonium molybdate with hydrogen. The molybdenum is produced as a fine powder, and can be of very high purity.

Table 2.6 Typical Composition of Technical Molybdic Oxide

Component	Percentage
Molybdic Oxide (molybdenum trioxide)	80-90%
Molybdenum	54-60%
Copper (max.)	0.5%
Sulphur (max.)	0.25%

2.4 SYNTHESIS OF MOLYBDENUM DISULPHIDE

Because of the abundance of naturally-occurring molybdenite, there is little real incentive for the synthesis of molybdenum disulphide, but it has been synthesised in small quantities. In most earlier work syntheses have been carried out only for research purposes, either to investigate the synthesis reactions themselves or to compare the properties of natural and synthetic material. Larger quantities seem to have been synthesized only when a country with insufficient natural sources wanted to ensure a reliable indigenous supply.

Several different processes have been used, the simplest being by the reaction of hydrogen sulphide with molybdenum pentachloride, or the reaction of sulphur vapour with molybdic oxide or molybdenum metal. The last of these processes has been called the SHS process (Self-Propagating High-Temperature Synthesis) and Russian workers have reported[37] that the product is less contaminated with impurities and has almost identical lubricating properties to natural molybdenum disulphide. The crystal structure is considered in more detail later, but it seems probable that the initial product of syntheses has a disordered or rhombohedral structure and that it can be converted by heat into the same hexagonal structure as the natural product.

In the early nineteen-eighties there was a surge of interest in photoelectrochemical cells for solar energy conversion, and molybdenum disulphide was extensively studied for this purpose[38-41]. It attracted attention initially because of its ready availability as a natural product, but it was found that the polycrystalline material had reduced efficiency and was more susceptible to degradation due to electrically-induced chemical reactions. There was therefore renewed interest in synthesis as a means of obtaining purer material and single crystals. A 96% yield of stoichiometric product was obtained[42] by reduction of molybdenum trisulphide with hydrogen at 200°C and 5.6 MPa, and the average crystal size was increased from 10 - 20 μm to 50μm by heating in argon at 900°C. Larger single crystals and polycrystalline films were prepared[43] by electrodeposition from a sodium tetraborate melt at temperatures over 800°C. Other synthesis procedures have been used to produce *in situ* films on bearing surfaces, and these are described in Chapter 9.

The differences between synthetic and naturally-occurring molybdenum disulphide are considered later. Unless otherwise specified, information in this book relates to the hexagonal crystal form obtained from natural molybdenite.

~ ~ ~ ~ ~ ~ ~ ~ ~ ~ ~ ~ ~ ~ ~ ~

CHAPTER 3.

MOLYBDENUM AND ITS COMPOUNDS

3.1 MOLYBDENUM

Molybdenum is the element of atomic number 42, and its atomic weight is 95.95. It is a shiny grey metal, resembling steel in appearance, but with an unusually high melting point of 2610°C. It has seven known natural stable isotopes whose mass numbers are 98, 96, 92, 95, 100, 97 and 94 in decreasing order of abundance. It was first isolated by P.J. Hjelm in 1782 from molybdenite, which he converted to oxide and then reduced by heating with charcoal[44]

Molybdenum is in Sub-Group VI A/B of the Periodic Table, and in the second series of transition elements. Transition elements are those which have an incomplete inner orbit in their atomic structure (see Table 3.1), and such an incomplete orbit is less stable than a filled orbit. The result is that the transition elements, and their compounds, show resemblances to each other and peculiarities in comparison with non-transition elements. It is therefore interesting that a number of compounds of other transition elements have been studied for solid lubricant use, and some of them have been found to be very effective, but no-one has yet shown any particular relationship between transition element structures and lubricating performance. The electron orbital assignments for these various elements are shown in Table 3.1.

Like all the other transition elements, molybdenum is a metal, and it is widely used as an alloying element and as a metallic coating. Some of its physical properties are listed in Table 3.2. In chemical reactions it shows little tendency to form cations, which is usually a characteristic of metals. In fact its salt-forming properties resemble those of non-metals, in that it has the ability to form salts (molybdates, sulphomolybdates, etc) or other complex compounds with another metal and a non-metal such as oxygen, sulphur or a halogen. The chemistry of molybdenum is very complicated, and apart from the oxides, halides and chalcogenides, very few simple compounds are known.

Table 3.1 Electron Orbital Assignments for Some Transition Elements

Shell	K	L	M	N	O	P
Subgroup	1s	2s 2p	3s 3p 3d	4s 4p 4d 4f	5s 5p 5d 5f	6s
Titanium	2	2 6	2 6 2	2		
Vanadium	2	2 6	2 6 3	2		
Chromium	2	2 6	2 6 5	1		
Zirconium	2	2 6	2 6 10	2 6 2	2	
Niobium	2	2 6	2 6 10	2 6 4	1	
Molybdenum	2	2 6	2 6 10	2 6 5	1	
Hafnium	2	2 6	2 6 10	2 6 10 14	2 6 2	2
Tantalum	2	2 6	2 6 10	2 6 10 14	2 6 3	2
Tungsten	2	2 6	2 6 10	2 6 10 14	2 6 4	2
Rhenium	2	2 6	2 6 10	2 6 10 14	2 6 5	2

Molybdenum oxidises slowly in air at 350°C and readily at temperatures above 500°C, so that it cannot be used unprotected at such temperatures in air or other oxidising environments such as water vapour or carbon dioxide. The rate of oxidation depends on the temperature and the availability of the oxygen. It is rapid enough to cause ignition of very fine powder but not of the bulk metal. There is no intergranular penetration or diffusion of oxygen into the metal, and the surface film of oxide produced may be acceptable in some cases for components which are only required to operate for short periods at high temperatures.

At high temperatures the metal will react slowly with certain gases. With carbon monoxide it produces a surface film of carbide, with nitrogen it produces a nitride film, and with hydrogen sulphide it reacts to form molybdenum disulphide. All of these films presumably interfere with the flow of gas to the metal surface, and in each case only a thin film of the product arises. Molybdenum is also very resistant to corrosive attack by mineral acids except for those such as nitric acid or chromic

Table 3.2 Some Physical Properties of Molybdenum (Ref.45)

Property	Value
Melting point	2610°C
Heat of fusion	28 KJ/mole
Boiling point	5560°C
Heat of vaporization	491.5 KJ/mole
Heat capacity (298.16-1800°K)	$22.9+5.4\times10^{-3}T$ J/°mole
Specific heat (20°C)	268 J/kg.°C
Thermal conductivity	
200°C	125 W/m.°C
1100°	100 W/m.°C
2200°C	86.2 W/m.°C
Mean linear expansion coefficient	
20-150°C	5.43×10^{-6}/°C
20-1600°C	6.65×10^{-6}/°C
Electrical conductivity (0°C)	34% IACS
Electrical resistivity	
0°C	5.2 microhm-cm.
725°C	23.9 microhm-cm.
1525°C	47.2 microhm-cm.
2525°C	78.2 microhm-cm.
Magnetic susceptibility	
25°C	0.93×10^{-6} cmu/g.
1825°C	1.11×10^{-6} cmu/g.
Lattice	
Type	Body-centred cubic
Parameter (25°C)	3.14767Å
Density	10.22 g./cm^3
Modulus of elasticity	324×10^3 MPa

acid which are oxidising. In oxygen-deficient or inert gases it can give useful lives at temperatures over 1250°C.

Table 3.3 (a) lists the principal uses of molybdenum in 1968 and it can be seen that lubricants represented only about 2% of the total. Table 3.3 (b) shows an estimated breakdown in 1994[31], and although the categories are somewhat different the major change is that the usage for chemicals has more than doubled, from 7 - 8% to 17 - 19%. The figures in both tables represent Western utilisation, as figures for Eastern Europe and Asia are not readily available. The total consumption represented by Table 3.3 (b) is about 100,000 tonnes per year. In the manufacture of the various ferrous alloys, molybdenum is normally used in the form of ferromolybdenum or of technical grade molybdic oxide. The figures in Table 3.3 show that less than 10% of molybdenum production is reduced for use in the form of the elemental metal.

There are five principal reasons for the use of molybdenum as an alloying element in steels and cast irons.

(i) To improve hardenability, or in other words to enable the same hardness to be achieved with less rapid quenching.

(ii) To improve toughness, (i.e. help to prevent brittleness)

(iii) To improve hot hardness of tool steels, that is to enable the tools to cut steel at temperatures up to red heat.

(iv) To improve the corrosion resistance of certain stainless steels.

(v) To increase the strength of steels at high temperatures.

A smaller proportion of molybdenum production is used in the form of metallic molybdenum or in non-ferrous alloys. The metallic molybdenum is produced by reduction of molybdic oxide or ammonium molybdate with hydrogen. The resulting powder is used as such for the manufacture of non-ferrous alloys, but can be converted into massive metal either by arc-casting or by pressing at up to 20 tons per square inch and sintering at high temperatures (1,100° to 2,200°C). The small commercial use of molybdenum metal is due to its exceptionally high melting point, chemical resistance, strength at high temperatures, high electrical and thermal conductivity, high modulus of elasticity and low thermal expansion.

3.2 OXIDES OF MOLYBDENUM

At least nine oxides of molybdenum have been reported[46], but of these the monoclinic dioxide MoO_2 and the orthorhombic trioxide are the most common and most stable. The others, with formulae lying between MoO_2 and MoO_3 are metastable mixed valency oxides with intense colour and metallic lustre, formed from one or both

Table 3.3(a) Approximate Breakdown of Molybdenum Utilisation in 1968

Use	%
Stainless steel	20
Tool and high-speed steels	11
Other alloy steels	44
Cast iron and steel mill rolls	8
Chemicals	5
Superalloys	5
Molybdenum metal and alloys	4
Lubricants	2
Miscellaneous	1

Table 3.3(b) Approximate Breakdown of Molybdenum Utilisation in 1994

Use	%
Stainless steel	20-25
Tool steels	7-10
Alloy steels	28-40
Cast irons	4-6
Chemicals	17-19
Special alloys	6
Molybdenum metal	6

of the stable oxides. The important oxide is the trioxide, particularly in relation to the use of molybdenum disulphide, since the trioxide is usually the final product formed on oxidation of the disulphide. In addition, the trioxide is the main intermediate in the conversion of molybdenite to other commercial products.

Molybdenum trioxide, or "molybdic oxide", melts at 795°C and boils at 1155°C at normal atmospheric pressure. It has a fairly high hardness and in many early publications was described as a harmful abrasive product from the oxidation of molybdenum disulphide in service. In fact the trioxide is not highly abrasive, being possibly less abrasive in some circumstances than molybdenum disulphide itself[46], and it has been recommended for use as a lubricant at elevated temperatures to 700°C.

Molybdenum dioxide, although generally less important than the trioxide, is more abrasive[46]. It has been suggested that abrasion associated with synthetic or oxidised[47] molybdenum disulphide may be due to the presence of the dioxide.

3.3 SULPHIDES OF MOLYBDENUM

The only important sulphide of molybdenum is the disulphide MoS_2, whose properties are described in the next chapter. Apart from the disulphide, three other sulphides have been reported[48], and some of their salient properties are listed in Table 3.4.

The sesquisulphide Mo_2S_3 is said to have been prepared[49]by rapid heating of the disulphide in the absence of air, and extraction with cold dilute aqua regia, or by combination of molybdenum and sulphur at 1300°C[50].

The pentasulphide Mo_2S_5 is said to be formed as a dark brown amorphous precipitate of the trihydrate when hydrogen sulphide is passed through an aqueous solution of a pentavalent molybdenum coompound, but like all pentavalent molybdenum compounds it is unstable.

The trisulphide MoS_3 is precipitated when hydrogen sulphide is passed through weakly acid solutions of molybdates, or can be obtained by thermal decomposition of thiomolybdates, but it has also been identified[51] in Chilean ores. It decomposes on heating above 1000°C to give free sulphur and molybdenum disulphide.

Table 3.4 Properties of Less Common Molybdenum Sulphides

Compound	Formula	Colour	Crystal Form	Remarks
Sesquisulphide	Mo_2S_3	Steel grey	Needles	Harder, denser than MoS_2
Pentasulphide	Mo_2S_5	Dark brown	Amorphous	Only as hydrates
Trisulphide	MoS_3	Brown/black	Amorphous	Decomposes to MoS_2 on heating

3.4 OTHER COMPOUNDS OF MOLYBDENUM

The instability of the electronic structure caused by the unfilled inner orbit of a transition metal leads to a very variable valency. Molybdenum exhibits valencies of 2,3,4,5 or 6 in different compounds, and it is considered that it has zero valency in its hexacarbonyl $Mo(CO)_6$. Because of this variability in its valency, many of its reaction products are mixtures of compounds in which it has different valencies. In solid form such products may be quite homogeneous in composition, and best represented by a nonstoichiometric molecular formula. The hypothetical simple compounds present in such products cannot be separated readily from one another because valency shifts occur in processing.

Apart from the oxides and sulphides, other groups of simple compounds are the halides, chalcogenides and molybdates, and there are a number of well characterised organic compounds. Simple halides can be prepared by direct reaction between molybdenum powder and chlorine, fluorine or bromine, and the pentachloride $MoCl_5$ has been used as a chlorination catalyst.

The important chalcogenides are the disulphide, diselenide and ditelluride, and their use in lubrication is described later. Other selenides and tellurides have also been identified and studied to a limited extent, but as a class these materials tend not to be stoichiometric. One group of relatively simple compounds is that of the molybdates. These are salts of ammonia or metals with molybdic acid H_2MoO_4.

3.5 MOLYBDENUM COMPOUNDS IN LUBRICATION

Apart from molybdenum disulphide, diselenide and ditelluride, several molybdenum compounds have been used as additives in lubricating oils and greases. A paper by Braithwaite and Green[52] in 1978 described the results of testing two soluble organo-molybdenum materials as additives in automotive engine and transmission lubricants. The first of these was a commercial product Moly van L, marketed by R T Vanderbilt Co. Inc., consisting of a sulphurized oxymolybdenum organophosphorodithiolate. The second was a reaction product of a molybdate and 3,4-dimercaptotoluene containing a mixture of tris(toluene-3,4-dithiolato) molybdenum (VI) and pentakis (toluene-3,4-dithiolato) dimolybdenum (V), and was called molybdenum dithiolate.

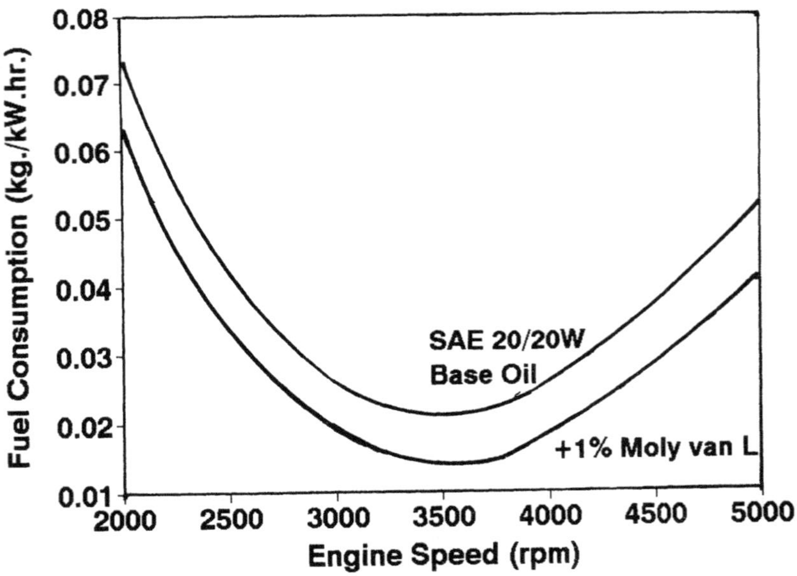

Figure 3.1 Reduction in Fuel Consumption with an Oil Containing an Oil-Soluble Molybdenum Compound (Ref.52)

Moly van L was effective in reducing fuel consumption, as shown in Figure 3.1, and both Moly van L and molybdenum dithiolate improved transmission efficiency at higher temperatures, by 4% and 2% respectively. Scott[53] showed that the molybdenum dithiolate was effective in increasing the rolling contact fatigue life of EN31 steel balls in the rolling four-ball test when added to mineral oil or ester lubricants, but Molyvan L was not effective in this respect in a diester base oil. Several Russian papers[54-57] also showed the effectiveness of certain oil-soluble molybdenum compounds in automotive engine oils in reducing both friction and wear. Some of the additives, including Moly van L, contained both molybdenum and sulphur, and it was implied that all of them did.

These organo-molybdenum compounds are in commercial use, for example as friction-modifiers. Their mechanism of action is considered in Chapter 9, in connection with the *in situ* production of molybdenum disulphide films.

3.6 CHEMICAL USES OF MOLYBDENUM

Although eighty percent of molybdenum production is used in the metallurgical industries, the fastest-growing sector of use is in chemicals, which has more than doubled in the past thirty years. The most widely-used compound is probably molybdenum disulphide. Apart from its use in lubrication, it is used as an additive to thermoplastics, where it improves the mechanical and thermal properties. It also has a number of potential applications in high density electric batteries, although the extent of commercial use is not clear.

Molybdates are used in a variety of industries. Sodium molybdate is used for the synthesis of pigments such as molybdate chrome orange, which is a homogeneous mixture of lead molybdate, lead chromate and lead sulphate. This use is likely to decline because of concerns about health hazards associated with lead, but phosphomolybdates and phosphotungstomolybdates are used to complex certain dyestuffs to produce pigments. A few of the best-known of these are Malachite Green, Rhodamine B and Methyl Violet, also used as indicators in analytical chemistry.

Molybdenum is important in agriculture, and plays a vital part in the fixation of atmospheric nitrogen. However, the concentration present in the soil is critical in relation to copper metabolism. If the molybdenum intake by animals is too high, especially with ruminants, then a copper-deficiency problem called "molybdenosis" can occur. On the other hand, too low an intake of molybdenum can lead to excessive copper metabolism and copper poisoning. The total use of molybdenum in agriculture is about 500 tonnes per year. Most of this is in the form of sodium molybdate, but technical molybdic oxide is also used to provide slow release into the soil.

Molybdic oxide is used as a fire retardant and smoke suppressant in plastics, and molybdenum compounds are used as corrosion inhibitors, especially in large cooling towers and similar systems. Finally, molybdenum compounds play an important and increasing part as catalysts in chemical and petroleum processing, both as homogeneous and heterogeneous catalysts. One of the major applications is in desulphurisation of petroleum. Others are in the single-stage conversion of methanol to formaldehyde, conversion of propene to acrylonitrile, liquefaction of coal, and denitrification.

~ ~ ~ ~ ~ ~ ~ ~ ~ ~ ~ ~ ~ ~ ~ ~

CHAPTER 4.

PROPERTIES OF MOLYBDENUM DISULPHIDE

4.1 PHYSICAL PROPERTIES

Molybdenum disulphide is a dark blue-grey or black solid which feels slippery, or greasy, to the touch. Because of its ready transfer to almost any solid surface, and the difficulty in removing it, it is a "dirty" material to handle. It exists in two crystal forms, hexagonal and rhombohedral, and the crystal structure is discussed in detail later. By far the most common form is the hexagonal, and the following data refer to this form.

The easy transfer to surfaces is probably the reason for the early names "plumbago" and "molybdaena", meaning lead-like, since lead also produces dark marks on paper and fabric. Lead rods were used in ancient times for marking out parchments, and this has led to the expressions "lead pencil" and "black-lead" which have been common until the twentieth century. Both terms now commonly refer to graphite, which resembles molybdenum disulphide in many ways, but the latter was almost certainly used in the same way until the late eighteenth century. The two materials can be easily distinguished by the lower density of graphite.

Molybdenum disulphide can be cleaved like mica, and thin sheets several centimetres square can be separated from a large crystal. These thin plates resemble lead foil in appearance, but are less malleable.

Some of the most important physical properties are listed in Table 4.1.
The crystal structure of natural molybdenite has been shown[8] to be hexagonal, with six-fold symmetry, two molecules per unit cell, and a laminar, or layer-lattice

structure, as shown in Figure 4.1. Each sulphur atom is equidistant from three molybdenum atoms, and each molybdenum atom is equidistant from six sulphur atoms, the interatomic spacing being 2.41 ± 0.06Å*.

Table 4.1 Physical Properties of Molybdenum Disulphide

Property	Value
Melting point	1700°C under pressure
Molecular weight	160.08
Density	4.9 +
Crystal form	Hexagonal, rhombohedral
Hardness (basal planes)	1.0-1.5 Moh's scale; 12-60 Knoop
Hardness (crystal edges)	7-8 Moh's scale; 800-1000 VPN
Colour	Blue-gray to black
Magnetic properties	Diamagnetic (but see Chapter 4)
Electrical conductivity	Low but variable (see Chapter 4)
Sublimation temperature	1050°C in high vacuum
Dissociation temperature	1370°C +

Each molybdenum atom is thus at the centre of a right triangular prism whose corners are the six sulphur atoms, the height being 3.17 ± 0.1Å and the triangular edge being 3.15 ± 0.02Å. Since the unit cell contains two molecules, the lattice parameters are a = 3.15Å, b = 12.39Å. The distance between adjacent sulphur layers is 3.49Å, which is greater than the overall thickness of a molybdenum disulphide layer, and Dickinson and Pauling[8] inferred that the excellent basal cleavage of molybdenite referred to earlier was caused by this great distance between the sulphur atoms.

The layer-lattice structure has often been compared with that of graphite, but in fact there are important differences. All the atoms in graphite are identical, and there is a relatively large inherent interlayer attraction caused by the interplanar π electron pairs. In molybdenum disulphide there are two different atomic species and the attraction between molybdenum and sulphur is powerful covalent bonding, but between lattice layers there is only very weak van der Waals attraction. Thus in any comparison, the units which must be compared are the molecule of molybdenum disulphide and the atom of graphite. Once this is understood the difference between the coulombic attractions of the graphite and the van der Waals forces of the molybdenum disulphide is understandable and the similarities between the two materials are not over-stated.

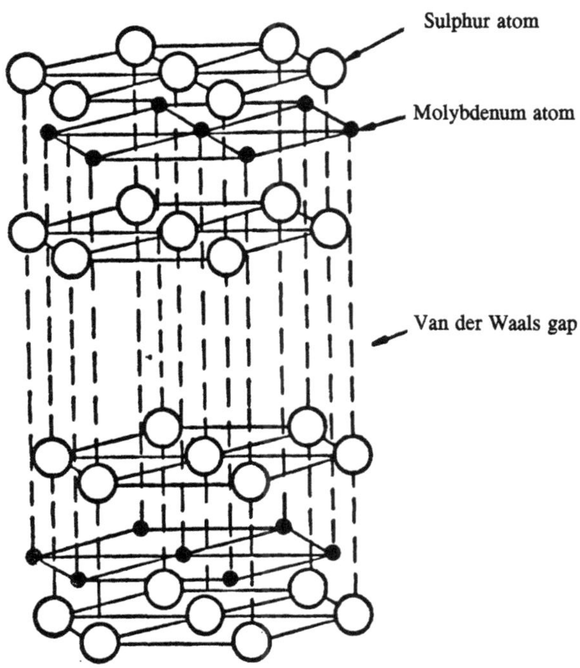

Figure 4.1 Crystal Structure of Molybdenum Disulphide (Courtesy of E W Roberts)

The most detailed study of the crystallography to date has been done by Takeuchi and Nowacki[58]. They found by a theoretical analysis that there should be four simple polytypes, all having the same coordination of the molybdenum atom between the sulphur layers in a right trigonal prism. The different polytypes result from different stacking of the simple laminae. The four polytypes are rhombohedral (3R), two hexagonal ($2H_1$, $2H_2$) and trigonal (2T). Apart from the common hexagonal form of natural molybdenite, only one of these polytypes has been found. This is the rhombohedral, which was first identified in a synthetic material[59] and subsequently found in several natural sources[60,61]. The hexagonal form has also been found in synthetic molybdenum disulphide.

A poorly crystalline "rag" structure has been described[62] for synthetic product obtained by the reaction between molybdic chloride and lithium sulphide in tetrahydrofuran. The product was purified by repeated washing with tetrahydrofuran to remove the lithium chloride. Heat treatment of the amorphous powder gave a low degree of crystallization.

In general it seems that the usual first product in the synthesis of molybdenum disulphide is highly disordered but is dominantly rhombohedral in crystal structure. However, the rhombohedral structure is less stable and on prolonged heating at temperatures ranging from 400°C to 1200°C it is converted to the hexagonal. The hexagonal form is stable at all temperatures up to decomposition temperatures, and heating in argon at 900°C has been used to produce crystals up to 50μm in size.

There is some disagreement as to whether the two types of crystalline material differ in their lubricating properties. There is some practical evidence that their frictional behaviour is similar, but this could be at least partly due to conversion of rhombohedral to hexagonal by frictional heating and shear. Differences between the lubricating properties of natural and synthetic material have also been reported by some workers[63] and denied by others[64]. In general this problem is not of great practical importance because of the dominant use of natural hexagonal molybdenite in lubrication, but it is significant in connection with *in situ* processes, and is considered further in Chapter 9.

4.2 INTERCALATION COMPOUNDS

It is possible to insert additional atoms or molecules into the inter-lamellar gap of many layer-lattice materials, including molybdenum disulphide, creating what are called intercalation compounds. The intercalated substances may be alkali[65] or alkalyne-earth metals (sodium, potassium, rubidium, caesium, calcium, strontium), salts or organic bases such as ethylene diamine or pyridine[66].

Many layer-lattice compounds can intercalate additional metal atoms of the same element as comprised in the original structure (e.g. niobium in niobium diselenide), but molybdenum disulphide will not do so. The behaviour may be determined[67] by the availability of electrons suitably oriented to form bonds with the additional metal atoms, although it seems unlikely that this single factor applies to all intercalation effects.

The effect of intercalating like metal atoms is of course to change the atomic ratios, and for example it has been reported[68] that niobium diselenide can intercalate additional niobium atoms to a composition of $Nb_{1.3}Se_2$ There will also be corresponding changes in the crystal lattice parameters, and these are discussed in relation to lubrication properties in Chapter 14.

The physics and chemistry of molybdenum disulphide intercalation compounds have been reviewed by Woollam and Somoano[69]. Perhaps the most interesting of these properties is superconductivity below 6.9°K,[70] obtained with either organic bases or alkali metals. Some of the intercalation compounds show high alkali ion diffusivity, and this has led to them being considered for use in electrodes for high energy-density batteries[71].

What might perhaps be considered as an extreme intercalation is the storage of hydrogen in atomic form in a strong magnetic field in exfoliated molybdenum disulphide[72].

4.3 ELECTRICAL PROPERTIES

The application of molybdenum disulphide and other dichalcogenides has become important in electrical brushes, especially in spacecraft, and its electrical properties are of considerable interest. It is therefore surprising to find that there is no clear agreement about its electrical conductivity.

It is usual to state[73,74] that molybdenum disulphide is a 'p' type semiconductor, while niobium diselenide is a conductor. However Mikhailov[75] has shown that pure molybdenum disulphide is a conductor and that only specimens having a developed film of oxidised material on the surface of the lamellae show semiconductor properties. Correspondingly a composite containing 15% was found[76] to have a specific contact resistance of only 0.4 $m.ohm.cm^2$. compared with 0.7 $m.ohm.cm^2$

for otherwise identical material containing 15% of niobium diselenide. On the other hand, the studies which have been made of its potential for use in photoelectrochemical cells show clearly that under the conditions of study it acts as a semi-conductor.

Table 4.2 Electrical Resistance of Molybdenum Disulphide at Various Temperatures (Data from Ref.20)

Temperature °C	Specific Resistance(Ω)
-60	8.33
+20	0.79
+73	0.47
+92	0.41

If we accept the general view that it is a semi-conductor under normal conditions of purity, temperature and environment, then it is certainly clear that with increasing temperature it becomes a conductor. Table 4.2 lists values[20] of specific resistance at different temperatures, which show a gradual decrease in resistivity with increasing temperature. Figure 4.2 shows a similar relationship, although the absolute values are very different. It has also been reported[29] that as the temperature approaches red heat in an inert atmosphere it becomes a good conductor, but in general the actual values quoted for resistivity are completely erratic, varying by factors of 10^9.

The resistance depends to some extent on the direction of the current flow in relation to the crystal structure[77]. At 70°C interpolated results were approximately 17 ohms parallel and 10 ohms perpendicular to the C-axis, but at 19°C the value was 29 ohms in both directions.

The resistance also varies with the applied potential, with pressure, and with light[48]. No detailed study has been made of the interactions between these various influences, but none of them seems sufficient to account for the wide range of measured values. It is probable that impurities have a dominant influence on the conductivity, and this is supported by the effect of intercalated molecules in producing superconducting derivatives.

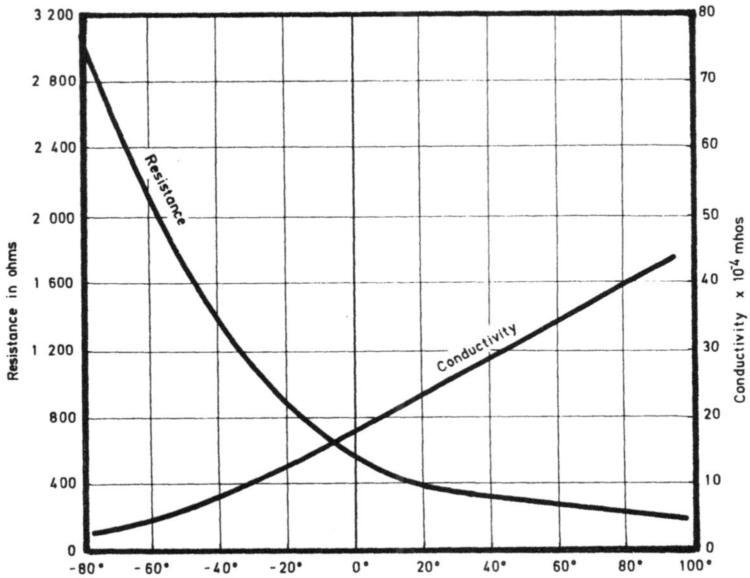

Figure 4.2 Change of Electrical Resistance and Conductivity of Molybdenite with Temperature (Ref.77)

The composition and performance of compacts and composites used in electrical brushes are considered in some detail in Chapter 12.

The effect of light is not simply to change the resistance, but to generate voltages by a photoelectrochemical process. This is true of all the lubricating dichalcogenides. They have high stability in use because the light irradiation produces d - d band transmissions which do not weaken the crystal bonding.

Overall the electrical properties of molybdenum disulphide are obviously both interesting and complex. The influences of anisotropy, heat, light, contaminants and intercalation have already been shown to be associated with a range of properties from semiconductivity to superconductivity, as well as power generation.

4.4 CHEMICAL PROPERTIES

In general molybdenum disulphide is chemically very inert. It is resistant to attack by most acids, except aqua regia and hot concentrated sulphuric, nitric and possibly hydrochloric acids. Whereas most metals form salts when attacked by acids, molybdenum has no such tendency, and the product of acid attack is normally molybdenum trioxide. The same appears to be true of the disulphide, and the limited attack by acids can be considered more as a form of oxidation. There is considerable variation in the resistance of different samples to acid attack, and the reactions involved may therefore be primarily those of contaminants rather than of the molybdenum disulphide itself.

It is attacked by fluorine but there is no reaction with dry hydrogen fluoride, and only a slow reaction with hydrofluoric acid. Reaction with chlorine produces molybdenum pentachloride. Heating in hydrogen reduces the disulphide directly to molybdenum metal.

Reduction with some of the traditional reducing agents for metallic ores is less effective[78]. When heated with mixtures of graphite and PTFE there was some change in the chemical composition and the crystal lattice parameters, but neither metallic molybdenum nor its oxides were formed. With graphite and coal-tar pitch in air there was no change at temperatures below 150°C. Between 150°C and 180°C there was partial degradation, up to 10% of metallic molybdenum being formed. In an inert environment there was no change up to 1100°C, and this suggests that oxidation was an intermediate phase in the reaction mechanism. Normally no oxidation of molybdenum disulphide itself would take place at temperatures as low as 180°C, and it is possible that some initial oxidation of the coal-tar pitch initiates a peroxy or other free radicle attack.

Adsorption on molybdenum disulphide is important because of its effect on lubrication, and Kalamazov and co-workers[79], studied the adsorption of oxygen, hydrogen, nitrogen and water vapour. They found that after desorption at 900°C and 10^{-4}Pa (10^{-6} Torr) subsequent re-adsorption was at a lower level, and inferred that active adsorption sites had been destroyed by the vacuum and high temperature. They found that at 700°C adsorbed water vapour was dissociated, causing oxidation and the liberation of hydrogen.

Matsunaga studied adsorption of n-amylamine on cleavage faces and edge sites of the crystals by the use of Auger spectrometry[80]. He confirmed the easy adsorption and slow desorption on crystal edges, and the very slow adsorption and very easy desorption on cleavage faces. This behaviour is discussed later in relation to the effects of contaminants on friction.

At temperatures above 300°C Holinski found[81] that molybdenum disulphide produced embrittlement of stainless steel. He suggested that free sulphur released at these temperatures reacted with nickel in austenitic alloys to deposit nickel sulphide preferentially at grain boundaries, thus leading to a form of stress corrosion. Knappwost similarly reported[82] that molybdenum disulphide reacted with iron at 700°C to produce ferric sulphide and free molybdenum, and Tsuya et al showed[83] that it reacted more rapidly with iron and nickel than with silver or copper in a vacuum of 10^{-5} Torr above 500°C. The reaction with copper was in fact slow above 500°C but very rapid about 700°C.

4.5 EFFECTS OF TEMPERATURE

The effect of temperature in an oxidising environment is discussed in the next section. In an inert gas or in vacuum molybdenum disulphide has very good thermal stability. Figure 4.3 shows the loss in weight as the powder at 10^{-8} to 10^{-9} Torr (0.13 to 1.3μPa) was heated[84] in stepwise fashion to 1260°C. Weight loss can be seen to begin at 930°C, and beyond that point the weight loss increased with temperature.

Free sulphur was detected at 1090°C, indicating dissociation, although this was believed to have begun at lower temperatures. On the other hand this temperature of 1090°C is interesting because it is very close to the temperature of 1060°C at which Cannon[85] reported the commencement of sublimation. It seems certain that this temperature is the lowest at which significant thermal degradation begins, and weight losses reported at lower temperatures are probably caused by volatilisation of molybdenum trioxide contamination.

At 1600°C it decomposes readily to give gaseous sulphur and molybdenum[86], but if the temperature is raised to 1700°C it melts with decomposition[87]. In the absence of other reactive substances it should therefore be theoretically possible to use it as a lubricant for extended periods to 1000°C and for shorter periods to perhaps 1500°C. In practice there is an increase in weight loss when rubbing takes place, and effective lubrication has not yet been reported at temperatures higher than 700°C.

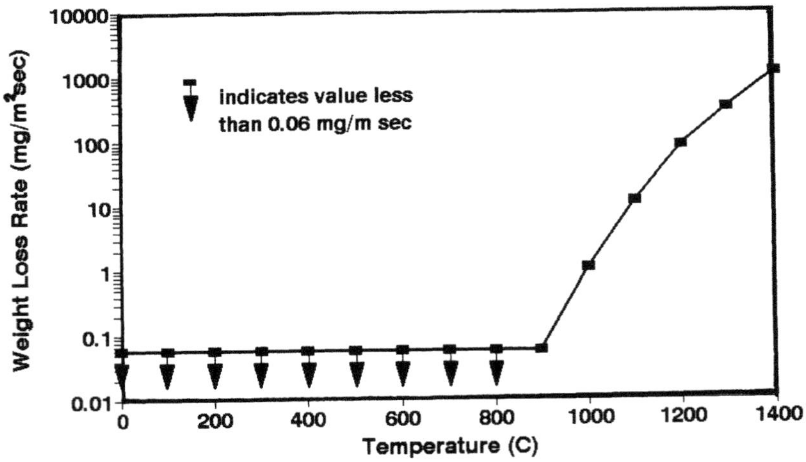

Figure 4.3 Loss of Weight of Molybdenum Disulphide with Temperature in a Vacuum of 0.13 to 1.3μPa (10^{-8} to 10^{-9} Torr) (Ref.84)

At very low temperatures it is apparently completely stable, both chemically and physically. Its reactivity is reduced to the extent that it has been satisfactorily tested at -182°C in liquid oxygen for use in ball bearings[88].

4.6 OXIDATION

The oxidation behaviour of molybdenum disulphide is of considerable practical importance. The presence of oxidation products causes an increase in friction, and the life of a burnished or bonded film in air may be largely determined by oxidation. The maximum temperature for satisfactory use in air or any other oxidising medium is therefore also controlled by oxidation.

The presence of moisture increases the tendency to oxidation, just as it encourages so many other chemical reactions. Slight oxidation occurs in moist air even at room temperature in long storage[89] but the rate of oxidation is extremely low.

Ross and Sussman showed that even after 100 hours at 85°C only about half of the surface layer is oxidised[90]. The oxidation is confined to the outermost layer of a crystal, and the oxidised layer appears to protect the remainder against further oxidation, so that at temperatures up to 300°C further oxidation is very slight.

It has often been assumed that any molybdenum disulphide surface will begin to oxidise immediately after cleavage, and that small amounts of oxide are probably even present within the crystal lattice, but Buckley[91] was unable to detect any oxide by Auger spectrometry on the surface of a crystal cleaved in air. It is clear therefore that at normal temperatures and in the absence of high concentrations of moisture the extent of oxidation is extremely small.

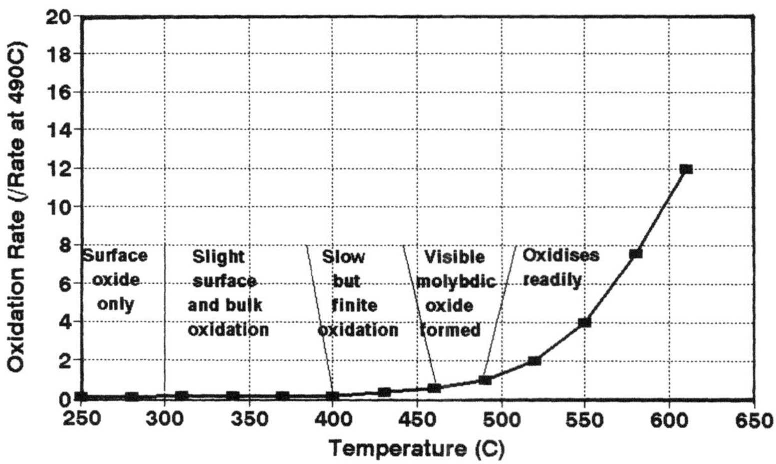

Figure 4.4 Oxidation Characteristics of Molybdenum Disulphide (Refs.89-93)

Figure 4.4 summarises the oxidation characteristics of molybdenum disulphide from room temperature to 610°C, the oxidation rates being compared with the rate at 490°C. It is clear that no arbitrary temperature can be defined below which oxidation does not occur, but that the oxidation rate is extremely small below 400°C, and Sliney[93] found that the rate constant at 370°C was only 6×10^{-4} for surface-layer oxidation and 2×10^{-4} for bulk oxidation.

The specific surface area of the molybdenum disulphide particles affects the rate of oxidation[94]. Finely divided powders have a high specific surface, and tend to be more rapidly oxidised, but larger particles are not necessarily resistant to oxidation, since they may consist of porous agglomerations of fine particles. Ducas found[21] that a sample ground to give a powder with a specific surface of 30 m^2/g gave sulphuric acid in 10% yield when extracted with water after exposure to air of 80% relative humidity for one month. The actual particle size was not stated, but a specific surface of 30 m^2/g represents a particle size of the order of 0.1μm. A more representative powder for lubricant use would have a particle size of about 1-3μm, and oxidation of such a powder would be far slower, probably far less than 1% oxidation under the same period and conditions of exposure.

While studying the influence of molybdenum disulphide on the wettability behaviour of steel, Braithwaite and Greene[95] found that the pH of a powder fell from 6.07 to 3.55 in one hour and to 2.60 in two hours when heated at 350°C in air. This powder was more typical of a fine lubricant grade, having a BET specific surface of 3 m^2/g, but the temperature used was very much higher and it is difficult to compare their results with those of Ducas because of the different technique used to assess the acidity produced.

Table 4.3 Variation of pH with Surface Area of Powders

Powder	pH	Surface Area m^2/g
A	6.1	3
B	5.5	8
C	3.9	12
D	3.2	16.5

Braithwaite and Greene also gave some figures for the pH of some powders of different particle size, and these are shown in Table 4.3. The acidities of these powders are probably due to the grinding process and to subsequent atmospheric oxidation, but they confirm the general tendency for a higher level of oxidation and acidity with finer powders.

The normal stable product of oxidation is molybdic oxide or molybdenum trioxide MoO_3, and this oxide is not abrasive, so that satisfactory lubrication can be obtained with molybdenum disulphide even after considerable oxidation. The cohesion of molybdic oxide and its adhesion to metal surfaces are inferior, however, and ultimately films will fail for these reasons. The presence of the trioxide causes no increase in wear[96].

It is possible under certain circumstances for lower oxides of molybdenum to be formed when the disulphide is oxidised, especially in the early stages of oxidation or where there is no local excess of oxygen present, as in carbon dioxide. Kalamazov reported[79] that oxidation in oxygen at 450°C took place through a number of intermediate oxides, including the sesquioxide Mo_2O_3 and dioxide MoO_2. Wyn-Roberts[97] also stated that the exposure of molybdenum disulphide to atomic oxygen, which can occur in space, can lead to the formation of the sesquioxide, which he described as being an abrasive.

Molybdenum dioxide is highly abrasive, and its formation would be a serious disadvantage to lubrication performance, but in practice it seems clear that the formation of the dioxide is transitory or exceptional, and that the oxidation process normally produces the more harmless trioxide.

There is considerable evidence that oxidation and other reactions take place preferentially at the crystallite edges rather than on the flat faces of the lamellae. The scope for such reactions would be small except with very finely divided powders because the proportion of edge surface to total surface area is small, and this theory is consistent with the general stability of the disulphide.

4.7 EFFECT OF MOISTURE

The presence of water has considerable influence on the lubrication properties of molybdenum disulphide, and their interaction has been studied in some detail. Nevertheless there is still conflict and confusion about this aspect of behaviour, as about many others.

Molybdenum disulphide has been shown to adsorb[98] or chemisorb[99] water, but the work of Johnston and Moore[100] has proved that the behaviour is quite complex. They found that in its normal commercial form it contains molybdenum trioxide, adsorbed water, chemisorbed water and adsorbed sulphuric acid as surface contaminants. When they removed sulphuric acid the adsorption of water was inhibited, and when they removed the molybdenum trioxide the chemisorption of water was markedly reduced.

The presence of moisture induces oxidation even at room temperature in long storage[89]. The ultimate product of this oxidation is almost certainly molybdenum trioxide, but it has been suggested[101] that the primary product at elevated temperatures in the absence of air is an oxysulphide $MoOS_2$, with the release of gaseous hydrogen.

$$MoS_2 + H_2O \longrightarrow MoOS_2 + H_2$$

According to this work no hydrogen sulphide is formed at any stage, but at temperatures above 300°C sulphur dioxide is released. Such a process requires the release of considerable amounts of gaseous hydrogen if we assume that the second stage is on the following lines:-

$$MoOS_2 + 6H_2O \longrightarrow MoO_3 + 2SO_2 + 6H_2$$

The liberation of gaseous hydrogen was shown by Kalamazov[79] to occur when water vapour was dissociated on the surface of molybdenum disulphide at 700°C. It has also been shown that hydrogen sulphide was produced during sliding of molybdenum disulphide in moist nitrogen, presumably by the reaction

$$MoS_2 + 2H_2O \longrightarrow 2H_2S + MoO_2$$

whereas in moist air the gaseous product was sulphur dioxide.

These complex and differing mechanisms are obviously due partly to the fact that the reactions are heterogeneous, occurring at solid surfaces, and partly to the variety of static and dynamic conditions which have been studied. Lancaster[102] has pointed out more simply that the thermodynamically favourable oxidation route is the one which liberates sulphuric acid.

$$2MoS_2 + 4H_2O + 9O_2 \longrightarrow 2MoO_3 + 4H_2SO_4$$

On the basis of these various reports the effect of water can take place in accordance with the following processes.

1.

$$MoS_2 + H_2O \xrightarrow[\text{or oxygen}]{\text{absence of air}} MoOS_2$$

$$MoOS_2 + H_2O \xrightarrow[300^\circ C]{\text{water}} MoO_3 + H_2$$

2.

$$MoS_2 + H_2O \xrightarrow[\text{rubbing}]{\text{metal}} MoO_2 + H_2S$$

$$MoO_2 + O_2 \longrightarrow MoO_3$$

3.

$$MoS_2 + H_2O \xrightarrow[\text{vapour}]{\text{water}} MoO_3 + H_2$$

4.

$$MoS_2 + MoO_3 \xrightarrow{\text{chemisorption}} MoS_2 + MoO_3 + H_2SO_4$$

$$MoS_2 + MoO_3 + H_2SO_4 \longrightarrow \text{adsorption of water}$$

The location of these reactions on the crystal surfaces has not been proved. Cannon and Norton[101] quoted evidence that reaction was monomolecular over the whole surface, but more recent work tends to confirm the general theoretical view that polar molecules such as water are adsorbed at crystallite edges while non-polar molecules such as hydrocarbons are adsorbed on the lamellar faces.

~~~~~~~~~~~~~~~~
~~~~~~~~~~~~~~~~

CHAPTER 5.

MECHANISM OF LUBRICATION

5.1 FUNDAMENTALS OF FRICTION

It is a curious fact that the English word "friction" has no direct equivalent in many other modern languages. It means simply the force which opposes movement between two surfaces in contact. (The apparently-equivalent words in certain other languages, such as "reibung" in German, "frottement" in French, and "treniye" (трение) in Russian, can all refer more generally to the rubbing process as a whole, as well as more restrictively to the friction force.)

In a very wide variety of situations friction closely follows two laws generally known as Amontons' Laws. These state that:-

(i) The frictional force is independent of the area of contact;

(ii) The frictional force is proportional to the contact load.

In fact both laws were recognised by Leonardo da Vinci about the end of the 15th century, but like most of his scientific work remained unknown until the 19th century. They were only re-discovered by Amontons about two hundred years later, and were then proved by Coulomb a further eighty years later, in 1781.

The second law leads to the concept of a coefficient of friction, since, if the friction force is proportional to the load, then

$$\text{friction} = \text{constant} \times \text{load}$$

or, rearranging

$$\frac{\text{friction}}{\text{load}} = \text{constant}$$

and this constant is called the coefficient of friction, usually written as μ.

Several different processes can contribute to the friction between two surfaces, but for most materials the only important process is adhesion. The surfaces of most objects are rough on a microscopic scale, and the load between two surfaces is supported only at the points where the peaks of the asperities on one surface are in contact with those on the other. The real area of contact between the surfaces is therefore extremely small. As a result the pressures generated at the contacts are very high, often exceeding the yield stress of one or both of the contact materials. The contacting asperities therefore deform elastically and plastically, thus increasing the area of contact. In addition the high pressures cause significant adhesion at the contacts. When a lateral force is applied to one of the bodies in contact, there is a resistance to sliding as a result of the adhesion at the contact points, and this resistance is the adhesive friction.

Where one or both of the contacting surfaces becomes permanently deformed during sliding, the energy required to produce the deformation represents an additional component of the friction force. For engineering surfaces the amount of permanent deformation which can be tolerated is very limited so that the deformation friction is small in comparison with the adhesive friction. Elastic deformation only makes a significant contribution to the total friction when there is a high level of hysteresis in the elastic recovery, such as in vehicle tyres, and this is not normally a consideration when molybdenum disulphide is used. For practical purposes it can therefore be assumed that adhesive friction is the only type of friction which needs to be considered.

The frictional force F between two solids is approximately equal[103] to the product of the critical shear stress of the softer solid and the real area of contact A_n. In the absence of tangential forces the real area of contact with a ductile material is equal to the normal force W divided by the yield stress p. When a tangential force is applied, the force to be supported at the contacts is the resultant of the normal and tangential forces, and the real area of contact increases. This is known as the phenomenon of junction growth. The magnitude of the increase is commonly small, but with clean ductile materials junction growth can in theory continue until the real area of contact becomes equal to the apparent area of contact. In practice junction growth is much more limited because the shear strength at the interface is reduced by such factors as embrittlement and the presence of contaminants, and the limiting real area of contact A_l will often be between $2A_n$ and $6A_n$.

The yield stress p or hardness is typically about five times the critical shear stress S. The coefficient of friction will therefore be given by

$$\mu = \frac{F}{W} = \frac{A_l S}{A_n p} = 0.4 \ to \ 1.2 \qquad \text{Equation 1}$$

The above discussion is relevant to isotropic bulk materials. Where a thin film is deposited on a harder substrate, there is a general tendency for the area of contact to be determined by the yield stress (or approximately by the hardness) of the substrate, while the shear stress is determined by the surface film. In the ideal case

$$\mu = \frac{A_l S_f}{A_n p_s} \qquad \text{Equation 2}$$

where S_f is the shear strength of the film material and p_s is the yield stress of the substrate. If we consider a hypothetical system in which the hardness of the film material is one tenth of that of the substrate, then combining Equations 1 and 2 we will have

$$\mu = 0.04 \text{ to } 0.12$$

and this is the general basis for the use of soft films for lubrication.

The ideal situation represented by Equation 2 will be modified in practice depending on the thickness of the film and the magnitude of the contact load. At low loads or high film thicknesses, the friction will increase because an increasing proportion of the load will be carried by the softer film material, and the real area of contact will increase. Conversely, at high loads or low film thicknesses, the friction will increase because an increasing proportion of the asperities on the harder substrate will interact. The result is shown in Figure 5.1, which shows the variation of friction with film thickness for a steel rider sliding against a tool steel substrate coated with a film of indium. The system shows a minimum coefficient of friction of 0.075 at a film thickness of 0.7μm.

This discussion assumes uniform pressure distribution over the whole apparent area of contact. In the case of elastic contact between non-conformal surfaces, the contact pressure varies over the apparent contact area in accordance with a Herzian pressure distribution. For an elastic contact between a spherical surface and a flat surface the relationship becomes:

$$\mu = \frac{S}{W^{\frac{1}{3}}}\left[\frac{3R}{4E}\right]^{\frac{2}{3}}$$

where R is the radius of the sphere and E the combined elastic modulus. The effect in both situations is that the friction coefficient decreases with increasing contact load, and is proportional to the shear stress of the film material. With anisotropic materials such as molybdenum disulphide the situation is further affected by the orientation of the material.

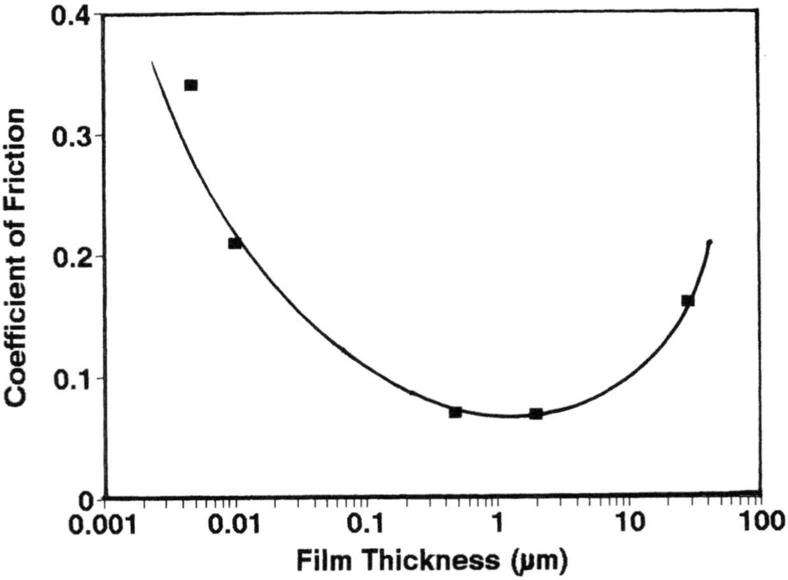

Figure 5.1 Variation of Friction with Film Thickness for a Coating of Indium on Steel (Ref.103)

5.2 FRICTION OF MOLYBDENUM DISULPHIDE

Molybdenum disulphide adheres readily to most substrates. As a result, when sliding takes place between molybdenum disulphide and a solid surface, the phenomena of adhesion and possibly junction growth will take place, and high frictional forces will be generated. This adhesion will be augmented by the action of burnishing (see Chapter 6). However, while adhesive forces between molybdenum disulphide and solid substrates are usually high, the cohesive forces between lamellae of molybdenum disulphide are low. It follows that the coefficient of friction between lamellae will be much lower than that between a lamella and a ductile substrate, and slip will take place preferentially between lamellae.

The same is not true of all layer-lattice materials[104]. In some the bond energies between layers are very high, for example ionic bonding in the case of mica and Π-bonding in the case of graphite. For these it is only when the bond energies are reduced that the shear strength and therefore the coefficient of friction are low. Bond energies can be reduced by the presence of contaminants such as water, or by the intercalation of other atoms or groups into the crystal structure.

The low inter-lamellar attractive forces in molybdenum disulphide consist only of weak Van der Waals forces. In addition the separation distance between the sulphur layers of adjacent lamellae is 3.49Å, and is larger than the 3.17Å thickness of an individual lamella. Cleavage or shear of molybdenum disulphide crystals between adjacent lamellae is therefore inherently likely to be easy.

However, Jamison[67,105] has intensively studied the relationship between the crystal and electronic structures of layer-lattice solid lubricants and their frictional properties, and has shown that other aspects of its electron distribution give a particularly favourable structure to molybdenum disulphide. In its structure the molybdenum atoms in one layer do not lie directly above or below the molybdenum atoms in an adjacent layer, but are opposite holes in that layer. The sulphur atoms are directly opposite other sulphur atoms, but do not have any unpaired electrons to provide strong bonding. It is this lack of electronic interactions which leads to the high interlamellar spacing, and low interlamellar attraction.

With some of the other layer-lattice solid lubricants, the natural electronic structure does not provide the same benefit but a favourable structure can be brought about by intercalation of metallic atoms into their crystal structures, and this is described in more detail in Chapter 14.

There has been some discussion over many years as to whether cleavage or shear is the dominant mechanism which results in low friction in the sliding of molybdenum disulphide. The argument in favour of a cleavage-dominated mechanism is in fact not easy to understand. For cleavage to take place, a component of force at right angles to the basal planes of the crystals would have to arise. It is now well established that the lowest coefficients of friction with molybdenum disulphide occur when both surfaces consist of fully-oriented basal planes and sliding takes place parallel to the basal planes. While slight deviations from pure parallel sliding may arise, the resulting normal stress component would be very small compared with any practical applied load, and the applied load would inhibit any tendency for cleavage to occur. Experimental evidence also generally supports the view that shear is the important mechanism[106].

5.3 EFFECT OF CONTACT LOAD ON FRICTION

A number of investigations have been made into the influence of contact load on the frictional properties of molybdenum disulphide. Puchkov and Pashkov[107] used a technique which they claimed to differentiate between shear stress and surface friction. They studied the effect of varying compressive stress on the resistance to shear in an epoxy-bonded film. The results, in Figure 5.2, show that there is a linear relationship between compressive stress and shear stress, but that there is a finite shear stress in the uncompressed state. As a result, the coefficient of friction decreases as the contact pressure increases as shown in Figure 5.3.

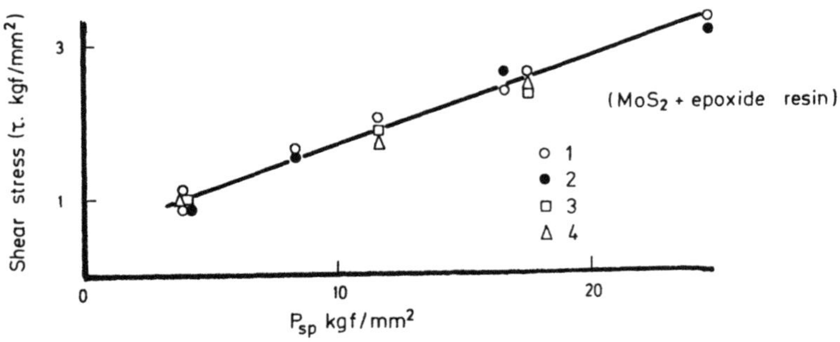

Figure 5.2 Change of Shear Stress with Load for Bonded Molybdenum Disulphide Film (Ref.107)

The form of this relationship follows that established more generally for inorganic compounds by Bridgeman[108] in 1936. Briscoe and Smith found a similar linear relationship[109] between shear strength and contact pressure for unbonded molybdenum disulphide, and their results provide some evidence that those of Puchkov and Pashkov were not due in any way to the presence of the epoxy binder.

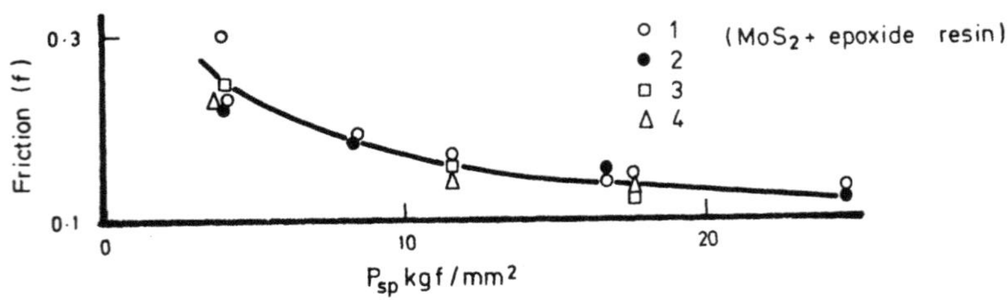

Figure 5.3 Change of Friction with Load for Bonded Molybdenum Disulphide Film (Ref.107)

Akaoka and Nitanai[110] carried out similar studies, but used a thicker film and obtained much lower shear strengths. They found a non-linear relationship

$$\tau = aP^{b}_{mean}$$

but the spread of their experimental results prevents reliable determination of a best-fit equation.

The equation established by Bridgeman was

$$S = S_0 + \alpha P$$

where S is the shear stress at pressure P, S_0 the shear stress at zero applied pressure, and α a constant. On the assumption that the friction is determined by easy shear, and dividing throughout by P, the coefficient of friction μ is given by

$$\mu = \frac{S_0}{P} + \alpha$$

This exactly follows the relationship found experimentally by several investigators, and provides strong support for the generally-accepted view that the low friction of molybdenum disulphide is due to easy shear between lamellae.

The discussion so far has considered only the problem of slip taking place between adjacent lamellae in a single crystal, or in some other form in which the interaction between lamellae simulates their behaviour in a single crystal. The same arguments will not necessarily apply to the practical case of sliding which takes place between two components lubricated with molybdenum disulphide.

The general subject of film formation is considered in the next chapter, but at this point it will be useful to mention a few aspects of film behaviour in order to clarify the nature of friction between lubricated components. In the first place, if two surfaces slide against one another with only free molybdenum disulphide powder present as a lubricant, then initially the coefficient of friction is quite high. It is only when a smooth adherent film has formed on at least one of the surfaces that lower friction occurs. In the second place, if a smooth adherent film of molybdenum disulphide is present on only one of the surfaces, then the lowest possible friction will still not be obtained. It is only when a useful film is also present on the second surface, either formed in advance or formed by transfer from the film on the first surface, that the lowest values of friction will be found.

In other words, effective lubrication by molybdenum disulphide requires the presence of a smooth adherent film on both interacting surfaces. The practical situation of interest is therefore that of two effective lubricating films of molybdenum disulphide sliding against one another. In such a situation any significant degree of slip within one of the coatings (intracrystalline slip) would inevitably lead to depletion of the film, and a limited service life. It is much more likely that slip will take place between the surfaces of the two films (interfacial or interfilm slip.)

As will be explained later, it is considered that the surface of such a film normally consists of a thin layer of fully-ordered crystalline material with the basal planes oriented parallel to the plane of the substrate surface. Conformal contact between two such films will then be similar to the contact between two adjacent lamellae within a crystal. As a first approximation it might therefore be assumed that interfacial slip will resemble intracrystalline slip. However each surface may be degraded by the presence of contaminants, surface defects, and deviations from planarity, and it cannot be assumed that interfacial friction will be completely governed by the same considerations as intracrystalline friction.

A study carried out by Masao Uemura and colleagues[111] was specifically designed to distinguish between the occurrence of cleavage, shear and interfacial slip (which they referred to as intercrystalline slip.) They concluded that cleavage took place when surface material was not fully oriented parallel to the basal planes, and that the friction coefficient was then of the order of 0.1. When shear was taking place the coefficient of friction was about 0.06, whereas when interfacial slip between fully basal-plane oriented surfaces was occurring the coefficient of friction was as low as 0.025.

This result is interesting and unexpected, since it suggests that inter-lamellar friction within a crystal is higher due to some form of bonding, and that this bonding is reduced by the presence of some contaminant material when the lamellae are at the surfaces. Most evidence suggests that the inter-lamellar bonding within a crystal is at an irreducible minimum and is increased by any known contaminants.

A very interesting approach to establishing the nature of interfacial friction was taken by Kanakia and Peterson[112], who assembled data from a number of sources and plotted the results as a single graph of coefficient of friction against contact pressure. This is reproduced as Figure 5.4, except that only their points representing molybdenum disulphide films have been included. The dashed lines are based on a theoretical analysis of shear behaviour, and the authors interpret the results as showing:-

(i) In the horizontal part of the curve, that interfacial slip is taking place and the load is being carried on the asperities of the films at low pressures. The real area of contact is directly proportional to the apparent contact pressure, so that the coefficient of friction is constant.

(ii) The pressure P* represents the film surface hardness, so that above this pressure surface contact is complete. Slip then takes place between fully conformal lamellae, whether interfacially or within a film, and the friction follows the form of Bridgeman's equation.

A further implication is that the coefficient of friction of a fully-burnished film is probably determined more by the burnishing pressure than by the subsequent operating pressure, since the films represented by the horizontal line apparently have similar surfaces while the operating pressures vary from 1.5 to 1000MPa.

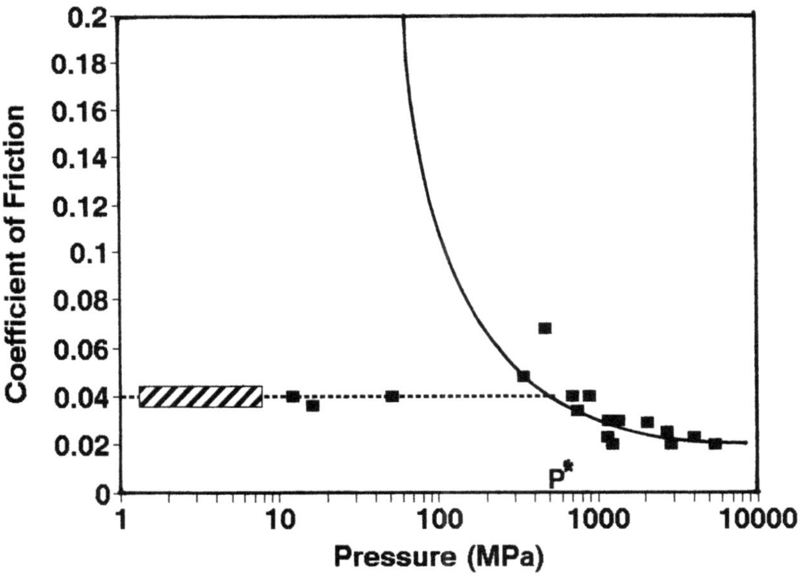

Figure 5.4 Friction of Molybdenum Disulphide Films Over a Wide Range of Pressures (Ref.112) (Data from several different sources)

If this interpretation is correct, then the figure provides strong support for the argument that interfacial friction between molybdenum disulphide films and intracrystalline friction are both determined by the same factors. This is because at the point of intersection between the horizontal and decreasing portions of the curve, where interfacial friction changes to intracrystalline friction, there is no discontinuity, and the same friction value is given for both types of slip.

It can therefore be accepted that whether it is in the natural crystalline form or in a consolidated film, the friction of molybdenum disulphide is adhesive friction, and its low magnitude is due to the easy shear between adjacent lamellae which is made possible by the unusually favourable crystal and electronic structure. However, it must be remembered that the correct orientation of the crystallites is essential for the maintenance of low friction. The shear strength is low only parallel to the basal plane of the lamellae. In other directions the shear strength is high, so that the coefficient of friction will also be high[113].

5.4 EFFECTS OF VAPOURS AND OTHER CONTAMINANTS

The work of Jamison has been referred to earlier[67,105], in which he showed how the electronic structure of the hexagonal molybdenum disulphide crystal is uniquely favourable for producing low sliding friction. Any contamination by vapours or reagents is therefore inherently likely to affect this favourable structure adversely rather than beneficially, and to produce an increase in friction rather than a decrease. This has in fact been shown experimentally to be the case, and a wide range of publications over many years reported increases in sliding friction parallel to the basal planes of the crystallites, resulting from the presence of moisture[114] or other vapours and liquids[115]. Conversely, it was found that in pure dry air or in vacuum[116] the friction was very low.

By the nineteen sixties there was therefore general agreement that the sliding friction of molybdenum disulphide is an inherent property which does not depend on the presence of vapours or other contaminants. As a result there was considerable surprise when in 1976 a group under Matsunaga published a paper[117] showing that the friction of clean molybdenum disulphide was reduced by the presence of a variety of organic vapours. This group carried out an intensive re-examination of the general frictional behaviour of molybdenum disulphide and the accepted friction theory between 1974 and 1982. These studies began with an investigation of the so-called "stop time effect".

After a shut-down in high vacuum of 10^{-6} Torr or better it was sometimes found that there is a brief increase in the friction. This disappears after a short period of operation[116]. In other cases there may be a decrease in friction during a shut-down. The effect is known as the "stop time effect". Matsunaga's first investigation confirmed that the presence of contaminants was involved in this phenomenon, and showed that the friction increase on re-starting could be described by an equation based on a simple model of contaminant diffusion within the lubricant film.

Further investigations[118-120] confirmed that with a variety of contaminants and with several different types of molybdenum disulphide film the presence of contaminant caused a decrease in friction and its removal an increase in friction. The contaminants, apart from water vapour, were all organic, and included propane, butanol, propionic acid, caprylic acid, stearic acid, propyl chloride, n-amyl chloride, a hexyl chloride and n-amylamine.

The films used consisted of fine particles deposited by sputtering, by electrophoretic deposition, or by flotation onto a plate from the surface of a liquid. However, all of these techniques will give a randomly-oriented deposit, the use of fine powder gives a high proportion of crystal edges, and the films were not burnished or consolidated in any way. It follows that the sliding tests involved a high proportion of crystal edges, and this is supported by the high measured coefficients of friction, which were sometimes as high as 0.18. The group eventually concluded[78] that the beneficial effects of contaminants in reducing friction are restricted to the crystal edges. The low friction of clean cleavage faces was confirmed as an inherent property which is not improved by the presence of any contaminants which have yet been studied.

The mechanism by which water vapour increases the coefficient of friction has not been established. The effect can arise with well run-in and burnished films in which the exposed surfaces consist for practical purposes entirely of crystallite basal planes, and can typically result in an increase in the coefficient of friction from 0.05 to 0.15. Lancaster[99A] pointed out that the higher friction is comparable with that which occurs between a molybdenum disulphide film and a metal substrate during the initial formation of a transferred film. He therefore inferred that the increased friction on exposure to moisture must be due to the replacement of interfacial sliding by subsurface shear. He postulated that this could only be due to one of the following mechanisms:-

(1) Vapour penetrating the porosity within the films, leading to reduced adhesion (to the metal substrate), film disruption and a greater proportion of metal-to-metal contact.
(2) Adsorption of vapour onto the (surface) basal planes, increasing the adhesion between them to a level exceeding the subsurface shear strength.

He went on to point out that both mechanisms are inconsistent with other evidence, such as (1) that the increase in friction on vapour admission is virtually instantaneous, and occurs with compacts in the same way as with films on metals, and (2) that vapours adsorb more readily on crystallite edges than on basal planes.

In fact, the first of Lancaster's suggested mechanisms does not in itself provide a sufficient explanation for the increase in friction. Unless the limiting interfacial shear stress increases, any reduction in shear stress or adhesion within the subsurface regions can only result in a reduction in friction. It follows that the presence of water vapour or other contaminants must lead to an increase in the interfacial shear stress. As is shown elsewhere, the interface between fully oriented basal planes in a contact is virtually indistinguishable from the interface between adjacent lamellae within a crystal. It therefore appears incontrovertible that water and many other contaminants alter the bond energies between the crystal lamellae in such a way as to increase interlamellar attraction.

Once the limiting interfacial shear stress has increased, shear will subsequently take place at the point at which the limiting shear stress is first exceeded, and the friction will be determined by the value of the limiting shear stress at that point. This may well be influenced by the first of Lancaster's two mechanisms, or by any one or more of the many other mechanisms which have been proposed, and which will be discussed in more detail in Chapter 7.

5.5 LOAD-CARRYING CAPACITY

Apart from its low-friction properties, the other attribute of molybdenum disulphide which is important in lubrication is its very high load-carrying capacity. Having said that, it is then impossible to give a specific value for the load-carrying capacity, because it depends entirely on the form and conditions in which it is used.

The most dramatic demonstrations of this property are in testing a grease containing 35% or more of molybdenum disulphide in a Seta-Shell Four-Ball Test (IP239) or the Falex Test (ASTM D-3233). In neither case can a weld be produced. In the Four-Ball Test the steel top ball will ultimately be extruded through the gap between the three bottom balls. In the Falex test the steel journal will ultimately be reduced in diameter and extruded from the ends of the V-blocks. If we assume a yield stress for the steel of about 700 MPa, this gives some indication of the load-carrying capacity of the molybdenum disulphide grease. However, other evidence suggests that the load-carrying capacity would be higher in the absence of the grease.

This high load-carrying capacity is a result of three separate properties of the lubricant. The high structural strength normal to the plane of the lamellae resists collapse under high load. The strong adhesion to steel, and to many other metals, resists removal under shear. The low friction reduces frictional heating, and thus reduces the tendency for the metal surfaces to soften and weld together.

5.6 SUMMARY

It is now generally accepted that the very low sliding friction of molybdenum disulphide is due to the very low shear strength parallel to the basal plane of the crystal lamellae, compared with the high strength or hardness perpendicular to the basal plane. The low shear strength is caused by the wide separation distance between adjacent lamellae, which is related to the low inter-lamellar bond strength. This is in turn caused by the uniquely-favourable electron distribution in the hexagonal molybdenum disulphide crystals. These properties are inherent in the molybdenum disulphide structure, and are not improved by the presence of any contaminant investigated so far.

The actual coefficient of friction of a molybdenum disulphide film will depend on the integrity of the film, contact pressure, temperature, humidity, film thickness and presence of contaminants. For a pure, smooth, dense, properly-oriented film at high contact pressure in a clean, dry atmosphere in unidirectional sliding, coefficients of friction as low as 0.02 have been reported. With impurities, poor orientation, humidity and low pressure, the coefficient of friction may be as high as 0.3.

~ ~ ~ ~ ~ ~ ~ ~ ~ ~ ~ ~ ~ ~ ~ ~

CHAPTER 6.

FORMATION OF MOLYBDENUM DISULPHIDE FILMS FILMS

6.1 FILM FORMATION

For effective solid lubrication, it is not enough to have a material with low internal or external friction. It is also necessary for it to form films with sufficient adhesion to a substrate, and internal cohesion, to withstand rubbing under high loads. Molybdenum disulphide has this ability to a very high degree. It can be made to adhere readily and firmly to a substrate, forming a strong, cohesive film.

Because of this ready adherence to a substrate, molybdenum disulphide films can be produced in a wide variety of different ways, including flotation from the surface of a liquid, spraying, brushing or dipping in a volatile dispersant, bonding with adhesive or polymeric compounds, rubbing with powder, transfer, and vacuum sputtering. The nature of the initial film produced depends on the way in which it is applied, and all the important types will be discussed in subsequent chapters.

However, the films can be broadly divided into two types. The harder types, including burnished and sputtered films and some bonded films, are not significantly altered during their service life except by wear and oxidation. The other types, including those produced by application of a dispersion, and many of the bonded films, are initially softer and must undergo consolidation by means of a running-in or burnishing process in order to attain adequate film integrity and low friction.

This chapter will be mainly concerned with the type of consolidated film produced by burnishing or running-in, and consisting mainly or entirely of molybdenum disulphide. The processes occurring in the production of such films have been most clearly established for the case of burnished films produced directly from powder, and that case will be described first.

6.2 BURNISHED FILMS FROM POWDER

In general use the word "burnish" means to produce a shiny or glossy surface on a material by rubbing. This may be achieved by a variety of techniques from the use of a soft pad to the use of a hard burnishing tool, depending on the nature and material of the surface.

Figure 6.1 Machine Used to Apply Burnished Coatings to Rings or Cylinders (Ref.121)

In the simplest case, burnished molybdenum disulphide films are produced by applying a smooth sliding pressure to molybdenum disulphide powder against the hard surface which is to be coated. This can be done by means of a pad of soft material such as fabric or cotton waste under hand pressure, but a variety of mechanical devices has been used in order to produce more consistent films. Figure 6.1 shows an example of a device used to produce burnished films on rings or cylinders[121].

It is obvious from simple geometrical considerations that in any process for applying molybdenum disulphide to a solid substrate, the first contact is likely to be at the peaks of the asperities on the substrate. However, Johnston and Moore[122] were the first to study the burnishing process in detail, using a cylinder covered with fabric to apply molybdenum disulphide powder to a flat copper substrate. They found that in their tests the first hundred traverses of the burnishing device filled the low spots on the substrate so as to produce a smooth surface. Subsequent traverses built up further layers of molybdenum disulphide onto the film, and the film thickness appeared to increase indefinitely without any significant subsequent change in the texture of the surface.

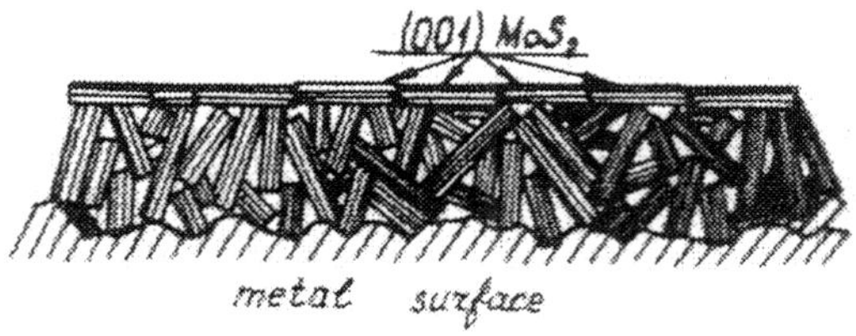

Figure 6.2 Structure of a Burnished Molybdenum Disulphide Film (Ref.112)

Bartz and Müller[123] also concluded that during the early stages of film formation under low load the crystallites first fill the low spots in the surface texture. The result of this initial infilling of low spots is that the surface finish of the film is generally much smoother than that of the substrate. Once this smoother film has been formed with the basal planes of the upper crystallites more or less parallel to the mean orientation of the substrate surface, further crystallites continue to add with their basal planes in the same orientation[124,125]. Brudnyi and Karmadonov[126] showed by X-Ray diffraction of films on copper that the surface consisted of an apparent single-crystal layer 2 to 5μm thick with the basal planes oriented parallel to the sliding surface, but this highly-oriented layer was on top of a randomly-oriented layer, as shown in Figure 6.2.

It is now generally accepted that effective lubricating films of molybdenum disulphide, after burnishing or running-in, have the type of structure shown in Figure 6.2, although the thickness of the randomly-oriented layer will vary depending on the way in which the film was produced. At its lowest level, the random layer may be only a few nanometres thick, as can be seen in High-Resolution Transmission Electon

Micrographs obtained by Takahashi and Kashiwaya[127] of films produced by transfer from bulk solid.

There are believed to be three different mechanisms by which the first layer of crystallites becomes attached, although none of the three has been proved beyond doubt to take place, nor their relative importance.

The first of these mechanisms is a simple infilling of low spots on the surface. There is no doubt that such an infilling takes place, as shown by the work of Johnston and Moore[122] and Bartz and Muller[123] mentioned previously. Further support for this process is provided by the fact that optimum film formation is strongly influenced by the surface texture of the substrate. It is difficult to accept that this geometrical process alone can account for any effective attachment of crystals, since loose molybdenum disulphide powder applied to a machined surface shows little tendency to form an attached film unless some pressure is applied to it. Compression of powder into a low spot would almost certainly be needed to interlock particles with each other and with the slopes of the cavity, and this interlocking could be expected to be retained when the applied pressure is removed.

The second mechanism believed to occur is embedment of crystallites in the surface[128-130]. Molybdenum disulphide is highly anisotropic, and although the mean hardness is only about 1 to 1.5 on the Moh scale, the crystallite edges can be as high as 8 Mohs, roughly equivalent to 1000VPN. This high edge hardness may encourage adhesion in two ways, first by direct embedment of crystallite edges in a softer substrate, and second by abrasively producing scratches which form keying sites for attachment of crystallites.

Bowden and Tabor[103] have described two phenomena which may have important influences on this embedment process. They showed that when a copper slider passed over a harder steel surface, fragments of copper adhered to the steel. In addition the steel surface was ruptured and copper penetrated the fissures created. In view of the very high hardness of the crystal edges in molybdenum disulphide, such a rupturing and penetration process seems even more likely to take place than in the case of copper on steel.

Secondly, they showed that under the high hydrostatic pressures generated in contact of a crystalline material (rock-salt) and a steel slider the rock-salt ceases to be brittle, and can undergo marked plastic deformation. Such plastic deformation in the case of molybdenum disulphide would facilitate the formation of intimate penetration into fissures created in a substrate surface. Furthermore, during plastic deformation cracks in the crystalline material will heal, creating strong cohesion.

The third proposed mechanism is chemical bonding between molybdenum disulphide and the material of the surface. Both chemical and hardness effects will be discussed in more detail later, but at this stage it is important to note that both crystal hardness and chemical reactivity are greatest at crystallite edges, and relatively low on the basal planes.

Whichever of these three mechanisms is most significant in the initial attachment of molybdenum disulphide to the substrate, the result will be a layer of crystals which are either randomly oriented (with simple infilling) or preferentially oriented at a relatively high angle to the surface (with embedding.)

As the burnishing process continues, further crystals will tend to attach to those in the initial film on the surface. They will attach preferentially at their edges, experiencing high adhesive forces due to the hardness, abrasiveness and high free energy at those edge sites. Silin and Aparin pointed out[131] that they will then experience a couple which will tend to rotate them until they become oriented parallel to the plane of sliding. At this stage the adhesive forces abruptly diminish, and the tendency to rotate ceases. As a result, crystallites which have attained this parallel orientation will remain in position, and the result is a fully oriented surface.

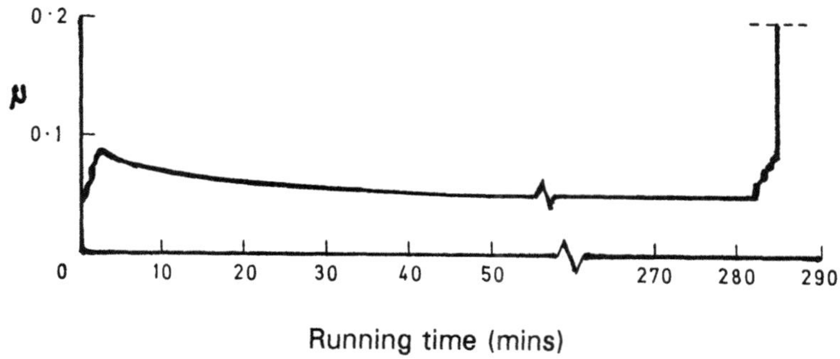

Figure 6.3 Change in Friction with Running Time for a Rubbed Film of Molybdenum Disulphide (Ref.132)

Kinner, Pippett and Anderson monitored the change in friction during the running-in process[132] and a typical graph of friction against time is shown in Figure 6.3. The initial increase in friction to a peak was found to be characteristic, and may represent the energy required in re-orienting the crystallites, which is likely to involve crystal fracture and the making and breaking of large numbers of edge-site junctions.

Where the surface of the film is not too highly burnished (that is, not highly reflective) it is possible to continue to add to the depth and density of the film by adding more powder and continuing to rub or burnish it onto the existing film[122,133]. Where films have been burnished to a high level of reflectivity, it may be difficult to add to the film thickness[134], and Finkin[135] reported difficulty in resupplying a film by means of loose powder. It seems probable that where the film is not too highly burnished, there are stacking defects and discontinuities present which leave edge sites available for further attachment, and this was presumably the case in the work of Johnston and Moore[122]. With a highly reflective film, as will be shown later, only crystal basal planes are exposed, which presumably provide no attachment points.

In any case, there seems to be little or no advantage in producing burnished films thicker than about 20μm.

6.3 BURNISHING OF SOFT FILMS

Loose molybdenum disulphide powder has only a limited tendency to adhere to solid surfaces. Very fine powders will attach loosely to metal surfaces, but the coarser grades commonly used for lubrication will not. However, there are several techniques which will produce soft adherent coatings.

The flotation process used by Matsunaga[118] and Tsuya[136] involves floating fine molybdenum disulphide powder on the surface of a liquid, and lifting it off onto the surface of a flat metal plate. After the liquid is removed by draining and drying, a weakly-adherent thin uniform film of the powder is left on the metal surface. The film appears to consist of randomly-oriented crystals, and has been extensively used for research purposes, but not for use in practical machinery.

Much thicker films can be produced by the use of dispersions of molybdenum disulphide powder in volatile liquids. The dispersion can be applied to a solid surface by dipping, brushing or spraying, and the liquid is then allowed to evaporate, either at room temperature or with additional heating. Although the dry films are much thicker and more strongly adherent than those produced by the flotation process, they are still soft and the crystals are randomly oriented. The subject of dispersions is described more fully in Chapter 9.

Even more strongly adherent films can be formed by the use of bonding agents. A wide variety of bonding agents has been investigated and many of them have been sold commercially and used in service equipment. The subject of bonded coatings is described in some detail in Chapter 11. At this stage the important factors are that the molybdenum disulphide in a bonded film as applied is randomly oriented, and that the films vary considerably in hardness. The softest films, even when heat-cured, are soft enough to be scraped off with a finger-nail, whereas the hardest will retain their integrity even under the highest service loads.

All the soft, randomly-oriented films, whether produced by flotation, from dispersions, or with bonding agents, initially show high friction and wear. In order for such films to give satisfactory low friction and wear in service, the initial period of operation must permit satisfactory running-in ("breaking-in") during which the film is consolidated and oriented so as to improve the friction, load-carrying capacity and wear rate. Alternatively the film can be burnished before use. This has important practical advantages in ensuring efficient consistent operation from the beginning of service, and has therefore been studied by several workers.

The soft films, whether bonded, floated or deposited from dispersions, obviously differ in one important respect from those produced by burnishing of loose powder. That is, that infilling of low spots in the surface texture clearly takes place during the initial film formation. On the other hand it is far less likely that embedding and chemical bonding play a significant part in the initial film formation, and the extent to which they occur during burnishing will depend critically on the way in which stress is applied to the film and transmitted through it to the substrate. With bonded films the bonding agent is of course assisting the adhesion of the film to the surface, and even with films applied in dispersions or by flotation there appears to be some factor improving the adhesion and cohesion of the films.

Three separate processes will take place during running-in or burnishing of a soft film. These processes are compression, shear and crystallite re-orientation, and the nature of the burnished film will depend on the relative extents of these processes.

The coefficient of friction of the randomly-oriented film is relatively high, possibly as high as 0.3 under light loads, so that in the first stages of the running-in process a high shear stress is applied to the soft film. The limiting shear stress of a soft film is correspondingly low, so that it is easy in the early stages for a high proportion of the film to be removed.

The compressive yield stress is also low, so that pressure applied normal to the surface results in compression and densification of the film, with a resulting increase in the limiting shear stress and the compressive yield stress. Burnishing without excessive loss of film material is therefore best achieved if compression takes place simultaneously with sliding, in other words if the counterface is curved or slightly inclined to the film surface.

Re-orientation of surface crystallites, and transfer to the counterface, take place quickly under the influence of sliding so that the coefficient of friction and the shear stress decrease. At the same time the compression of the film under the normal component of the applied stress forces the film material into the low spots of the substrate surface, and it is at this stage that embedding and chemical bonding are likely to become more significant.

The early stages of running in or burnishing will therefore result in some loss of film material, improved orientation of the surface crystallites with reduced friction, densification of the film and improved film adhesion. Further running-in or an increase in the applied stress will result in further reduction in film thickness, improved adhesion and possibly further reduction in friction.

The final burnished film will have the same general structure as those produced from burnished powder, as shown in Figure 6.2, but the thickness of the randomly-oriented layer will vary more widely. If the original soft film was relatively thick, and the running-in process took place under light loads or was brief, then the surface of the film will not be highly reflective, the friction will be high, the random layer will be thick and attachment to the substrate relatively weak. If the running-in process took place under high loads and was sufficiently prolonged, then the surface will be highly reflective, the friction low, the random layer thin or virtually non-existent, and attachment to the surface strong. However, with a thick initial film and high loads, there is a tendency for the coating to flake. There is an optimum thickness for long life, and this is discussed further in Chapter 7.

6.4 FILM FORMATION BY TRANSFER

Films of molybdenum disulphide can also be formed on solid surfaces by direct transfer from many types of source, including single crystals, composites, and other films. The general subject of transfer is considered in detail in Chapter 8.

In general high contact loads are required for effective transfer, so that crystals can attach firmly enough to the counterface to be detached from the source. The stresses are such that transferred films tend to be strongly attached, and rapidly become consolidated and highly oriented, with low friction and high load-carrying capacity. They are relatively thin with little or no randomly-oriented layer above the peaks of the substrate asperities. They thus resemble films produced by the burnishing of powder, which is in fact a form of transfer.

6.5 STRUCTURE OF BURNISHED OR RUN-IN FILMS

Brudnyi and Karmadonov[126] described the degree of crystal orientation at the surface of a burnished film in terms of a reflection or texture coefficient C, such that

$$C = K\frac{J_1}{J_2}$$

where J_1 is the intensity of interference of MoS_2 (100), J_2 the intensity of interference of MoS_2 (001) and K a proportionality constant. They established the value of K by making C equal to 1 for randomly-oriented crystals and zero for a completely uniform crystal orientation.

The reflection coefficient of a fully burnished film was found to be the same as that of the MoS_2 (001) plane and their reflectivities were also the same. They therefore concluded that the surface of the film consisted of an assembly of (001) basal planes. They also inferred that a highly reflective surface occurs when the coefficient of friction is a minimum.

Yuko Tsuya also used electron diffraction[136] in studying the progress of running-in with an unconsolidated "floated" film initially about 0.5μm thick, using an oscillatory friction tester with a stroke of 20 mm and load 8 kg. She found a gradual transition from a layer of randomly-oriented particles with little adhesion or cohesion to a smooth cohesive layer with the molybdenum disulphide crystallites almost fully oriented with the (001) basal planes parallel to the substrate surface. The coefficient of friction was fairly steady at about 0.1, which is high for a fully oriented surface, and may be associated with the use of a reciprocating tester (see below).

Gamulya and co-workers[137] also concluded on the basis of electron microscopy and micro-electron diffraction that the production of a highly reflective surface occurs when the coefficient of friction reaches a minimum level. The full orientation appears to be limited to a very thin surface layer, which they found to be about 0.1 μm thick in their tests, while Brudnyi and Karmadonov described it as being only one crystal thick, regardless of the force used for burnishing. On the other hand, a fully densified, fully-oriented layer without discontinuities may not be distinguishable in practice from an extended single crystal.

This highly-oriented "surface" film will adhere strongly to a metallic counterface, and will readily shear from the disordered substrate material, thus forming a transfer film on the counterface. A new highly oriented film will then reform rapidly from the disordered subsurface crystallites as sliding continues. The repetition of this process when sliding takes place against a bare metal surface represents continuing depletion and wear of the primary lubricant film. Where the counterface already carries a transfer film, wear may be negligible, and the life of the lubricant film will be determined by other factors.

One interesting aspect of these and other studies which may be worth emphasising is that the same ultimate surface condition may arise with either burnished powder or certain bonded films. It appears[122] that the same orientation mechanism can occur with or without the presence of a binder. On that basis the function of at least the softer binders would be simply to retain the molybdenum disulphide in position while the running-in or burnishing is taking place. This subject is further discussed later.

There are indications that when the molybdenum disulphide film is formed and run in under very high loads (55MN/m^2) the tendency for it to fill the grooves and cavities of the surface is actually reduced. Adhesion then appears to have taken place on the summits of the asperities[138], and although a cohesive film is formed it bridges the low areas between asperities, leaving voids between film and substrate. A possible reason for this bridging phenomenon is that under high running-in loads the asperities are elastically flattened. A cohesive contact film is formed in the usual way, but on elastic recovery of the asperities the film is lifted clear of the low points on the surface.

It has been reported that running-in is ineffective under oscillating sliding motion. A suggested mechanism for this[138] is that during the formation of a lubricating film the lamellae are laid down with their exposed edges all pointing in the direction of counterface movement (Figure 6.4), as with the overlapping of tiles or slates of a roof. Reversal of motion will then cause geometrical interaction between the step edges, tending to disrupt the film and presumably increase friction. If this proposed mechanism is correct it implies that in the formation of burnished films it is important to avoid fully reversed directions of sliding by using a curved sweeping action or a continuous single-direction sliding.

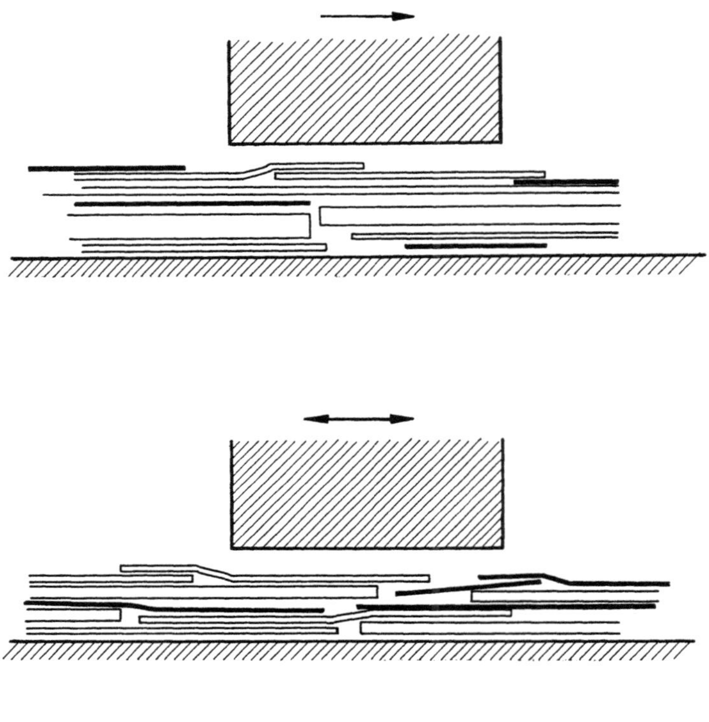

Figure 6.4 Arrangement of the Layers During Rotational (Above) and Oscillational (Below) Movement (Ref.138)

The same proposed phenomenon may be responsible for the higher friction which is reported when a surface is run in under unidirectional sliding and the direction of operation is subsequently reversed. Unless the load or speed are extreme it would seem reasonable to expect that the surfaces will again run themselves in the new sliding direction, but there must inevitably be some reduction in life. In critical applications where this situation can arise it might therefore be useful to carry out tests to assess the effect on wear life.

6.6 EFFECTS OF THE SUBSTRATE ON FILM FORMATION

The importance of substrate hardness in the formation of effective lubricating films has been clearly established[129]. Where the substrate is significantly softer than the molybdenum disulphide (7-8 Mohs at the crystallite edges, or 800-1000 VPN) satisfactory adhesion can be obtained, but for harder materials, such as tool steel, an initial roughening by grinding or grit-blasting helps to provide effective keying.

No information seems to exist about the formation and performance of films on very soft materials, but down to the lowest hardnesses which are of interest for bearing use, i.e. about 200 VPN, the wear life of the molybdenum disulphide tends to improve as substrate hardness decreases, although Tsuya's findings were less clear-cut[136].

The evidence for more specific physical and chemical influences of substrate composition is less clear. Chemically, molybdenum disulphide is very inert. The sulphur atoms which form the surface layer of a lamella are strongly bonded to the molybdenum atoms, and their valency electrons are fully occupied in those bonds. Although molybdenum disulphide is highly polarised in its hexagonal crystals, the free energy at the lamellar surfaces is very small.

Nevertheless, in spite of the very low attractive forces between lamellae, some workers have suggested that they account for the strong adhesion which occurs between molybdenum disulphide and a metal substrate. This argument is unconvincing, because strong adhesion would require fairly accurate spacial matching of the weak surface charge distributions. Such matching does not generally occur, although there is an approximate matching with ferric oxide.

Fleischauer[140] has carried out a detailed analysis of the electronic structure of molybdenum disulphide. This analysis showed that all the accessible orbital electrons for both molybdenum and sulphur are used in intralayer bonding, leaving only high-energy antibonding orbitals available for bonding between layers or for basal surface adhesion to substrates. There are no accessible orbital electrons on either molybdenum or sulphur surface atoms. The lone pairs of sulphur 3s electrons occupy very stable orbitals and cannot interact with external atoms.

He concluded that the undisturbed (001) basal surface of molybdenum disulphide has no ability to form bonds or to react unless its molecular orbital structure is altered by physical or chemical means.

It is of course far more likely that any chemical adhesive effects occur at crystallite edges or at stacking faults or other defects rather than at undisturbed crystal faces. Apart from the mechanical factors involved in embedding of crystallite edges, there is far greater free energy at edge sites. This is a normal consequence of the unbalanced energy distribution which occurs at fractured edges of plate-like crystals. It is also indicated by the high adhesive friction at crystal edges, and its reduction by absorbed contaminant molecules.

Practical metal surfaces, at least those generally used in engineering, are of course almost always oxidised to some extent. Any discussion of chemical effects of the substrate must therefore usually be considered as referring to the oxides on a metal surface rather than to unoxidised or unreacted metal. To clarify this aspect, Stupian and Chase[133] studied the effect of the surface oxide on molybdenum disulphide adhesion, using a series of metal surfaces which had been deliberately oxidised. They found, in confirmation of earlier investigations[141], that the strength of the bonds between the sulphur of the molybdenum disulphide and the metal of the substrate was a major factor in determining the strength of adhesion.

The role of the surface oxide was mainly in influencing the accessibility of the metal to the molybdenum disulphide. With copper, for example, the oxygen is not strongly bonded to the metal and can be displaced. On the other hand, with titanium the oxide is strongly held, but lattice vacancies are present which expose titanium metal to the molybdenum disulphide.

Clearly, any abrasion of surface oxide by the hard crystal edges of the molybdenum disulphide is likely to be particularly important in exposing free metal to reaction with the sulphur atoms. In addition, the depletion of surface oxide in sliding in high vacuum should make it easier for molybdenum disulphide to attach to a worn surface, but the potential of this for re-supply of a molybdenum disulphide film in high vacuum applications has apparently not been studied.

Gänsheimer[142] showed that chemical reactions take place between the sulphur of molybdenum disulphide and metallic surfaces during sliding contact, but established no direct correlation with friction, adhesion or wear life.

Reid and Schey[143] studied the role of substrate composition and other factors in the formation and performance of films on various metal substrates, including copper, aluminium, titanium and mild steel, tested against themselves and against an alloy steel. They used a twist-compression test to assess performance, and concluded that substrate hardness and composition had the greatest influence on film formation and life. They believed that film formation and especially durability are improved by chemical reaction if a substrate, such as copper or iron, has a strong tendency to react to form a sulphide, provided that the reaction kinetics are favourable. However, they found no direct evidence of reaction or of sulphide formation. Their conclusions were based on the fact that the durability of the films was found to be in the sequence aluminium, titanium, iron, copper, which is the same as the order of the free energies of formation of their sulphides.

M J Devine[144] demonstrated a general relationship between the wear life of a molybdenum disulphide film and the chemical composition of a metal substrate. He carried out a detailed study using eighteen different substrate metals with a bonded film consisting of 71% molybdenum disulphide, 7% graphite, and 22% of sodium silicate as the binder. All tests were performed under identical running conditions, and the results are shown in Table 6.1. They show that with a molybdenum substrate the wear life was approximately twice as long as with the best of the other substrates, while the third longest life was for a molybdenum tool steel. The specific benefit of molybdenum as a substrate was confirmed by a test with a sprayed film of molybdenum on steel as the substrate. The bonded film gave the same life on the sprayed molybdenum coating as on the original molybdenum substrates.

Similar tests were then carried out with a variety of metal sulphides on six different metal substrates, and the results are shown in Table 6.2. Again they showed that the lives were consistently better with molybdenum substrates, for all the sulphides tested. This strongly suggested that some chemical interaction was taking place, possibly between free sulphur or sulphur compounds and the molybdenum in the substrate. Where the original sulphide under test was itself molybdenum disulphide, this reaction would represent a re-supply mechanism. Where some other sulphide was under test, the reaction would again enable molybdenum disulphide to be formed which would supplement the life of the original sulphide.

Table 6.1 Wear Life of Different Alloys with a Bonded Molybdenum Disulphide Film (Ref.144)

Pin Designation	Pin Hardness	Major Metal	Alloy Constituents	Running Time (mins)	
				Run 1	Run 2
Titanium Alloy	R_C31	Ti	Al,Va	1	1
Titanium Alloy	R_C29	Ti	Mn	1	1
AISI 302	R_C32	Fe	Cr,Ni	6	5
Inconel X	R_C27	Ni	Cr,Fe,Ti	18	10
Hastelloy C	R_B52	Ni	Mo,Cr,Fe,W	19	16
AISI 3135	R_B78	Fe	Ni	21	18
AISI 440C	R_C57	Fe	Cr	20	36
Tungsten	R_C36	W	-	31	-
AISI 1095	R_B90	Fe	-	29	32
AISI M2	R_B92	Fe	W,MO,Cr,V	28	36
Tenelon	R_C42	Fe	Cr,Mn	37	32
AISI 52100	R_C61	Fe	Cr	38	44
AISI 4130	R_B88	Fe	Cr	47	39
AISI T1	R_C60	Fe	W,Cr,V	47	39
Ta-782	R_B97	Ta	W	53	-
AISI M10	R_C60	Fe	Mo,Cr,V	73	50
Molybdenum	R_B93	Mo	-	160	123
Molybdenum, 0.5Ti,0.08Zr	R_B97	Mo	-	130	114

There was therefore strong, but not conclusive, evidence of some chemical action taking place. This was supported[145] by correlating the reactivity of the various test sulphides with the beneficial effect of a molybdenum substrate. It should also be mentioned that the tests were carried out at high load and speed, where frictional heating would be expected to encourage chemical reaction. In other studies at lower load and speed molybdenum showed no advantage over other substrates[129]. Devine pointed out[146] that there is a high degree of lattice matching between molybdenum disulphide and molybdenum, so that electronic effects could not be ruled out, but that would apparently not explain the good performance of tungsten disulphide or titanium disulphide on the molybdenum substrates. Thus the beneficial effect of a molybdenum substrate on molybdenum disulphide performance has been clearly demonstrated, but the reason is not fully understood.

Table 6.2 Wear Lives in Minutes for Metal Sulphides on Different Metal Substrates (Ref.144)

Metal Sulphide	Molybdenum	Molybdenum .5Ti,.08Zr	SAE M-10	AISI-C3135	Inconel X	Ti-6Z1 -4V
MoS_2	160,123	130,114	73	21	18	1
WS_2	146	-	<1	2	10	1
Ti_2S_3	94	92	<1	<1	<1	<1
Cr_2S_3	51	31	<1	<1	<1	<1
HgS	72	103	<1	<1	<1	<1
ZnS	84,86	-	<1	<1	<1	<1
Fe_2S	50	128	<1	<1	<1	<1
CaS	130	120	-	<1	-	-
FeS	70	64	-	<1	-	-
CdS	6	8	-	<1	-	-
TiS_2	135,120	-	-	<1	-	-
Ag_2S	7,8	-	<1	<1	<1	<1
Bi_2S_3	27	-	<1	<1	<1	<1
Sb_2S_3	30	-	<1	<1	<1	<1

Another chemical aspect which seems to have been clearly established is the generally unsatisfactory performance of molybdenum disulphide on a copper or copper alloy substrate. There is some evidence[147] that sputtered molybdenum disulphide will not adhere properly to copper either in the initial formation of a film or in transfer to a counterface. Reid and Schey[143] also found indications that with their sprayed and burnished films copper became intermingled with the film and transferred to the counterface.

The use of certain chemical conversion treatments on a substrate prior to application of a bonded molybdenum disulphide film can give very considerable improvement in the load-carrying capacity and wear life. Phosphating of a steel substrate[148,149] increased the seizure load and the time to seizure at low load. Similar improvement on steel was given[150] by the commercial Sulfinuz sulphiding process. Although these conversions were developed and proved for bonded coatings, they have also been used for burnished films. Another surface coating specifically recommended[151] for burnished films is flame-sprayed zirconium silicate, but such a high technology substrate seems curious in conjunction with an essentially low-technology process like powder burnishing.

Overall, the evidence for specific chemical effects of the substrate is strong, but limited to only a few substrate materials, such as molybdenum and copper. The effects of the pre-treatments mentioned above have been clearly documented, but it is less clear whether these effects are chemical or physical. Several of them are well known as treatments to improve the running-in of the appropriate metals in oil-lubricated systems by preventing severe adhesive wear and ensuring that surface damage is limited to mild wear. Their mechanism of action is related to the modification of surface texture and the production of surface films which are readily abraded, and not to any particular chemical activity of the surface. It is therefore probable that their action in connection with molybdenum disulphide films is also physical, in improving embedding of crystallites rather than directly encouraging stronger chemical bonding.

The most important aspect of pre-treatment is to provide the optimum surface roughness to give mechanical "keying" of the film to the surface. It is generally accepted that the optimum surface finish is between 0.5 and 2.0μm c.l.a. Most of the published work on surface finish effects has been related to bonded films, and the subject is described in detail in Chapter 11, but similar surface finishes are also desirable for burnished films, especially on hard substrates. However, for sputtered films, which may be only 1μm thick, much smoother substrate surfaces give the maximum life.

It is interesting to note here that the beneficial effect of phosphating on film life is quite small if the substrate has an optimum surface finish of 0.5μm, and this supports the suggestion that the benefit of phosphating is at least partly in providing a suitable surface texture.

6.7 EFFECTS OF MOISTURE AND OTHER VAPOURS ON FILM FORMATION

The presence of moisture during burnishing of powder increases the rate of film formation and the maximum film thickness. It was shown in Chapter 5 that moisture increases the friction in inter-lamellar sliding, but may reduce friction at crystal edges. In view of the mechanism of film formation described in Section 6.2, it seems probable that a reduction in friction at crystallite edges would delay the formation of a fully-ordered film by reducing the rotary moment about crystal edges. Increased inter-lamellar shear strength would tend to cause greater adhesion of fresh lamellae to the material already deposited. Both mechanisms would increase the eventual film thickness, but only the higher interlamellar shear stress would seem likely to increase the rate of deposition, so that this may in fact be the dominant effect of moisture.

There is some evidence that running-in of a film is improved in the presence of moisture, and this may again be due to the effect of increased friction in removing high spots in an unburnished film. This would in fact be a specific case of the general effect of moisture on film wear, which is to increase wear rate[152] and reduce wear life[114]. Both effects are consistent with the effect of moisture in increasing friction, since in general any factor which increases adhesive friction will also tend to increase adhesive wear.

There seems to be little specific information on the effect of other vapours on film formation or running in.

~ ~ ~ ~ ~ ~ ~ ~ ~ ~ ~ ~ ~ ~ ~ ~

CHAPTER 7

PROPERTIES OF MOLYBDENUM DISULPHIDE FILMS

7.1 FRICTION

The frictional properties of molybdenum disulphide films have been discussed in the previous two chapters, and it is not necessary to repeat the same information here. However, before proceeding to discuss the influence of various factors on the magnitude of the friction, it may be worth emphasizing the fact that friction varies with the gaseous environment, humidity, temperature, load, purity and the state of orientation and consolidation of a film. No-one has ever attempted the huge task of carrying out a parametric study of all these factors together, and most of the published work has failed to define one or more of the influential conditions. As a result it is very difficult to establish absolute values of the coefficient of friction in any particular situation.

It also follows that it is impossible in most cases to compare the results published by different investigators with any reliability. Where a report shows a variation in the friction with change in one test condition, the relative values can usually be accepted with a great deal of confidence, but the absolute values will often not agree from one investigation to another. In order to predict the value of the friction in a situation, it is necessary to analyse a variety of reports and infer those conditions which are not completely defined. An attempt is made to do this later in this book, and Table 16.7 lists a range of typical coefficients of friction under different conditions.

In summary, the friction is at its lowest for fully ordered surface films in dry air or vacuum at high load and highest for randomly-oriented films in the presence of water vapour or certain other vapours at low load.

Except for the anomalous low friction reported for an ultra-pure sputtered film (see Section 10.5) and under radiation (see Section 7.4), typical minimum coefficients of friction in dry air or vacuum are 0.02 to 0.03. For fully disordered films in dry air the static coefficient of friction can be between 0.12 and 0.15, but because re-orientation begins rapidly when movement starts any figures for the kinetic friction of disordered films must be suspect.

The effects of moisture and other vapours will be discussed in detail in the next section, but the friction can be more than trebled in the presence of moisture.

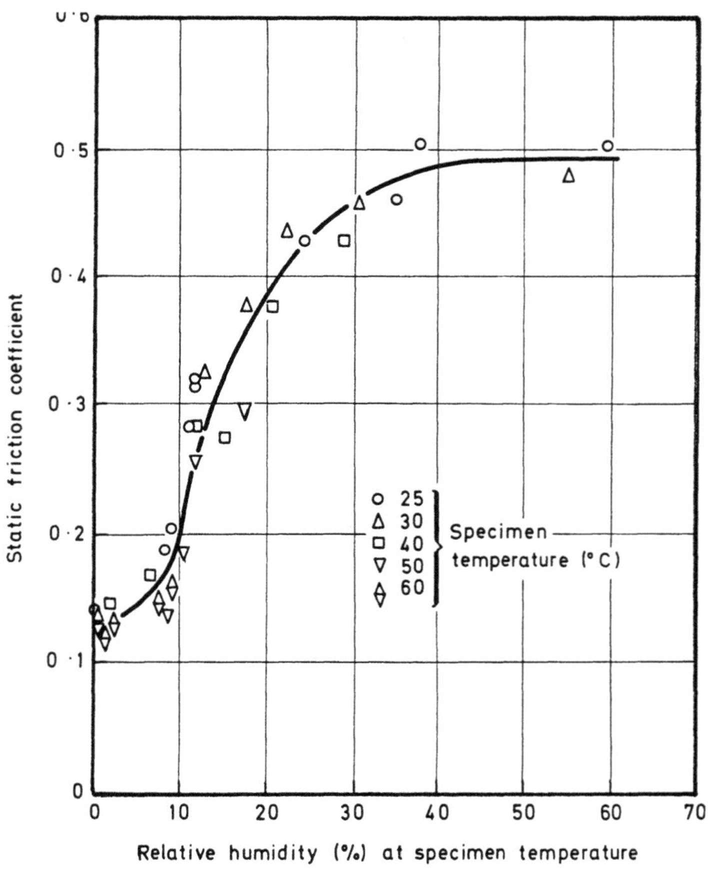

Figure 7.1 Effect of Humidity on the Coefficient of Static Friction of a Rubbed Film of Molybdenum Disulphide (Ref.114)

7.2 EFFECTS OF MOISTURE AND OTHER VAPOURS

In general the effect of moisture on molybdenum disulphide is to increase both friction and wear rate. Figure 7.1 shows the effect of relative humidity on the coefficient of static friction[114] for annular washers coated with an unbonded molybdenum disulphide film. The actual values of the coefficient of friction are high, because the films were unburnished, but the increase from 0.12 to 0.49 is significant.

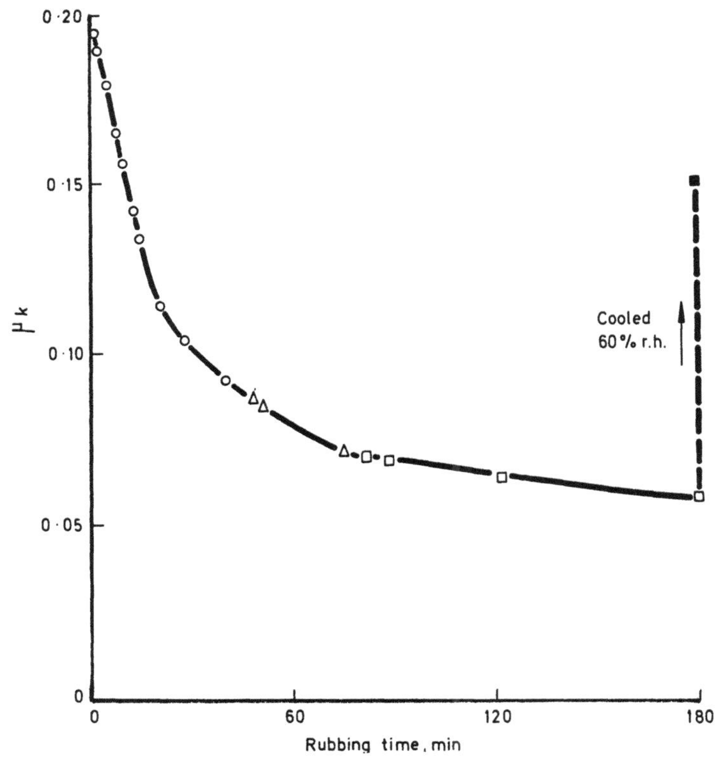

Figure 7.2 Change in Friction of a Rubbed Film of Molybdenum Disulphide with Time of Sliding (Ref.153)

The effect on kinetic friction is less straightforward. At constant humidity there is a tendency for the coefficient of kinetic friction to decrease with time of sliding[153], as shown in Figure 7.2, with speed, and with load[154], as shown in Figure 7.3. All these effects have been attributed to the influence of rubbing in raising the surface temperature and thus reducing the relative humidity. If the relative humidity is changed, there is a delay in reaching a new equilibrium value for the coefficient of kinetic friction[155], as can be seen in Figure 7.4.

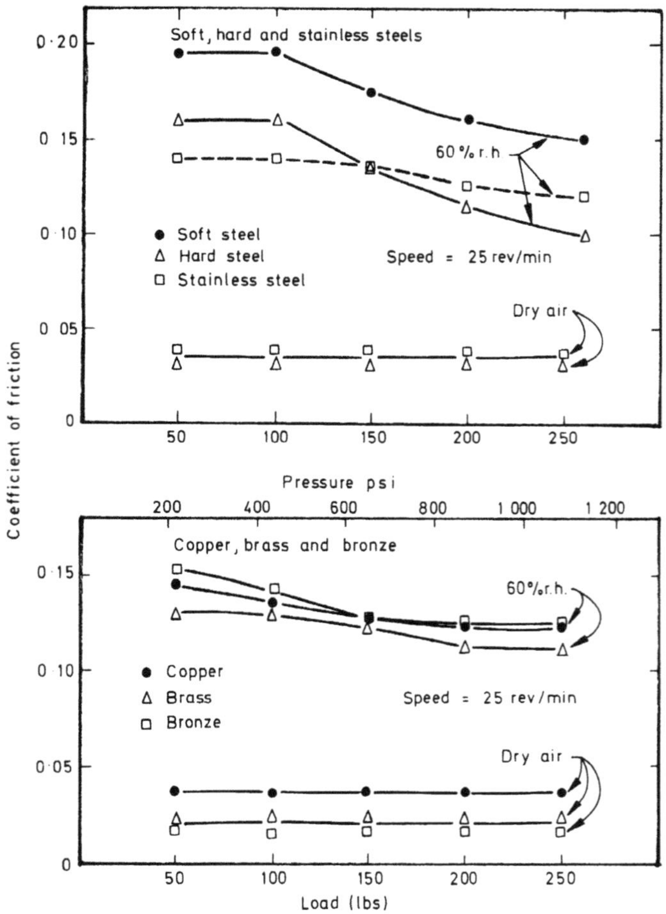

Figure 7.3 Effect of Load on the Dynamic Friction of a Rubbed Film of Molybdenum Disulphide (Ref.154)

The peaks in the curves of Figure 7.4 suggest that at humidities above about 65% the friction begins to decrease. Pardee[156] found some evidence of a similar effect for a bonded film on certain substrates, with a small reduction from the peak friction at very high humidities, above 80%. He suggested that the rise in friction with humidity was due to instantaneous oxidation of molybdenum disulphide at the rubbing surface, with the formation of molybdenum oxides. He also suggested that hydrogen sulphide formed in the oxidation process reacted with some substrate or counterface metals to produce a sulphide film which gave a reduction in friction at the highest humidities.

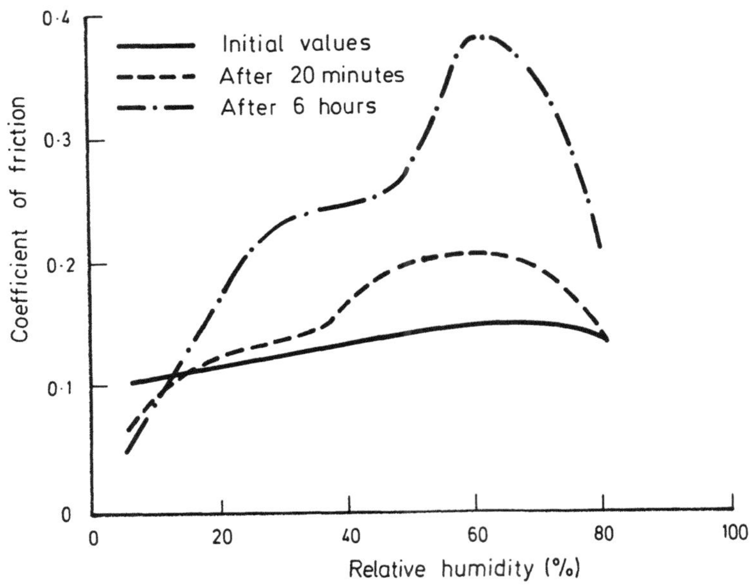

Figure 7.4 Variation of Molybdenum Disulphide Friction with Humidity (Ref.155)

Another theory of the reason for increased friction in the presence of moisture was proposed by Gao et al[157]. They found that in a humid environment molybdenum disulphide films were more readily thinned by sliding contact, 'due to increased ease of interlamellar slip.' They suggested that adsorption of water softened the films, and that resulting increased deformation by plowing in sliding contact led to a poorly oriented film and thus to increased friction. However, they considered that this was a short-term reversible effect which was not in conflict with theories of chemical breakdown. Gao et al also poiinted out the possibility that an increase in friction is caused by capillary pressure effects of moisture at asperity contacts.

The converse of these effects of moisture in increasing friction is that in a very dry atmosphere or in high vacuum, the friction is low. However, it is found that if sliding is stopped in a high-vacuum test, there is a transient increase in friction on restarting the test. This "stop time effect" has been described in Chapter 5. The effect appears to be caused by re-adsorption of water vapour during the shut-down and desorption within a short period of re-starting. It has been shown that at higher vacuum (e.g. 10^{-9}Torr) or after effective baking out[158], the transient friction rise does not occur.

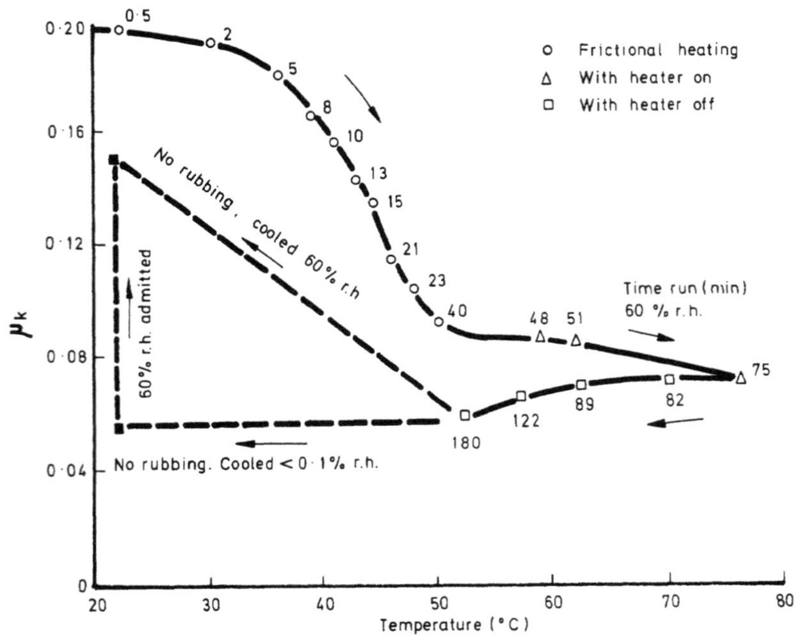

Figure 7.5 Effect of Humidity and Temperature on Friction of a Rubbed Film of Molybdenum Disulphide (Ref.153)

The effect of moisture on the coefficient of friction is generally reversible, as shown in Figure 7.5, and it is accepted that the effect is due to adsorption of water vapours[114]. The situation is complicated to some extent by the influence of moisture on oxidation, and the effect of surface oxide in promoting adsorption of water.

In one respect the effect of moisture is not reversible. Sliding in the presence of moisture has been shown to result in evolution of hydrogen sulphide gas[47], and this must be caused by irreversible chemical reaction, probably involving oxidation. It may therefore be the case that some part of the increase in friction in the presence of moisture is caused by an increase in the amount of surface oxide. If this is so, then it seems probable that this oxide is preferentially removed on further sliding, allowing the lubricant to revert to a low coefficient of friction in a dry atmosphere.

Other polar vapours also cause a significant increase in the coefficient of friction[115], but the increase due to the vapour of non-polar compounds is relatively small. There is a lack of specific information on the effects of other vapours on film formation or wear life. Analogy with the effects of water vapour would again suggest that polar compounds would increase wear rate while non-polar compounds have little effect.

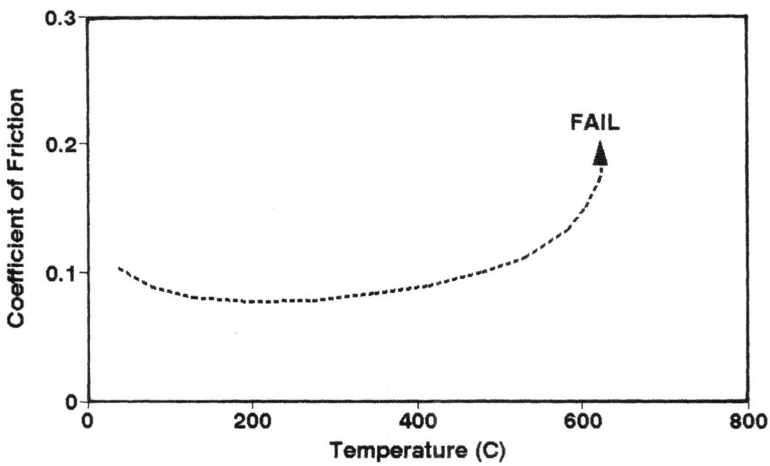

Figure 7.6 Variation of Friction of a Burnished Molybdenum Disulphide Film with Temperature (Ref.149)

7.3 EFFECTS OF TEMPERATURE

The general effect of increased temperature on molybdenum disulphide films is to reduce both friction and life. Most of the specific information available refers to bonded films, and at higher temperatures the behaviour is complicated by the effects on the bonding agents and other additives. Figure 7.6 shows a typical increase in friction with increasing temperature for a burnished film, and Figure 7.7 shows a typical variation in wear life with temperature.

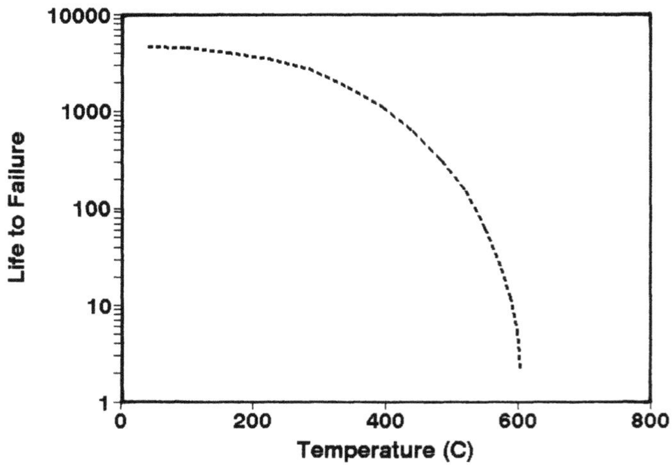

Figure 7.7 Effect of Temperature on Life of a Burnished Film (Ref.149)

The upper temperature limit for the use of molybdenum disulphide in air or oxygen has been quoted as 350°C for extended service, and up to 500°C for brief periods. This limitation is imposed by oxidation, which has been discussed in some detail in Chapter 4. Operation at much higher temperatures is possible in an inert atmosphere or vacuum. The thermal stability is excellent to over 1000°C, and theoretically lubrication should be possible to similar high temperatures. In practice the highest temperatures achieved successfully seem to have been about 700°C, even for short periods. Early workers quoted a limiting temperature in vacuum of about 650°C for long service, and this is probably too low. The reason for this low figure may be that early vacuum test environments still contained enough oxygen to cause significant deterioration.

However, there is one other factor which should be taken into account in discussing temperature limits, and that is the effect of frictional heating. Even with metals in an air atmosphere, frictional heating can cause local hot spots (flash temperatures) over 200°C hotter than the ambient temperature. With non-metallic surface films and the absence of any cooling liquid or gas, such temperature rises may easily account for the lower practical temperature limits in vacuum.

The coefficient of friction falls slightly with increasing temperature to a minimum at about 200°C and then rises as temperature increases further. With bonded films the effect of temperature on the binder will usually mask any effect on the molybdenum disulphide itself.

There seem to be relatively few publications about the effects of low temperature on performance, but the indications are that low temperature has no adverse effect on molybdenum disulphide itself. Hopkins and Campbell[159] tested twenty-two different bonded molybdenum disulphide coatings at -73°C in vacuum, and at room temperature and 204°C in nitrogen. Static friction values at -73°C differed from those at room temperature by between -33% and +114% except for one film for which test repeatability was poor. The average static friction for the remaining twenty-one coatings was 0.283 at room temperature and 0.277 at -73°C, so that the average change was not significant. This seems to suggest that the effect of low temperature was more on the other components than on the molybdenum disulphide. The average dynamic friction was 20% higher at -73°C than at room temperature, but this effect also seems likely to represent increased hardness of the various binders than any property of the molybdenum disulphide.

Gould and Roberts[160] carried out a test on a ball-bearing with a Duroid PTFE/glass fibre/molybdenum disulphide retainer and with a sputtered molybdenum disulphide film on the tracks. The bearing completed the 2×10^6 revolution test at 17°K (-256°C), but the sputtered film was then found to have been removed and replaced by PTFE. It is not clear whether this result represents any fundamental problem with sputtered films at low temperatures.

There have been a few reports of satisfactory operation while immersed in cryogenic liquids, including one by Devine et al[88], who reported that a sodium silicate bonded film was compatible with liquid oxygen. Rempe[161] found reasonable performance of a molybdenum disulphide compact in liquid hydrogen.

There have also been a number of satellite applications in which various molybdenum disulphide systems have performed satisfactorily with exposure to sub-zero temperatures, and overall it seems clear that low temperatures as such are not seriously detrimental to molybdenum disulphide performance.

7.4 EFFECTS OF RADIATION

Like most crystalline inorganic solids, molybdenum disulphide is highly resistant to nuclear radiation. Different studies have shown[162] little or no loss in wear life of inorganic-bonded films after a gamma-ray dosage of 10^9R or a total neutron irradiation of 2 x 10^{18} n/cm^2. On the other hand there was a significant reduction in wear life[163] at a neutron flux of 3 x 10^{12} n/cm^{2}sec and it seems possible that damage is related to dosage rate, or type of radiation, or both. Both the powder and inorganic-bonded or ceramic-bonded films have been used effectively in nuclear power stations[20], and it can be assumed, for example, that the radiation levels in space would have little effect.

Organic compounds are less resistant to nuclear radiation, and conventional oils and greases are therefore inferior to molybdenum disulphide in this respect. The same applies to the organic polymers which are often used as binders in bonded films. Inorganic binders should be used in preference to organic binders where whole-life radiation doses higher than 10^7R are predicted.

Dukhovskoi and co-workers reported a beneficial effect of irradiation in the occurrence of unusually low friction when an *in situ* coating of molybdenum disulphide was bombarded with a 2keV electron beam[164]. The tests were carried out in high vacuum with a pin-on-disc geometry. When the electron beam was switched on the coefficient of friction fell from about 0.04 - 0.05 at 10^{-7} Torr to less than 0.002, which was the limit of sensitivity for the measuring system. When the electron beam was switched off the friction rose again more slowly to 0.03. Similar behaviour had been reported previously[165,166] for irradiation with 2keV of helium atoms (or possibly α - particles), and the effect of both types of radiation is shown in Figure 7.8.

It was suggested[167] that greater efficiency of orientation of the crystallites was produced during sliding under irradiation because of enhanced removal of pollutants such as water vapour, which affect the surface energy distribution. This explanation is not entirely satisfactory for two reasons. The first is that the effect takes place

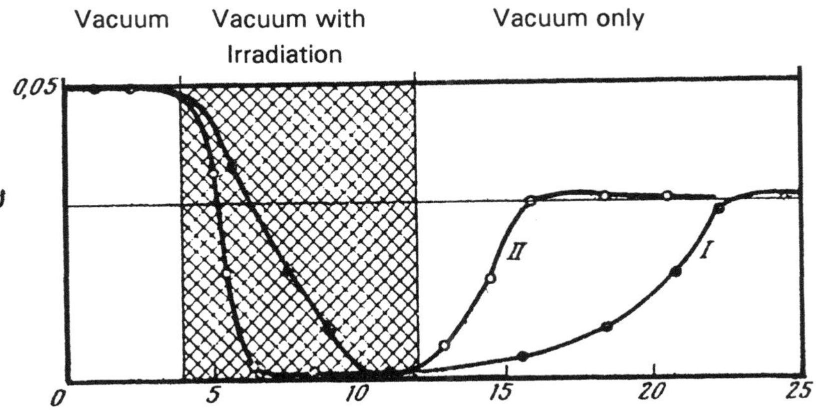

Figure 7.8 Reduction in Friction of an In Situ Molybdenum Disulphide with Irradiation (Ref.164 - 166)
(I) electrons (ß-particles)
(II) helium atoms (?α-particles)

more rapidly than the similar effect of sliding in high vacuum. The second is that the effect ceased while the specimens were still under vacuum, which suggests that re-contamination, and therefore loss of orientation, are unlikely to have occurred. On the other hand, a similar exceptionally low value was found for a sputtered film when extreme precautions were taken to exclude contaminants (see Section 10.5), so that removal of contaminants may be one relevant factor. It may also be that a more specific atomic structural effect is involved, possibly associated with the effect of ψ band electrons.

7.5 EFFECTS OF VACUUM

Vacuum in itself has no harmful effects on the lubrication performance of molybdenum disulphide. The effect of conventional atmospheres containing oxygen and water vapour is to increase the friction and decrease the wear life of molybdenum disulphide lubricants. It follows that in high vacuum, where such contaminants are absent, the friction and wear behaviour are generally improved.

The transient increase in friction in vacuum known as the "stop time effect" has been discussed earlier. The extent of the transient increase depends on the bearing geometry, the form of the lubricant and the duration of the rest period, but can be as high as 0.30, compared with a steady state value as low as 0.06. This is no greater than can be found in conventional atmospheres, and does not detract from the overall beneficial effect of vacuum on molybdenum disulphide lubrication, which has been confirmed by many studies, including tests on the Ranger spacecraft[168].

One effect of high vacuum which is deleterious for all materials is the loss of material by volatilisation. Under equilibrium conditions any material is in contact with its own vapour at a pressure, the vapour pressure, which is constant for the material at a given temperature. In high vacuum the equilibrium is disturbed because vapour atoms or molecules can escape readily from the vapour cloud, and the rate of loss of material then becomes finite.

The rate of loss is given by the Langmuir equation[169].

$$W = \frac{P}{17.14}\sqrt{\frac{M}{T}}$$

where W is the rate of evaporation, p is the vapour pressure of the material, M is its molecular weight in the gas phase, and T is the absolute temperature.

In practice the evaporation, or outgassing, of molybdenum disulphide has been found to be very low, and analysis of the gases evolved has shown them[170] to consist largely, if not entirely, of desorbed contaminants. These various published results are only at vacuums of 10^{-10} Torr. Detectable sublimation of molybdenum disulphide does not take place below 930°C[92] in 10^{-8} to 10^{-9} Torr.

It can therefore be accepted that the rate of outgassing at normal temperature is extremely low, although the actual rates are not known. In fact molybdenum disulphide has been used satisfactorily in many applications in space vacuum, some of which were listed in Table 1.1.

7.6 EFFECTS OF PARTICLE SIZE AND SHAPE

There is no single optimum particle size of molybdenum disulphide powder for lubrication, and the preferred size depends on the application. For example, the finest powders will be more easily dispersed and give more stable dispersions in light solvents, whereas larger particles are more effective for carrying heavy loads.

The largest particles give higher wear rates[46,171] especially in liquid carriers, and this is probably due to abrasion by the edges of large crystallites which are badly oriented. On the other hand the finest "micronized" powders can have increased corrosiveness, especially if size reduction is carried out in a humid atmosphere or for an unnecessarily long time[172]. This effect is probably due mainly to oxidation, but may also involve contamination during the grinding process, and can be reduced by careful control of the micronizing process. It follows that there is an intermediate range of particle sizes with which wear is minimised, and this usually lies between 0.7 and 7.0 μm[173], possibly with an optimum about 1-2 μm[174].

More important than the particle size are the purity and the particle shape. The effect of bad milling on a large lamellar particle is a particle having far smaller "shape factor" or aspect ratio than the original material. Since it is the basal plane surface which provides low friction and the edges which cause abrasion, it is therefore important in milling to maintain the highest possible aspect ratio consistent with the required particle size.

Ball milling in air tends to produce fine particles with an aspect ratio less than 15. Similar milling in a hydrocarbon gives material with aspect ratios between 15 and 100. The two types of product also differ in that the material ground in a hydrocarbon medium is readily wetted by organic liquids. It is known as oleophilic molybdenum disulphide and forms more stable dispersions in oils. Presumably the edge sites of the crystals adsorb organic molecules or radicals in the course of the grinding process and these promote wetting by organic liquids. Oleophilic molybdenum disulphide has far lower abrasiveness[175] than low-aspect-ratio air-ground material of the same average particle size, the improvement being as much as a hundred-fold. Much of this improvement is certainly due to the lower proportion of crystal edges, but the presence of adsorbed organic material on the edges would probably also help to reduce abrasiveness.

These effects of particle size and shape have all been described in connection with dispersions, pastes, greases or loose powders, in which particles remain separated. The effects of particle dimensions on the formation and properties of films are far less clear. The process of film formation as described in Sections 6.2 and 6.3, would seem likely to benefit from large particle size, for a number of reasons. Large particles would presumably embed more effectively in substrate surfaces. Subsequently large particles would experience higher rotational couples than small ones during the orientation process. Finally, once the surface is fully oriented, a surface formed from large particles might be expected to show fewer discontinuities or inter-crystalline faults than one formed from small particles, and would therefore give lower friction.

On the other hand, very large particles would cause heavier abrasion of the substrate, releasing wear debris. They would also tend to break up during running-in or burnishing, absorbing energy in doing so and rendering the running-in process less efficient. The optimum size may therefore be similar to the optimum for minimising wear, namely about 1-2μm.

7.7 EFFECT OF FILM THICKNESS

The initial thickness of most molybdenum disulphide coatings has an important influence on the performance and life of the coating[176]. The special case of sputtered films is considered in Chapter 10, and there is little information about thickness effects for *in situ* or transfer films. Many workers have investigated the effect of film thickness on bonded films, but, as was pointed out earlier, much of this work appears confusing, and sometimes contradictory, because of failure to understand and analyse the effects of running-in or burnishing on the consolidation and resulting structure of the films.

Soft films produced by deposition from dispersions, or with soft binders, may lose over half of the initial thickness during running-in or burnishing[177]. On the other hand, bonded films with hard binders will lose little or no thickness during running-in. Attempts to define a single optimum initial coating thickness for the whole range of coatings are obviously likely to be confusing without some attempt to establish the way in which the coatings behave. Nevertheless, once the problem of burnishing and consolidation is recognised, it is possible to analyse the various published reports and to establish the relationship between initial film thickness and performance.

It may be useful first to consider the minimum acceptable thickness for a film. Although the thinnest films are inherently unlikely to have long lives, there may well be cases in which other design considerations limit the film thickness which can be accommodated. The minimum desirable thickness will obviously depend on the roughness of the substrate and the counterface. Dayson[178] has shown that very thin films increase in friction as film thickness decreases due to an increasing degree of asperity penetration of the film.

An optimum surface finish of 0.5 μm c.l.a. is recommended for the substrate for a bonded film. The optimum roughness for a counterface for forming an effective transfer film has not been as well established, but if a similar surface finish is assumed, then the combined roughness parameter will be 0.7 μm c.l.a. Under normal circumstances a typical bonded film will have a thickness of between 5 and 10 μm, so initially at least there is no risk of asperity penetration.

If design considerations require it, a much thinner bonded film could be used, or a soft film could be burnished down to much reduced thickness. Under those circumstances it would be possible to use a smoother substrate, but coating performance deteriorates badly if the surface is too smooth, and a miminum acceptable surface roughness would probably be about 0.2 μm c.l.a., giving a combined roughness parameter of about 0.3 μm. It seems probable that in such a case an initial, or a fully-burnished, coating thickness of 1 to 2 μm would give a useful life before any serious problem of asperity penetration arises.

It is more difficult to establish the optimum initial film thickness for film performance and life, without considering the nature of the film. The structure of the film produced by burnishing a soft coating has been considered in Chapter 6, and takes the general form shown in Figure 6.2. With a hard coating the effect of burnishing or running-in will be to develop a thin smooth oriented layer on the surface, with the underlying greater part of the coating unchanged.

In practice there is a wide range of coating hardnesses, and the only practical way to define coating hardness is in relation to the way in which it will be used. If the stresses during running-in or burnishing produce a strong, highly-consolidated, well-oriented surface film over a softer under-layer, then the coating is behaving as a soft one. On the other hand, if the stresses during initial operation or burnishing are only sufficient to produce a very thin, low-friction surface film on a firmer unchanged underlayer, then the coating is behaving as a hard one. The various published results can be understood more clearly if we try to understand the category into which the test coatings fall.

Hopkins and Campbell[179] used Dual Rub-Shoe and Falex test machines to study the wear lives of a series of polyimide-bonded molybdenum disulphide coatings, with initial coating thicknesses varying from 2.5 μm to 25 μm. Both test machines use line contact, and they calculated the initial contact zone stresses as 276 MPa and 669MPa respectively, reducing to 103 MPa and 171 MPa by the end of testing. They

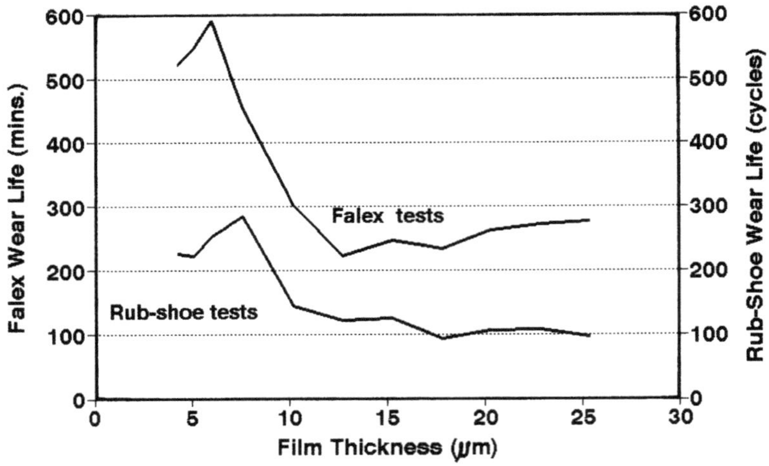

Figure 7.9 Effect of Initial Film Thickness on Life of a Bonded Molybdenum Disulphide Film Under High Contact Stress (Data from Ref.179)

found that with the Dual Rub-Shoe Tester the longest wear lives were obtained with initial thicknesses between 2.5 μm and 7.5 μm. At higher initial thicknesses the wear life reduced by over 50% for thicknesses greater than 10 μm. With the Falex machine the longest lives were obtained with initial thicknesses of 4.25 μm and 6 μm. At higher thicknesses the wear life was again reduced by over 50% for thicknesses greater than 9 μm. Both sets of results are plotted together in Figure 7.9.

Sauer et al[176] used the Alpha LFW-1 and Falex test machines to carry out similar tests on a resin-bonded coating containing molybdenum disulphide and graphite. The LFW-1 machine also used line contact. The contact zone stresses were not reported, but the test conditions suggest that they would be similar to those of Hopkins and Campbell. They studied a series of coatings with initial thicknesses from 3.5 μm to 24.5 μm, and found maximum lives with initial thicknesses between 6.25 μm and 7.75 μm with the LFW-1 tester and between 6.75 and 12.75 μm with the Falex machine. The reduction in life at higher initial thicknesses was not as dramatic as in Hopkins and Campbell's tests, but in both test series an increase in initial thickness of 8 μm from the optimum brought about a decrease in life of over 20%.

Sauer et al also found that with the LFW-1 tests the lowest friction occurred with the optimum initial thickness. They concluded that the optimum initial thickness was between 7 μm and 9 μm, whereas Hopkins and Campbell recommended that at heavy loadings in line contact the initial thickness should be between 5 μm and 7.5 μm. The specific figures will obviously vary with the type of coating and the loading conditions, but the important factor is that at higher thicknesses wear life is reduced.

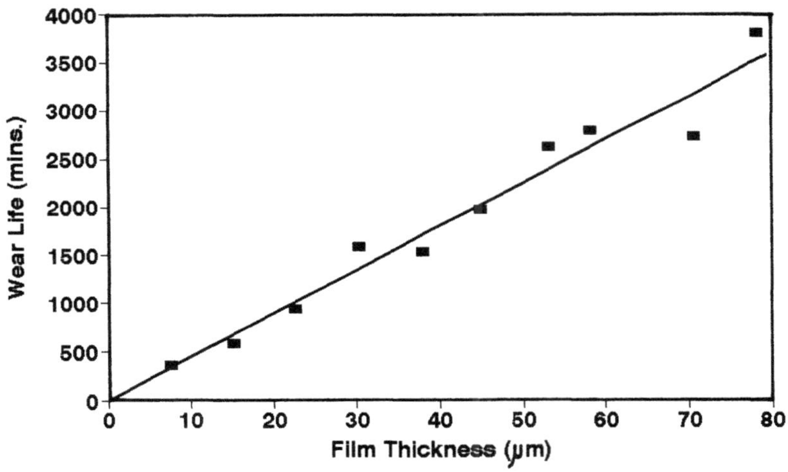

Figure 7.10 Effect of Initial Film Thickness on Wear Life of a Bonded Molybdenum Disulphide Film at Low Contact Stress in a Pin-on-Disc Test (Ref.179)

Hopkins and Campbell also carried out similar tests with a pin-on-disc tester in which the stress in the contact zone was calculated as 117 MPa, reducing to 1.3 MPa. The results of these tests were completely different, showing a linear increase in wear life with increasing initial film thickness over the whole range studied, as shown in Figure 7.10.

Bahun and Jones[180] performed a similar series of tests with a phenolic resin-bonded coating containing molybdenum disulphide and graphite. They used an Alpha LFW-1 tester modified to incorporate a conformal block rubbing against a cylindrical test ring, so that the contact zone stress was low, in the range 3 MPa to 29.7 MPa.

They also found that wear life increased linearly with initial coating thickness over the whole thickness range of 5 μm to 2.5 μm, as shown in Figure 7.11.

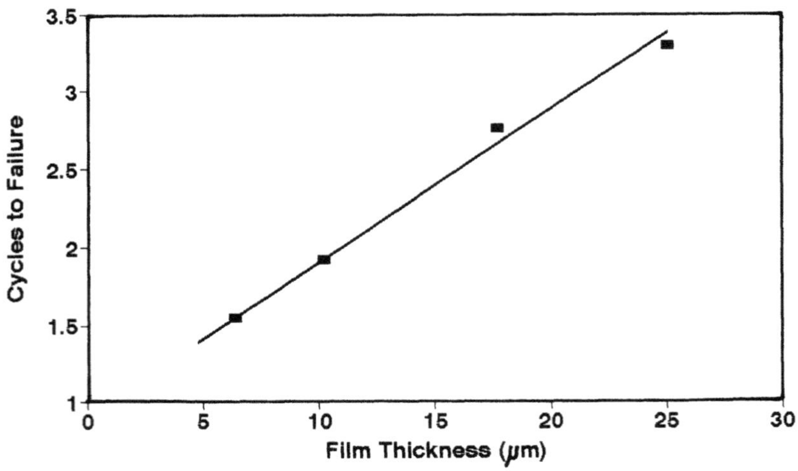

Figure 7.11 Effect of Initial Film Thickness on Wear Life of a Bonded Molybdenum Disulphide Film at Low Contact Stress in Conformal Contact (Ref.180)

It is possible to rationalise these results by assuming that under high stress all of the coatings tested behaved as relatively soft films, suffering consolidation and orientation to produce a run-in film with the structure shown in Figure 6.2. Such a structure would be stable as long as the soft underlayer is thin enough to provide a stable base for the highly-oriented and consolidated surface film. Above a certain critical thickness, depending on the hardness of the unconsolidated underlayer and the applied contact stress, the underlayer would no longer provide a firm base, and the surface film would then tend to break up. In confirmation of this theory, Hopkins and Campbell[179] reported that their thicker films did experience flaking.

On the other hand, under low operating stresses the coatings behaved as relatively hard films. Only a thin surface film would have been re-oriented, with little consolidation. With the low applied contact stress there would be little tendency for the underlayer to distort, and the thin oriented film would itself be more flexible. The surface would therefore wear gradually, and be replaced from the underlaying material, resulting in a steady rate of wear, and the life would be proportional to the film thickness. Hopkins had earlier[181] suggested that the phenomenon of a critical film thickness only occurs at high loads. It is in fact the contact stress in relation to coating hardness and thickness which determines whether the coating will run in or burnish to give a stable structure or break up.

In summary, if the operating contact stresses are high in relation to the hardness of the film, there will be an optimum film thickness for maximum wear life, and that optimum thickness is likely to be between 4 and 10 μm. If the operating contact stresses are low in relation to the hardness of the film, then the wear life will be directly proportional to the film thickness. These relationships have all been established by the study of bonded films, but similar effects seem likely to apply in the burnishing and use of films deposited from dispersions.

The thickness of a burnished film can be built up by adding further powder and re-burnishing to a total thickness of at least 15 μm. Such a film may be uniformly consolidated throughout its thickness, so that it will behave as a hard coating, and the wear life will increase with increasing film thickness[150].

The above discussion will only apply reliably where the life is determined by wear. If life is terminated by other effects, the same relationships may not apply.

The coefficient of friction will also vary with film thickness, as reported by Sauer et al, above, and this tends to confirm that the most effective consolidation and orientation occur at the optimum initial film thickness. Whitehouse et al also reported[182] a decrease in friction with film thickness to a value of 0.019 at 5 μm.

7.8 EFFECTS OF SLIDING SPEED

The effects of sliding speed on film performance are less clear than those of some of the other factors. In dry air or hard vacuum there seems to be little variation of friction with sliding speed[29,154,158]. In moist air a marked decrease in friction with increase in sliding speed has been reported, but this seems almost certain to be due to frictional heating at higher speeds causing a loss of moisture from the molybdenum disulphide films. If this is so, then the evidence suggests that the inherent friction is not affected by sliding speed. The effects are shown in Figure 7.12.

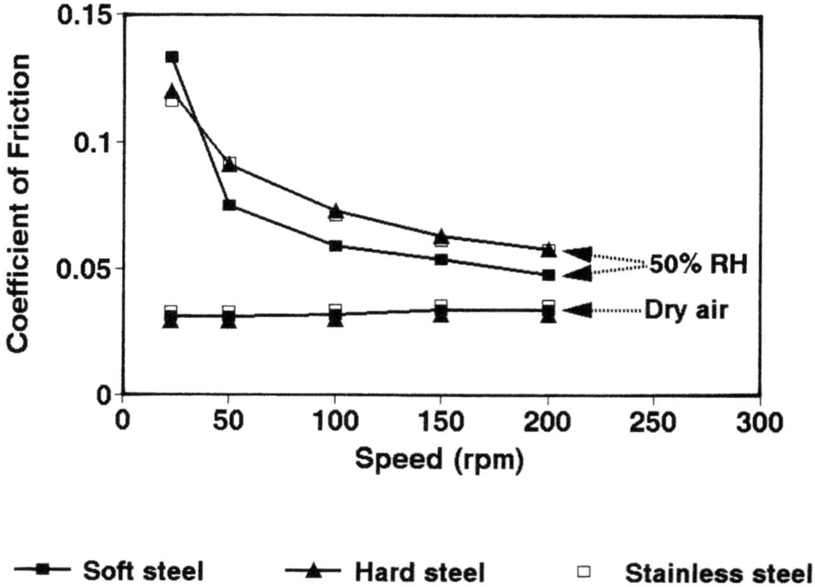

Figure 7.12 Effect of Speed and Humidity on Friction of Rubbed-On Molybdenum Disulphide Films (Ref.154)

Table 7.1 Variation of Wear Life with Sliding Speed (Ref.132)

Sliding speed (m/s)	Wear life (mins)	Wear life (m)
0.20	220±20	2.64×10^3
0.32	170	3.26×10^3
0.67	60±40	2.40×10^3
Test conditions: burnished film, conformal contact, specific load 1.64MPa, ambient air		

Several authors have reported a decrease in wear life with increasing sliding speed[132,183]. Table 7.1 shows the results obtained by Kinner[132], and Figure 7.13 those of Bartz[183]. Such results can be misleading because it would be normal for the life in minutes to be reduced in proportion to the increase in speed. Conversion of Kinner's figures to show life in sliding distance shows that in fact there was little change in life for a 235% increase in speed. With Bartz's results, however, the total sliding distance is also reduced at higher speed. This may be a result of higher frictional heating, and a consequent increase in oxidation rate.

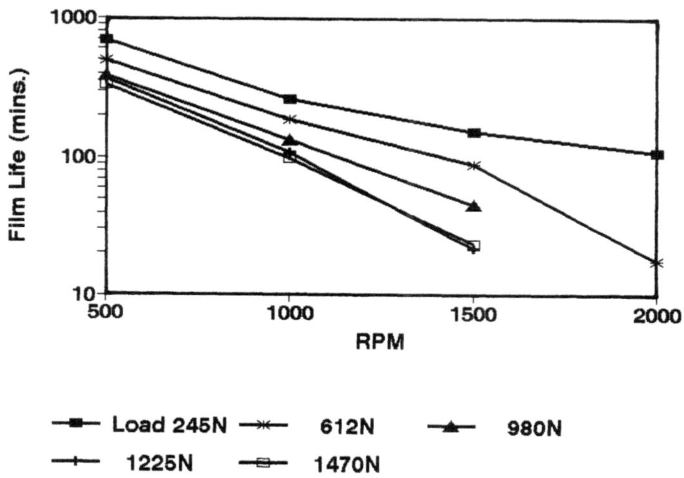

Figure 7.13 Variation of Bonded Molybdenum Disulphide Film Life with Sliding Speed (Ref.183)

7.9 FILM LIFE AND MECHANISM OF FAILURE

There are probably four important modes of failure of molybdenum disulphide films, namely loss of adhesion (or film break-up), wear, oxidation, and possibly fatigue.

One form of film break-up has been mentioned previously, in which the surface of a relatively thick film becomes consolidated over a soft unconsolidated layer. This can lead to shear in the soft layer, especially with high non-conformal loading, and the consolidated layer will break away. Loss of adhesion and film break-up can also occur, especially with a bonded film, if the surface pre-treatment has been badly performed, so that the coating simply fails to adhere and breaks away. These cases may be considered as premature failures caused by poor film preparation.

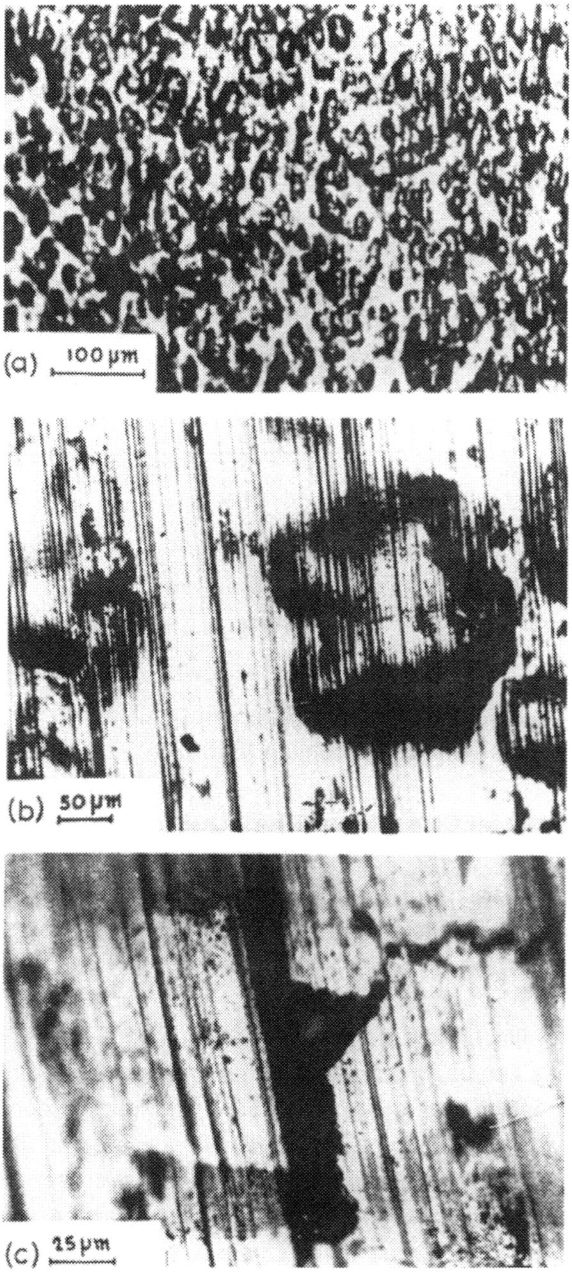

Figure 7.14 Three Stages in the Life and Failure of a Burnished Molybdenum Disulphide Film (Ref.184)

Where a film has been well prepared and burnished to a high degree of consolidation and reflectivity, and is run against a similar film on the counterface, sliding will take place almost entirely between the smooth surfaces. Under those circumstances the wear rate will be extremely low, and may for a period even be nil, so that the mean film thickness remains unchanged for the remainder of the film life[132,138,184].

The ultimate failure of such a film when oxygen is present, either in the atmosphere or in the form of water vapour, has been shown to take place by oxidation[184,185]. Towards the end of the period of smooth, zero-wear sliding, Salomon, De Gee and Zaat showed that blister formation starts to take place at the interface between film and substrate. The blistering increases to the stage where film adhesion is impaired, and the film eventually fails by flaking, causing rough running. The three phases of film life are shown in Figure 7.14, and the appearance of the blisters is shown in Figure 7.15.

Kinner showed by electron microprobe analysis[132] that rubbed films suffer progressive oxidation during sliding, and that this oxidation begins very early in the sliding process. The oxidation products include molybdic oxide and sulphate ion, which have higher friction than molybdenum disulphide, so that the film friction increases progressively. More importantly, the oxidation products lack the layer-lattice structure and the film-forming properties, so that both cohesion and adhesion of the film deteriorate. Like all irreversible chemical reactions the rate of oxidation increases with temperature. The rate of film breakdown will therefore be greater in high ambient temperatures, or where high loads or speeds cause a high level of frictional heating.

When lubrication failure finally occurs, the surface film contains large quantities of molybdic oxide and sulphate, and very little molybdenum disulphide. Presumably the disordered nature of the products causes a volume increase, which leads to the formation of the blisters if the nature of the contact permits it. Kinner considered that blister formation was unlikely to be a cause of failure in conformal contacts because the presence of uniform loading over the surface prevents any vertical development. The same oxidation process takes place in conformal sliding, except to the extent that conformal contact inhibits oxygen access, and even if blisters are physically prevented from forming, there must be a slow increase in friction and film break-up.

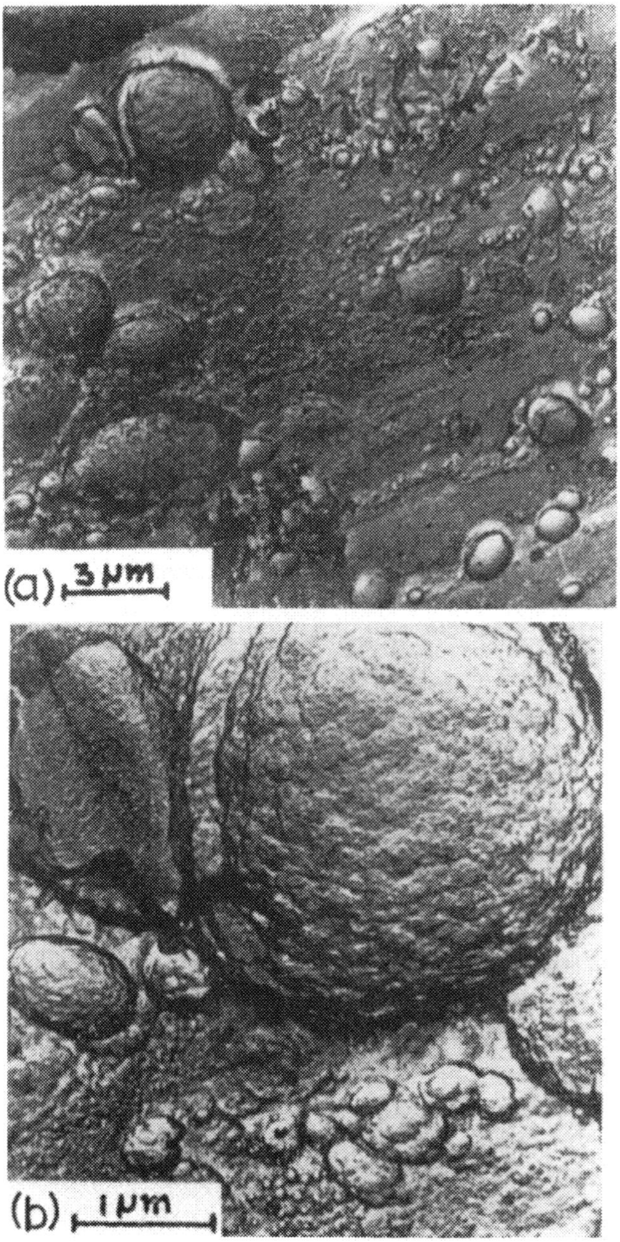

Figure 7.15 Blisters Developing in a Burnished Molybdenum Disulphide Film (Ref.184)

In the whole life cycle, a beneficial effect of humidity has been demonstrated, in that a relative humidity of 7% at the bearing surfaces (20% in the environment) gave an increase of over 100% in the duration of the smooth running period compared with a completely dry system. This may be due to improved running-in in a slightly humid atmosphere[122], with perhaps a thicker fully-ordered surface film.

Many workers have described situations, even in an oxygen-containing environment, where progressive wear takes place[126,186], and replenishment can occur from un-oriented sub surface material. It seems probable that this mechanism arises where contact conditions, either in operation or in prior burnishing, are such that the surface film is not perfectly ordered or highly reflective. In that situation there will be a proportion of discontinuities or dislocations in the surface film, which will provide sites for crystallite removal.

Fusaro[186] observed the behaviour of a rubbed film of molybdenum disulphide on a tool steel disc during sliding against an uncoated tool steel rider. He found that individual particles flowed plastically and coalesced, ultimately forming a smooth film. Loss of material during sliding was by plastic flow out of the contact region, with thinning of the film. Eventually the smooth film became black and powdery.

The way in which films fail in vacuum or in an inert atmosphere is less clearly understood, because the published results are less consistent. There are probably a number of reasons for this. There are fewer reports on vacuum, and it is particularly difficult to monitor the progressive deterioration of a film in vacuum. Some of the earlier investigations were in vacuum of only 10^{-5} or 10^{-6} Torr, and it is difficult to assess the effects of contaminants, residual oxygen, and adsorbents at crystal edges in those conditions.

It is clear that films which are not fully ordered, or whose surfaces are not highly burnished, can show progressive wear in inert gas or vacuum, in exactly the same way as in a normal atmosphere. Where both surfaces have highly-burnished coatings and the sliding is fully interfacial, it would appear that neither oxidation nor progressive wear can occur, so that film life would be indefinite. In practice this is not so, although very long lives can be achieved in vacuum.

Blistering and flaking have been reported in the deterioration of films in vacuum, but the nature and cause have not been adequately established. It is possible that in some cases oxidation took place due to the presence of residual oxygen or moisture, but otherwise the only likely causes are fatigue or some other chemical change such as loss of sulphur and conversion of the molybdenum disulphide to the sesquisulphide Mo_2S_3.

One factor which may be partly responsible for longer film life in inert gas or vacuum is an improvement in the formation of a transfer film on the sliding counterface as described by Fayeulle et al[187]. Formation of an effective transfer film on the counterface can have a significant effect in reducing wear of the primary film.

7.10 EFFECTS OF ADDITIVES

A very large variety of different chemicals have been used in conjunction with molybdenum disulphide to improve different aspects of performance of bonded films or composites, and these will be more suitably discussed in Chapters 10 and 11. A few have been used more generally to improve the life of molybdenum disulphide films, and it will be more appropriate to describe them here.

Haltner and Oliver[188] found that several metallic sulphides brought about an improvement in the load-carrying capacity when mixed with molybdenum disulphide. The sulphides included stannic and stannous sulphides, lead sulphide, ferrous sulphide and cuprous and cupric sulphides, and in a standard test procedure there was up to a ten-fold increase in load-carrying capacity. They speculated that the action of the added sulphides was similar to that of extreme-pressure additives in liquid lubricants. This would imply the formation of some protective film on the substrate surface. Pardee[156] later suggested that the effective mechanism was more likely to be oxidation inhibition. An alternative would seem to be the possibility that certain sulphides can act as an additional source of sulphur to form sulphide on the substrate surface, and thus improve adhesion of the molybdenum disulphide, as discussed in the previous chapter.

Probably the most important additive is antimony trioxide, which has been widely used in commercial formulations. Calhoun et al[189] studied a number of different substances as additives for bonded molybdenum disulphide films, but the only one which was effective in increasing wear life was antimony trioxide. Kinner[132] carried out a detailed study of the influence of antimony trioxide. He found that the improvement in wear life was particularly marked at high temperatures, and could be as great as 1000% at 200°C. The improvement was sometimes at the expense of some loss in room temperature performance.

He used spectrochemical techniques to study the composition of burnished molybdenum disulphide films from preparation to eventual failure. He found that the film surface at failure contained little molybdenum disulphide but contained molybdic oxide, sulphur, sulphate and iron compounds. In the presence of antimony trioxide, however, there was preferential oxidation of the antimony trioxide to the tetroxide Sb_2O_4. On heating a mixture of molybdenum disulphide and antimony trioxide in air at 500°C and 600°C, he found by X-ray diffraction that the products were mixtures of molybdenum disulphide and antimony tetroxide. Neither molybdic oxide MoO_3 nor antimony trioxide was present. When molybdenum disulphide alone was heated under the same conditions, it was almost completely converted to molybdic oxide.

These results strongly suggested that antimony trioxide improves the life of molybdenum disulphide films by acting as a sacrificial anti-oxidant and thus delaying the oxidative degradation of the molybdenum disulphide. Lavik and co-workers[190] confirmed the beneficial effect of antimony trioxide on film life with unbonded pellets and with a polybenzimidazole-bonded film. They also showed that the friction against a transfer film in air or against a clean track in air or vacuum was significantly reduced with antimony trioxide present.

On the basis of their results, they put forward an "oxide interaction concept", according to which the friction and wear of bonded films, composites and simple transfer films of molybdenum disulphide are improved by the presence of low-melting oxides which either combine easily or form desirable eutectics with molybdenum oxides. Such a concept would provide valuable guidance to the development of better solid lubricants, but, as Buckley[191] pointed out, the authors had in fact produced no evidence for the "oxide interaction" concept. The lower friction can in fact be explained by a reduction in the rate of oxidation, since the films exhibiting lower friction had all been run at some stage in air.

Centers[192] re-examined the published information about the performance of antimony trioxide, and rejected both the sacrificial oxidation theory and the oxide interaction concept. He concluded that antimony trioxide and certain other oxides and sulphides improve performance by providing a soft component which enables molybdenum disulphide more readily to acquire the optimal basal plane orientation for low friction and wear. However, he used compacts, and it is difficult to understand how his conclusion could apply to burnished films or well run-in films which already approach perfect crystallite orientation. His rejection of the sacrificial oxidation theory is based on a report by Gardos and McConnell[193], which does not include all of

Kinner's evidence. It is therefore possible that he did not have access to Kinner's work, which provided strong spectrochemical evidence for sacrificial oxidation.

Table 7.2 Synergistic Effect of Antimony Trioxide and Lead Monoxide on Wear Life (Ref.132)

Composition				Wear Life (mms)	
MoS_2	Sb_2O_3	PbO	Polyimide	Room temp.	200°C
2	0	0	1	670	140
1	1	0	1	155	390
1	1	0.025	4	1450	430
1	1	0.05	2	1550	280
1	1.5	0.05	2.5	1800	410
1	3	0.05	4	26	500

Many other substances have been tested for use as additives for molybdenum disulphide films, including silver, tin, lead, lead oxide, bismuth trioxide and boron nitride[189], alumina, arsenic oxide, cadmium oxide, cuprous oxide, molybdic oxide, titania, antimony trisulphide and antimony tetrasulphide[192], magnesia, silver oxide, zinc oxide and graphite[132], and boric oxide[194]. Of these, graphite has been shown to be beneficial[88] while antimony tetrasulphide and boric oxide were effective under unburnished conditions.

Lead oxide in conjunction with antimony trioxide showed a synergistic improvement in performance[132], as shown in Table 7.2. Bartz[183,195] showed a marked improvement in wear life for a three-component mixture of molybdenum disulphide, graphite and antimony tetrasulphide. Like those of Centers[192], his tests were carried out with unburnished/non-run-in films, against an uncoated block, and may not be representative of practical engineering situations.

~ ~ ~ ~ ~ ~ ~ ~ ~ ~ ~ ~ ~ ~ ~ ~

CHAPTER 8.

TRANSFER IN LUBRICATION

8.1 GENERAL PHENOMENON OF TRANSFER

Transfer of material from one surface to another is a common phenomenon when surfaces rub against one another, and Rabinowicz[196] suggests that it may be universal in dry contact. Some material combinations are of course much less prone to transfer than others, but Bowden and Tabor[197] describe transfer between a diverse variety of material pairs, including lead selenide and rock-salt, chromium and diamond, tin and platinum, and tungsten, lead or copper and steel.

Initial transfer at the atomic or molecular level is almost certainly universal in dry sliding, and transferred material has been detected down to the lowest limits possible with the analytical processes available. Transfer at such low levels can apparently take place without visible damage to the surface to which the transferred material becomes attached, so that the adhesion is a purely physical or chemical phenomenon, without any mechanical embedding. With transfer between metals there may also be diffusion of one metal into the other, or inter-diffusion of surface oxides. It is commonly found that with repeated sliding the quantity of transferred material increases to a maximum and it then becomes detached to form a wear fragment. In many cases this is a simple mechanical process, since a point at which the quantity of transferred material is greatest will experience the highest stress under a sliding contact, and will be most likely to rupture.

For many pairs the adhesion is weakened because of poor lattice matching between the transferred material and the substrate, or major differences in their mechanical properties, or by poor cohesion within the transferred material. Conversely, it follows that smooth, effective transfer is more likely to take place when the two materials have accurate matching of crystal lattices or electronic charge distributions, similar thermal and mechanical properties so that interfacial stresses are small when strained or heated, and where the transferred material has good cohesion.

Reference has also been made earlier (Section 6.2) to the possibility of rupture of the counterface and of plastic flow of transferred material into a fissure so formed. This may result in strong attachment of the transferred substance, even if other factors tend to weaken that attachment. Continued transfer and back-transfer between the surfaces of two different materials may lead ultimately to a similar composition on both surfaces. As this condition is approached, there is an increasing tendency for the surfaces to adhere together.

The factors which encourage material transfer are similar to those which lead to adhesive wear, since transfer is an essential first stage in the formation of an adhesive wear particle. The most severe form of adhesive wear is scuffing, and gross transfer commonly takes place under the influence of frictional heat when scuffing occurs.

8.2 TRANSFER OF MOLYBDENUM DISULPHIDE

Molybdenum disulphide will transfer readily to suitable clean solid counterfaces from a variety of different sources, including burnished films, bonded films, compacts, composites or single crystals. This ready transfer is of considerable importance in molybdenum disulphide lubrication, for two reasons.

An obvious disadvantage of easy transfer is that it results in depletion of the molybdenum disulphide in the primary source, and this may be a factor in determining the life of a lubricating film. However, once a smooth transfer film has been formed on a counterface, the transfer process either diminishes considerably or ceases altogether. The problem of depletion of the source is therefore only likely to be serious in systems in which the primary source is a film in continuous or repeated contact with fresh counterface, that is a counterface which is not yet carrying a transfer film. This is most likely to arise in processes such as metalworking, where molybdenum disulphide on a tool or die may be progressively transferred to fresh metal stock.

The second reason for the importance of transfer is that the lowest coefficients of friction in molybdenum disulphide lubrication are only obtained when a well-oriented film is present on both surfaces. If molybdenum disulphide is present initially on only one surface, then transfer is essential for building up a film on the counterface, and so achieving effective lubrication. It follows that in most mechanical systems in which molybdenum disulphide is used as a lubricant, the ability to transfer to a counterface is a beneficial, desirable, and often essential phenomenon. Lancaster[24] has in fact listed it as one of the most important features needed in a solid lubricant.

In some ways transfer of molybdenum disulphide to a metal surface resembles burnishing of loose powder. It follows that descriptions of the mechanism of transfer and of the factors which influence it will be similar in many respects to the description of the burnishing process in Chapter 6.

There have been many investigations of molybdenum disulphide transfer, but the majority of these have been concerned with finding practical solutions for operating rolling bearings or gears. Relatively few studies have been aimed at improving basic understanding of the transfer process, and because of the large number of factors which can be independently varied, the results of different studies are not easy to correlate. These factors include the nature of the molybdenum disulphide source, and the orientation of the crystallites in it, the material and condition of the substrate, the contact load and speed, the nature of the relative motion (i.e. linear, rotary, oscillatory or reciprocating), the temperature, and the nature of the gaseous environment, all of which appear to affect the transfer process.

As a result there is no clear detailed picture of the way in which transfer takes place, although there is a broad understanding of the nature of the process. It is generally accepted that transfer takes place by the movement of crystallites rather than at the molecular level. Where the source of the molybdenum disulphide is a single crystal, a crystallite in transferring to a counterface must be detached from the source crystal. This can take place by cleavage, fracture or shear, or a combination of one or more of these mechanisms.

Barton and Pepper[198] studied transfer from single crystals of molybdenum disulphide to surfaces of copper, nickel, gold or stainless steel, using Auger spectroscopy and sliding friction tests. They found in general that transfer took place readily with only a single pass and that the film thickness increased with the number of passes. They also found that when the single crystal was oriented with the basal plane parallel to the metal counterface, smooth transfer took place.

It was pointed out previously (Chapter 5) that the adhesive friction between the basal plane of molybdenum disulphide and a metal surface is higher than the friction between adjacent lamellae. It follows that when the basal plane of a crystal slides over a metal counterface, the adhesive stress developed at the interface is greater than the limiting shear stress between lamellae, and shear will take place in the crystal, with transfer of the surface lamella to the counterface. Even under ideal conditions, the first sliding contact will not produce complete uniform coverage of the counterface, but there will be individual transferred lamellae, more or less isolated from each other. Rabinowicz[196] has suggested that, for many systems, when the area covered by transferred fragments exceeds 10%, the cohesive energy of the transferred material begins to influence the rate of transfer, and adjacent transferred particles can work together to encourage further transfer.

In the specific case of molybdenum disulphide transfer it seems probable that the 10% figure would be exceeded very quickly. Furthermore, the cohesive energy at the edges of the transferred lamellae would be higher than the inter-lamellar cohesive energy, and further transfer of lamellae to fill the initial gaps would be expected to take place very readily to give complete coverage of the counterface. This mechanism would produce some irregularities and lattice defects at the junctions between transferred lamellae, and in conjunction with the original surface texture of the counterface, would result in a proportion of edge-site exposure in a generally well-oriented lamellar film. Such edge-sites would facilitate the build-up of further transferred lamellae. Continuation of the process would be in accordance with Barton and Pepper's finding that there is a tendency for further transferred material to build up smoothly on top of the original transferred lamellae with a parallel orientation.

They found, however, that when the single crystal was oriented with the basal plane inclined to the counterface, transfer of large particles could take place, sometimes with abrasion of the metal surface. This may be caused by the fact that when crystal edges are in contact with the counterface their hardness enables them to gouge the metal surface. This would facilitate embedding, while at the same time the higher stress at the contact compared with the parallel case, oriented at an angle to the lamellae, would tend to cause fracture across the lamellae and cleavage instead of shear along the basal planes. The end result would be the separation of a thick crystal fragment, firmly attached to the counterface by embedding and by the strong bond energies at the fracture faces, rather than a thin lamellar particle.

Barton and Pepper reported that the strength of adhesion was not directly related to the contact load between the single crystal and the counterface. This seems curious if mechanical embedding is a significant factor in the adhesion, but it is possible that the limiting factors in determining the extent of embedding are the cleavage and fracture stresses of the crystal. and the yield stress of the counterface material. In that case the effect of increased load might be to increase the rate of material transfer rather than the strength of adhesion.

Before leaving the subject of transfer from single crystals, it may be appropriate to point out that any film or compact which is highly burnished will have a surface which consists of fully-oriented material with basal planes parallel to the plane of the surface. This surface film will in fact be a sort of pseudo single crystal, and it would be reasonable to expect its transfer behaviour to resemble that of a true single crystal in parallel orientation.

There have apparently been no detailed studies of the mechanism of transfer from such a highly-burnished surface, in spite of the practical importance of that type of transfer. If we consider a typical coating in which a highly-oriented surface film overlies a softer and more randomly-oriented subsurface, then its initial contact with a metallic counterface will resemble the first contact in the parallel case as studied by Barton and Pepper. By analogy with their results it would be expected that in such a contact the surface lamellae of the coating would transfer readily and smoothly to the counterface. This would then expose softer and less highly oriented subsurface material. Further relative sliding would result in non-parallel, or edge-site, contact, and the crystallites on both surfaces would need to re-orient in order to provide efficient low-friction lubrication.

Lancaster[134] studied the transfer of molybdenum disulphide and graphite to low-carbon steel discs from compacts. The compacts were formed at relatively low pressures, up to 80 MPa, but later work[130] indicated that when compacts were formed at higher pressures up to 1500 MPa, there was no obvious difference in the nature of the transfer films produced from them. It can therefore be assumed that the crystallites in the compacts were mainly randomly oriented, although there is some evidence[199] that in unidirectional pressing of a lamellar solid a relatively high degree of orientation occurs in the surface layers.

Lancaster found that transfer of molybdenum disulphide to very smooth surfaces took place in large aggregated lumps up to 10μm thick on top of a smooth film about 0.05μm thick. On rougher surfaces, with surface finishes between 0.13 and 2.5μm C.L.A. the transfer film was more even, and the limiting volume of transferred lubricant was approximately equal to the volume of the surface depressions. This implies that the transferred films were very thin over the asperity peaks, but that the depressions were generally filled. The highest scuffing loads for the transfer films were obtained with a disc surface finish of 0.75μm C.L.A. Lancaster interpreted these results to mean that with the rougher surfaces transferred material was compacted into the surface depressions, giving strong mechanical attachment, whereas with smoother surfaces, embedding would be less effective, and only physical and chemical bonding would be significant.

It is interesting to try to correlate these findings with the later work of Barton and Pepper described previously. The transfer of large particles on smooth surfaces in Lancaster's work is similar to the nature of transfer found by Barton and Pepper when their single crystals were oriented with the lamellae at an angle to the counterface. Since in Lancaster's compacts many of the crystallites will also have been at an angle to the counterface, there is no contradiction in these results. On the other hand, it is relevant to consider why the same irregular transfer did not occur between the compacts and rough counterfaces.

In the description of the burnishing of powder in Chapter 6, it was shown that when crystallite edge-sites attach to a surface in a sliding contact, a couple is generated which rotates the crystallite until it achieves parallel basal plane orientation. In his work Lancaster found that the rate of loss of material from a compact was greater than the rate of transfer to the counterface, so that there was always a surplus of loose powder available. It is therefore probable that at least some of the film formation was not by direct transfer, but by attachment of loose powder, so that burnishing of powder was also making a contribution. In addition, with the rougher counterface, higher contact stresses would arise at the asperity peaks, which again might lead to higher aligning forces.

It is in fact interesting that in this and other publications, Lancaster appears to consider that transfer usually occurs by the formation and attachment of loose particles, whereas the work of Barton and Pepper suggests that, at least in some circumstances, direct transfer from source to counterface takes place.

The transfer films produced from compacts by Lancaster had high load-carrying capacities, over twice as high as a plain mineral oil, PTFE, or any of the graphite composites, and higher than 10% of graphite in a mineral oil. The coefficient of friction varied between 0.09 and 0.15 on mild steel, but when the steel was previously phosphated the lowest coefficient of friction was only 0.04 and the load-carrying capacity was unchanged. This suggests that the benefit of phosphating in this case was in improving the crystallite orientation in the bonded film.

The life of a transfer film without enrichment was 10,000 seconds, but when the compact remained in contact with the transfer film during life testing, the life was increased to 62,000 seconds. The continued presence of the compact did not maintain lubrication indefinitely. It appeared that in the early stages the rate of continued transfer balanced the depletion of the film by wear. Later, the surfaces became smoother, and the transfer rate decreased, so that the wear rate of the film was no longer balanced by continued transfer, and the film then failed by wear.

The decrease in transfer rate could be partly offset if the alignment of the compact was changed. This may indicate that the surface of the compact had become fully oriented, so that the shear forces decreased, and there were few if any edge-sites suitably positioned to favour adhesion to the steel counterface. A change in the alignment of the compact would then result in edge-sites being exposed to the sliding contact, improving adhesion and at the same time increasing friction and shear stress in the compact. He suggested that warming of the surfaces by continued sliding resulted in a loss of moisture, causing a decrease in inter-crystallite cohesion, so that a smaller proportion of the available lubricant attached to the disc. Some support for this explanation is provided by a reduction in load-carrying capacity for a transfer film formed at a higher temperature of 150°C.

An alternative explanation is suggested by some later work of Fleischauer and Bauer[200]. They found that the best performance of transfer films of molybdenum disulphide was obtained when molybdic oxide was present in the lowest layers of the film adjacent to the steel surface. Oxidation to molybdic oxide is increased in the presence of moisture, so that reduction of moisture content due to frictional heating may reduce the amount of molybdic oxide present, and thus have a direct adverse effect on transfer film life as well as reducing the rate of film formation. Fleischauer and Bauer also found indications that transfer film life was improved if a slight excess of sulphur was present at the interface between the film and the substrate. The presence of molybdic oxide or excess sulphur are undesirable in the bulk of the lubricant and especially on the sliding surface and they suggested that for optimum transfer film performance these factors need to be controlled independently at the substrate surface and in the bulk of the lubricant film.

Rabinowicz[196] concluded that materials having a high ratio of surface energy to hardness have a greater tendency to accept transferred material. This relationship has some validity, but does not explain the differences between steel and brass or nickel, which may have similar values of the ratio.

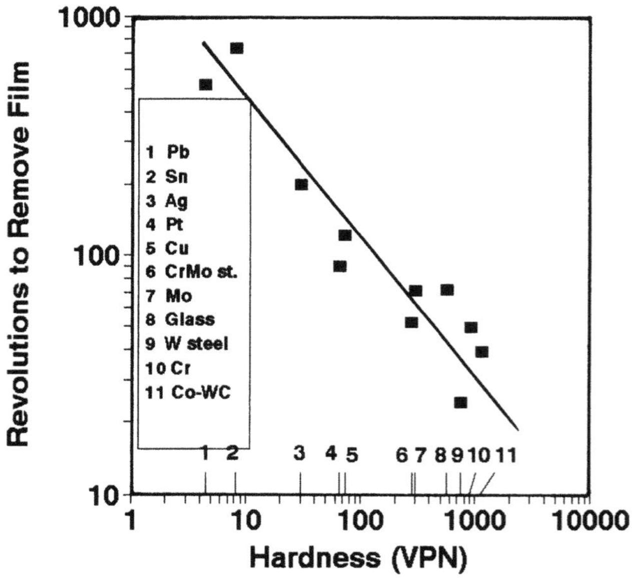

Figure 8.1 Effect of Substrate Hardness on the Life of a Transfer Film of Molybdenum Disulphide (Ref.130)

There is no general agreement about the effect of substrate material and conditions on the formation and performance of transfer films. Lancaster[130] showed an inverse correlation between transfer film life and counterface hardness, as shown in Figure 8.1, and suggested that the use of a soft metal plating on a hard substrate might give the optimum surface for transfer lubrication. On the other hand Barton and Pepper found that the strength of adhesion was not directly related to the ductility of the counterface material. Adhesion to 304 stainless steel took place more readily than to gold, but less readily than to copper or nickel, and this gives some support to the theory mentioned earlier that the strength of adhesion is related to the strength of the metal-sulphur bond.

8.3 APPLICATIONS OF TRANSFER

There are basically three ways in which transfer of molybdenum disulphide can be deliberately used for lubrication. These are the pre-coating of a bearing surface with a molybdenum disulphide film, transfer from one bearing surface to an uncoated counterface, and continuing replenishment from a reservoir during machine operation.

8.3.1 Pre-Coating of Bearing Surfaces In the early days of molybdenum disulphide lubrication, transfer from single crystals was one of the simplest techniques for creating a lubricating film on a bearing surface. It was more convenient and very much less messy than the alternative of using free powder.

Since about 1960, however, many different dispersions and bonded coatings have been commercially available. These are more convenient and generally cleaner to use, giving better control of the film-forming process and more predictable performance. As a result, the use of transfer for pre-coating bearing surfaces is now of little practical importance.

8.3.2 Transfer from Bearing Surface to Counterface This phenomenon is of major practical importance. The greatest system life with molybdenum disulphide films will usually be obtained when both of the interacting surfaces are pre-coated, but even in that situation transfer can be beneficial. Any initial gaps or flaws in the surface coatings can be repaired by transfer during the early stages of operation. In the same way, any deterioration of a coating caused by wear or flaking in the later stages of operation can also be repaired if there is enough surplus material present to transfer to the points of deterioration.

There are also situations in which the pre-coating of all the interacting bearing surfaces may be undesirable, inconvenient, or even impossible. It may be undesirable if the resulting sum of the coating thickness tolerances would be too great, or where loss of surplus material from a number of surfaces during running-in would create a contamination problem. It may be inconvenient if the necessary access to all the bearing surfaces would require excessive dismantling or more complex design and construction. Finally, it may be quite impossible if one of the surfaces to be coated is inaccessible, and an example of such a problem is the internal surface of a spline or thread.

In all these cases, effective lubrication may be obtained by coating only one of the bearing surfaces, and making use of transfer to create a film on the counterface. Two simple precautions need to be taken to make certain of satisfactory operation. The first is to ensure that the primary surface film on the one coated component is not too heavily burnished before assembly, since it is essential for enough molybdenum disulphide to be present to form two viable films. The second precaution is to run in the system under lightly loaded conditions, so that no surface damage or other fault develops before an effective transfer film is formed on the counterface.

In theory it should be possible to create transfer films on several successive surfaces from the one primary coating, such as in a gear train, but there are serious practical difficulties in doing so. In particular, such an arrangement would require one or more pairs of interacting surfaces to operate unlubricated initially.

In view of its practical importance, it is surprising that there has been relatively little detailed study of this form of transfer, even to the extent of defining the rate of formation of a transfer film, or the effects of such factors as counterface material, hardness and surface finish. The general design assumptions tend to be based on the requirements for transfer from composites, namely a surface roughness of 0.2 μm C.L.A., and the possible use of a soft plating or a chemical conversion coating on the counterface.

8.3.3 Lubrication by Transfer from a Reservoir is the most important practical application of transfer, as it provides a means for continually supplying molybdenum disulphide to a machine system during operation.

The general problem of resupplying a system with a solid lubricant is discussed elsewhere in this book, and various possible techniques are mentioned, but resupply by transfer from some form of reservoir is the most successful technique, and the only one which has been used commercially. It has been used in spacecraft, and in terrestrial applications for vacuum or for very high or very low temperatures, while rolling bearings lubricated by solid lubricant composite retainers or cages have been commercially available for over thirty years.

Some basic research studies, such as those by Lancaster, Barton and Pepper, Fleischauer and Bauer, have been performed in order to give a firm basis for the design and use of reservoirs for transfer lubrication. Far more projects have been carried out to evaluate specific practical applications. The two important variables in applying the technique are the composition of the reservoir material and its location in the system, both of which have to be related to the stresses and environmental conditions which will be experienced.

8.4 COMPOSITION OF THE TRANSFER SOURCE

The composition of the reservoir material is a compromise between structural strength and the availability of molybdenum disulphide for transfer. In general terms, when the concentration of molybdenum disulphide in the reservoir is high, the rate of supply of lubricant to the bearing surfaces is high but the structural strength is low. Conversely, the structural strength can be increased by incorporating the molybdenum disulphide in a strong matrix, but the lower the concentration of molybdenum disulphide, the lower will be the rate of transfer to the bearing surfaces. However, this is only a broad generalisation; the actual properties and performance will be affected by the nature of the matrix material, the presence of other components in the composite, and the rubbing conditions.

Molybdenum disulphide alone can be used as the reservoir material, either in the form of single crystal or as a compact. It is difficult to define the structural strength of single-crystal molybdenum disulphide. Because of its anisotropic nature, the ultimate stress in shear, tension or compression varies critically with the direction of the applied stress in relation to the crystal orientation, as discussed in Chapter 4, but some indication is given by the hardness values on the crystal faces and edges of 1.5 and 8 Mohs respectively.

Similar comments apply to compacts, with the additional complication that the crystallite orientation in compacts can vary from completely random to a high degree of orientation. Compacts can be successfully formed at pressures as low as 35 MPa, but the structural strength increases with compaction pressure, as shown in Figure 8.2, and pressures as high as 1350 MPa have been used[201] Another indication of the change in structural integrity with compaction pressure is given by the variation in wear rate, and an example is shown in Figure 8.3[130]. However, a high wear rate is not in itself necessarily a disadvantage, since transfer to the bearing surfaces requires wear of the reservoir material.

Overall, the structural strength of single crystals or simple compacts of molybdenum disulphide alone is not high enough to enable them to be used for the manufacture of bearing components, and they must be adequately supported to withstand any significant operating stresses. There are also difficulties in forming or machining them in any but the simplest shapes. The structural strength and forming problems are improved by the use of binders, or by incorporating the molybdenum disulphide in a metallic, polymeric or ceramic matrix. The concentration of molybdenum disulphide in such a composite can range from 3% to 90%.

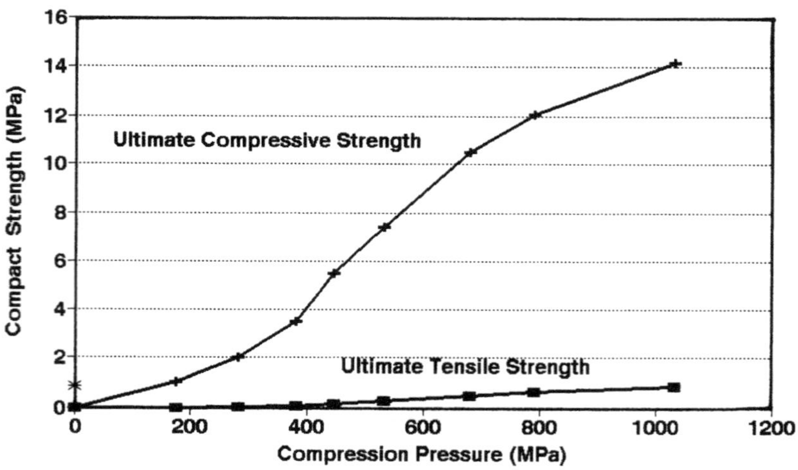

Figure 8.2 Variation of Structural Strength of a Molybdenum Disulphide Compact with Compaction Pressure (Based on data from Ref.201)

Lancaster[24] has suggested that because transfer is an inefficient process, the concentration of solid lubricant in a reservoir composite must be at least 25%. Successful results have been claimed for composites with much lower concentrations, but comparisons are difficult because different workers have used different criteria for successful operation. Certainly most of the successful composites which consist only of molybdenum disulphide in a strong matrix have contained at least 20% of the solid lubricant.

The upper limit for the concentration of molybdenum disulphide in such a simple composite is imposed by the low friction and low limiting shear stress of the molybdenum disulphide. This commonly results in poor structural integrity of polymer composites containing more than about 50% of molybdenum disulphide, so that they can only be used in situations where they are well supported and are not subjected to high stresses. Much higher concentrations can be used in metal composites.

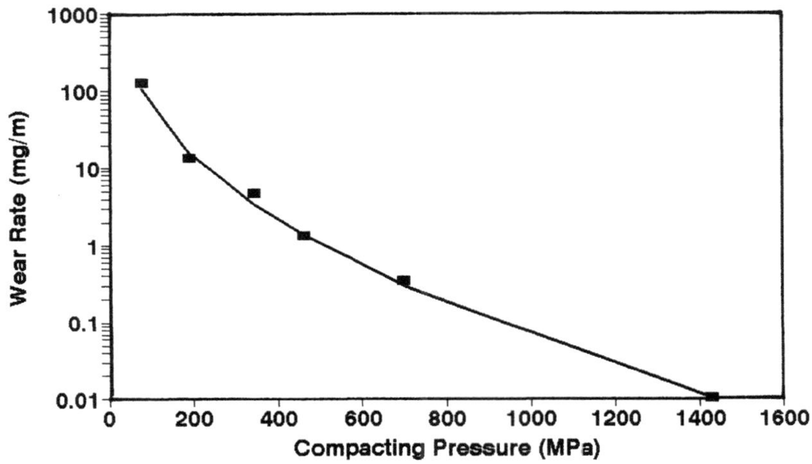

Figure 8.3 Effect of Compacting Pressure on Wear Rate of a Molybdenum Disulphide Compact (Ref.130)

Janes, Neumann and Sethna[202] reviewed the general subject of solid lubricant composites in polymers and metals. They pointed out that the reduction in mechanical properties with higher concentrations of solid lubricant can be offset by the use of fibre reinforcement. Glass fibre is probably the most commonly used reinforcing fibre, with carbon fibre as a second choice. Metal and ceramic fibres have been used experimentally to reinforce polymers, but have not apparently been used commercially. To some extent powders such as bronze, lead, silica, alumina, titanium oxide or calcium carbonate can be used to improve compressive modulus, hardness and wear rate.

In practice most composites for use in transfer applications have consisted of three or more components. The most widely-used composites usually contain PTFE as well as molybdenum disulphide, and these must also include a reinforcing material such as glass fibre to compensate for the low strength of unreinforced PTFE. PTFE is itself a useful solid lubricant in transfer applications, and Connelly and Rabinowicz[203] have shown that in continuous sliding it has a greater ability than molybdenum disulphide to migrate along the wear track and repair areas from which the lubricant has been worn away. In oscillatory motion, however, the reverse may be true, since PTFE has a tendency to migrate to unloaded regions, whereas molybdenum disulphide has been found to back-transfer from the point of reversal and to replenish depleted areas on the track[204].

The subject of molybdenum disulphide composites will be described more fully in Chapter 12. Most of them have not been developed specifically to provide transfer lubrication, but it can be assumed that in any situation where molybdenum disulphide is present in a sliding contact, it is capable of producing some transfer to a counterface.

8.5 NATURE AND LOCATION OF THE TRANSFER SOURCE

There are two fundamentally different ways in which the reservoir of transfer lubricant can be located. It can be a part of, or the whole of, one of the normal load-bearing machine components, and this has been variously described as direct, primary, or two-body transfer lubrication. Alternatively it can be a separate auxiliary component present only to provide a lubrication reservoir, whose sole function is to transfer lubricant to one of the other machine components. This has been described as indirect, secondary, or three-body transfer lubrication. In the Russian literature the latter is called "Rotaprint Lubrication" by analogy with the use of a separate inking roller to transfer ink to the cylinder in a rotary printing press[190-205-208].

8.5.1 Direct Transfer Lubrication Direct transfer lubrication has theoretical advantages in reducing the complexity of the overall design and in shortening the path from the reservoir to the lubricated bearing surface. However, a major limitation is that the lubricant reservoir must withstand the same load, speed, temperature and other conditions as are required by the function, load path, power, etc. of the machine design. This imposes severe constraints on the composition and method of incorporation of the reservoir material.

Where a load-bearing machine component is manufactured completely from the composite, the problem of material selection is the conventional one of relating material properties to the design requirements. It is sometimes possible to use the lubricating composite for the manufacture of a component which carries little or no load, or whose dimensional accuracy is less critical. The cage or retainer in a rolling bearing is the classical example of this situation, but another is to incorporate a lightly-loaded composite ring in the ring pack of a piston.

Where neither a load-carrying component nor a less critical component can be completely manufactured from the reservoir material, a useful alternative is to bond, rivet or press composite onto the surface of a metal component, or to incorporate inserts of the composite in a machine component. Many applications of this technique have been developed, and a few examples will serve to illustrate it.

One of the earliest, and still one of the most ambitious, applications of the use of solid lubricant inserts, was in a solid-lubricated piston engine constructed by M.J. Devine and co-workers at the U.S. Naval Air Engineering Center in Philadelphia in 1966[209]. They used a variety of inserts of different geometries to provide lubrication to different parts of a single-cylinder four-stroke engine. The composite used consisted of 71% molybdenum disulphide, 7% graphite and 22% sodium silicate as a binder, and had originally been developed as a bonded coating. It was filled into the various reservoir recesses or pockets in the form of a paste and air-dried. Some of the bearing surfaces were also sprayed with the same lubricant. The components treated are shown in Figure 8.4, and were as follows:-

(i) Connecting-rod gudgeon-pin (wrist pin) and big-end (crankshaft) bearings. Composite-filled reservoirs 0.125" in diameter and 0.035" deep in molybdenum alloy bearing inserts.

(ii) Crankshaft journal. A spiral groove, pitch 0.2", 0.096" wide and 0.032" deep machined in journal surface, filled with composite.

(iii) Piston. A spiral groove, pitch 0.25", 0.078" wide, 0.016" deep, filled with composite.

(iv) Gudgeon pin. Forty-five cylindrical reservoirs, 0.125" diameter, 0.048" to 0.063" deep, machined in surface of pin and filled with composite.

(v) Tappets. Molybdenum alloy face rivetted to tappets, rivet holes recessed 0.070", giving recesses 0.187" in diameter, filled with composite.

The test engine ran for 6 hours at 2500 - 3000 rpm, and failed due to cam wear. This compared with 10 seconds before seizure for a completely unlubricated engine, and 30 minutes for an engine with only a bonded coating.

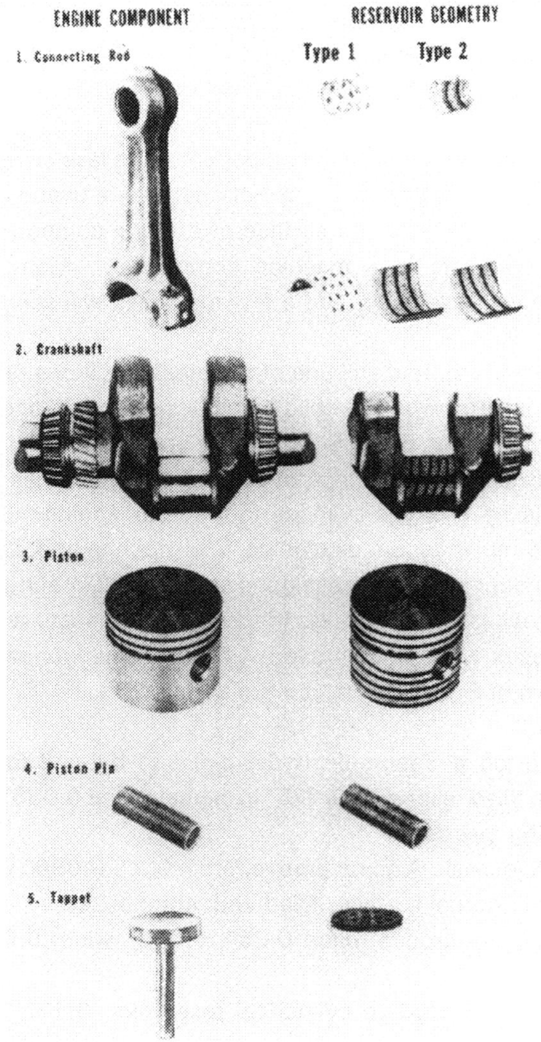

Figure 8.4 Some Solid Lubricant Reservoir Designs for a Small Piston Engine (Ref.209)

Van Wyk[210] used a similar geometry in developing plain spherical ceramic bearings for helicopter pitch control linkages. He tested several different composites of molybdenum disulphide in polyimide or metal, and these gave better performance than composites of PTFE or graphite. The best of them was a composite of 90% molybdenum disulphide, 8% molybdenum and 2% tantalum by weight. This was used for full-scale tests on spherical bearings similar to those used in Boeing Vertol H-21 helicopter rotor pitch linkages. The composite was filled into holes drilled in the ball and outer race surfaces, as shown in Figure 8.5. The hole size and spacing were not specifically stated but they appear to have been about 2.1 mm in diameter and 1.0 mm deep, spaced 1 mm apart in rows 2.5 mm apart. The composite reservoirs therefore covered about 17% of the bearing surface.

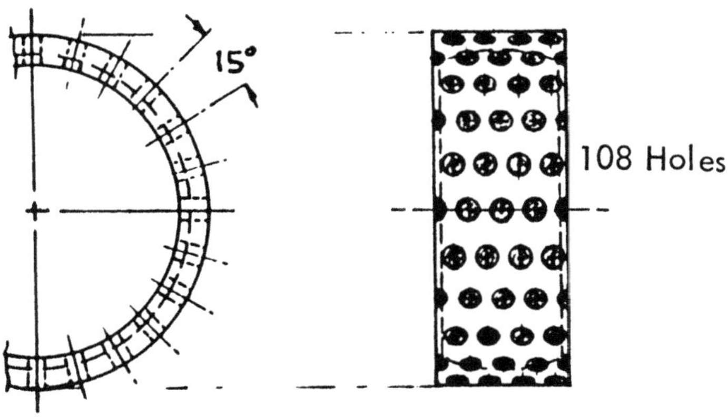

Figure 8.5 Lubricant Reservoir Pattern Used in a Helicopter Linkage Bearing (Ref.210)

After 50 hrs of testing, involving ±9° oscillation at 243 cpm with cyclic loading from 5.3kN to -2.9 kN, the bearing surfaces showed good lubricant transfer, with very little wear, comparable to the wear of conventional production bearings. A backing disc of polyethylene having a high coefficient of thermal expansion was fitted behind the lubricant composite, so that when friction increased the higher frictional heating caused the backing disc to expand. This pressed the lubricant composite against the counterface, thus increasing lubricant supply and reducing the friction.

This highlights one of the limitations in using the technique of filling a lubricant supply into pockets in one of the bearing surfaces. The limitation is that lubricant transfer to the counterface can only take place when the lubricant source is in contact with the counterface. It follows that, in the absence of some mechanism such as that used by Van Wyk, fresh lubricant can only become available as the surface of the bearing wears.

This has two important consequences for the design of the lubricant reservoirs. The first is that if the recesses containing the lubricant are deeper than the acceptible wear depth, then the deeper portion of the lubricant will be unusable. In Devine's work, the recesses, at 0.016" (0.4 mm) to 0.070" (1.8 mm), were much deeper than the permissible wear depth for the surfaces, presumably in order to give adequate lateral support for the composite material.

The second consequence is that it becomes important to make the concentration of lubricant over the bearing surface as high as possible. Because structural support is provided by the walls of the recesses, the structural strength of the composite or compact itself is less critical, and molybdenum disulphide concentrations from 50% to 95% have been used. The other factor which can be varied is the fraction of the total bearing surface which consists of lubricant composite, described by Lancaster[211] as the lubricant area fraction. Again a compromise is necessary, since a higher lubricant area fraction will give higher lubricant availability and lower structural strength, and vice versa. In his work Lancaster found that the optimum lubricant area fraction was 0.5.

Apart from the lubricant area fraction, the actual dimensions of the recesses are also important. Deep, narrow pockets are inherently likely to give strong support and retention of the lubricant material, but, as shown previously, deep pockets are wasteful of lubricant, while narrow pockets are susceptible to blocking with wear debris. On the other hand, wide shallow pockets are likely to provide poorer support and retention for the lubricant material. Wide recesses will also lead to a gross lack of uniformity in the surface strength of the bearing surface because the lubricant composite and the metal matrix are likely to have very different moduli.

Overall, there are therefore advantages in using a high areal concentration of small, shallow recesses. The ultimate in this respect may be the patterns of shallow circular recesses produced by Lancaster, using a photo-lithographic chemical etching process[211]. The patterns obtained are shown in Figure 8.6. The chemical etching technique also has advantages in avoiding the mechanical stressing and work-hardening which are likely to be inevitable with machining of recesses.

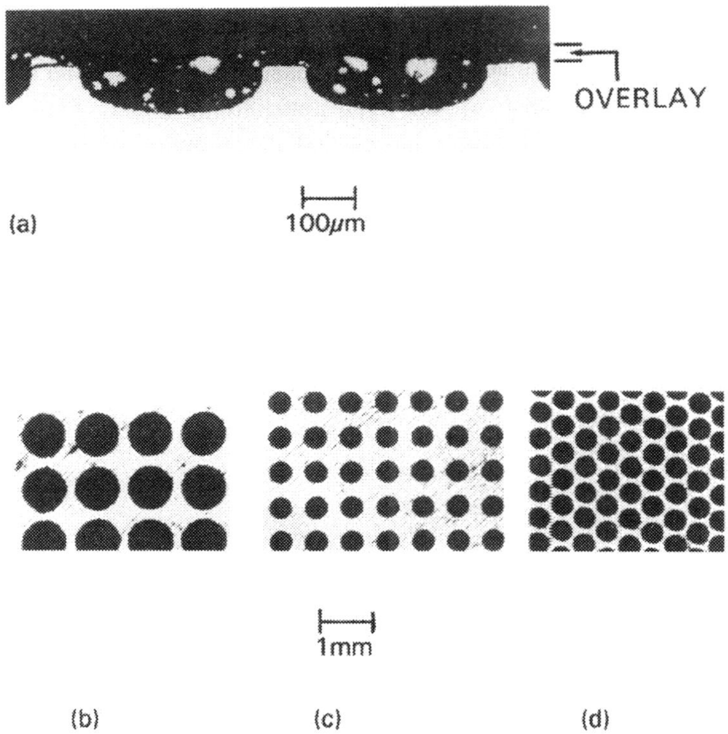

Figure 8.6 Etched-Pocket Lubricant Reservoirs:
(a) Pockets in Beryllium-Copper Filled with PTFE and Lead
(b),(c),(d) Plan Views of Alternative Patterns
(Ref.211, Courtesy of J.K.Lancaster)

technique also has advantages in avoiding the mechanical stressing and work-hardening which are likely to be inevitable with machining of recesses.

8.5.2 Indirect Transfer Lubrication The use of composite cages or retainers to lubricate the races and rolling elements of rolling bearings is in a way intermediate between direct and indirect transfer lubrication, having some features of both. The cage or retainer is itself an essential component of the bearing design, and the composite will transfer lubricant directly to the races or the rolling elements, but secondary transfer will then take place between the races and the rolling elements.

The clearest examples of indirect transfer lubrication relate to gears, in which a separate idler gear which is not part of the basic gear-train is used only to transfer lubricant to one or more of the load-transmitting gears. Paul H Bowen of Westinghouse Research Laboratories carried out a series of tests[212] on transfer lubrication of spur gears in 1963. The tests did not involve molybdenum disulphide, but a number of composites containing the similar dichalcogenide, tungsten diselenide. The results are therefore described in Chapter 14, but the test equipment provides an interesting example of the use of a lubricant composite idler gear for transfer lubrication, and is shown in Figure 8.7.

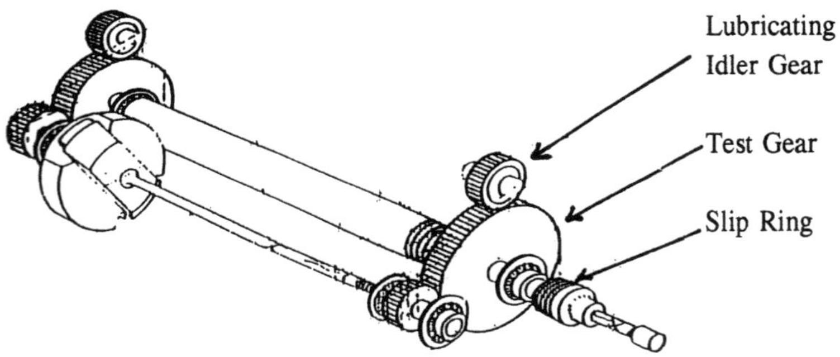

Figure 8.7 Use of Lubricating Idler Gears to Lubricate Gear Set (Ref.212)

A later example which used a molybdenum disulphide composite was described by Drozdov[207]. His gear arrangement is shown diagrammatically in Figure 8.8. By using two lubricating idler gears he was obviously able to reduce the distance over which lubricant transfer needed to be propagated.

The input speed to the gear train was varied from 1500 to 4500 rpm, the torque from 0.025 to 0.5 Nm, and the contact load on the idler gears from 0.5 to 30N. The peripheral speeds of the gears were up to 3.3 ms^{-1}, the relative slip 1 ms^{-1}, and the maximum contact stress 900 MPa. The test temperature was varied between 20°C and 250°C, and the chamber pressure was 10^{-8} Torr. 24 lubricating gears containing different proportions of copper, silver and molybdenum disulphide were tested, and the best performance was given by a composite of 87% copper, 5% silver and 8% molybdenum disulphide.

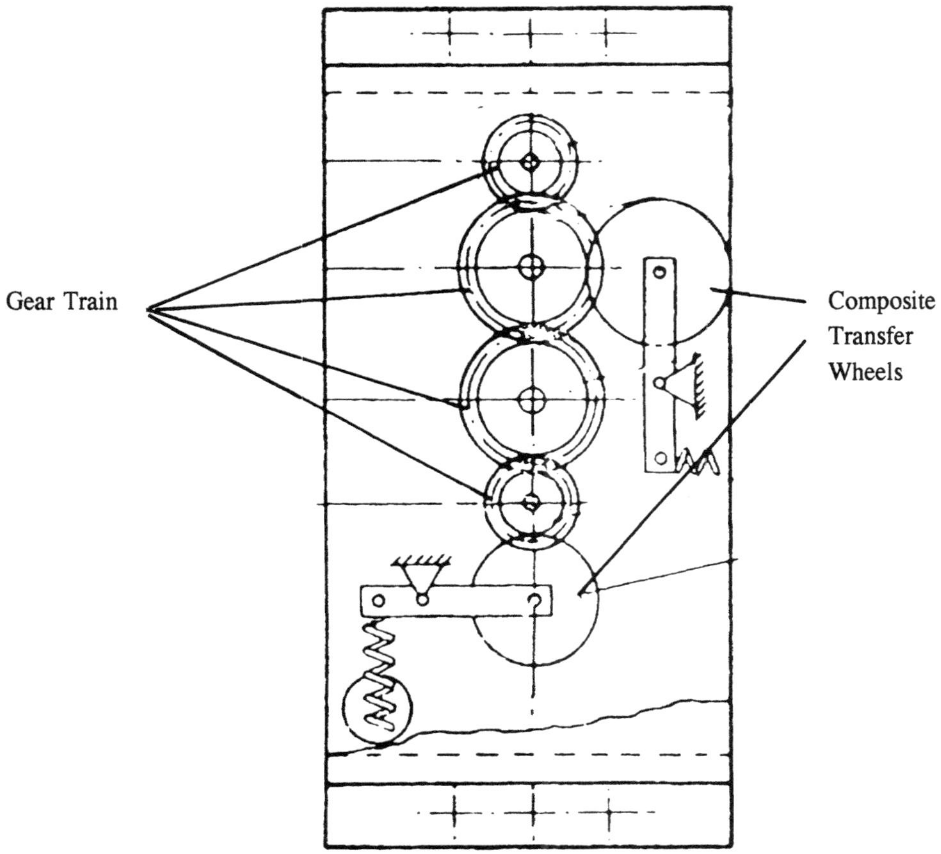

Figure 8.8 Transfer Lubrication of a Gear Train (Ref.207)

This result is interesting in relation to the comment by Lancaster, quoted previously, that because of the inefficiency of the transfer process, the concentration of solid lubricant needs to be at least 25%. It may be that in Drozdov's work the copper and silver also transferred and made a useful contribution to the lubrication, and this seems quite possible in high vacuum. Certainly the quality of the lubrication provided was quite good, since the life of the gear sets was generally over 100 hours at 4,500 rpm and 0.025 Nm torque. The lubricant effectiveness was not affected by the load applied to the lubricating gears. The life of the test gears was limited by wear of the lubricating idler gears, which was found to increase with the applied load and to be proportional to the number of load cycles.

There have been a few practical uses of the indirect transfer process for gears in aerospace applications, but such systems do not appear to have been produced commercially, or used in terrestrial applications.

~~~~~~~~~~~~~~~~
~~~~~~~~~~~~~~~~

CHAPTER 9.

LUBRICATION BY MOLYBDENUM DISULPHIDE ALONE

9.1 DIFFERENT TECHNIQUES OF USE

Molybdenum disulphide is intrinsically an excellent lubricant. No details exist about the way in which it was used in the distant past, but almost certainly it was first used as a free solid. While not deliberately mixed with any other material, it would have been fairly impure. In recent years it has become usual for it to be used in conjunction with other substances to improve some specific property or to overcome some specific problem. From such points of view as ease of application, re-supply, quality control, corrosion prevention or service life, there can be significant advantages in using it in combination with other materials. Nevertheless, there are still many situations for which its use unmixed with other substances can provide a satisfactory, or even optimum, solution.

Considered purely from the aspect of friction reduction, molybdenum disulphide in its fully oriented hexagonal crystal form is one of the two or three best lubricating substances yet recognised. It follows that when, in order to improve some other aspect, it is used in conjunction with an inferior substance, the frictional properties of the resulting mixture or composite are inherently likely to be degraded. In fact, even when used in conjunction with the other outstanding low-friction material, PTFE, the resulting composites are found, as will be shown later, to have frictional properties inferior to either of the two components separately. In the same way, the high thermal and chemical stability of molybdenum disulphide are likely to be diminished if it is mixed with less stable materials.

There are basically seven different ways in which molybdenum disulphide can be used alone, namely as a free powder, dispersed in a liquid, as a compact, by *in situ* formation, as a burnished film, as a transfer film or in a sputtered film. Some advantages and disadvantages of these various techniques are summarised in Table 9.1, and they will be described more thoroughly in this and the following chapters.

Table 9.1 Processes Using Molybdenum Disulphide Alone

Process	Advantages	Disadvantages	Applications
Free powder	Simplicity	Messy, not very effective	Open gears, screw threads
Dispersion in water	Convenience, fire resistance	Limited storage life	Metalworking
Dispersion in volatile liquid	Rapid evaporation of carrier	Limited storage life	Anti-seize, assembly
Dispersion in gas	Continuous feed	Complicated supply	Rolling bearings
Burnished film	High quality film and good	Laborious, limited to simple shapes	Shafts, bushes, slideways
Sputtered film	load-carrying capacity	Complex equipment required	Vacuum
In situ film		Complex process	Complex shapes
Compact	Mouldable to required shape	Low structural strength	Test work, transfer lubrication
Transfer film	Simple, possible to supply continuously	Quality control difficult	Bearings or gears

For clarity it may be worth mentioning at this point that in all these techniques other substances may be present in small quantities, either by accident or design, so that the description as "molybdenum disulphide alone" is not absolutely accurate. However, the phrase is useful in practice to distinguish these applications from those in which other materials, especially binders, metals or polymers, are present as significant, or even major, components and have a major effect on properties or performance.

9.2 USE IN FREE POWDER FORM

Many types of moving contact can be lubricated by simply feeding molybdenum disulphide powder into the contact zone either before operation or even during operation. This is probably the simplest of all techniques for using it, but there are several disadvantages. The powder is easily scattered and is black, so that the procedure can be very messy. It is also difficult to achieve uniform distribution, so that lubrication may be unsatisfactory where there is insufficient lubricant, or the powder may cause jamming where there is an excess present.

One useful way of applying the powder quite controllably is to use a vessel like a pepper-pot or sugar-sifter. The application can be controlled by the size and distribution of holes in the dispenser, and by applying a steady vibration to it. Some experimentation will always be necessary in order to achieve satisfactory application of powder. Any change in the particle size of the powder will significantly affect the distribution, so that once a suitable technique has been developed, care must be taken to maintain the particle size and the purity. Any major change in humidity will also affect particle flow and film formation.

The performance of molybdenum disulphide as a loose powder is limited, and the important factor is to convert it into a strong, adherent, well-oriented film. This is the practical significance of burnishing. Although such a film may be produced during operation of certain systems various techniques have been developed for producing consistent, effective burnished films on components from powder. These are described in Section 9.7, together with some practical applications. Use of the free powder directly in lubricated systems is a low-technology procedure, and there is a lack of detailed reports of such applications. In 1976 Yu N Drozdov et al[213] described a novel technique to improve the distribution of molybdenum disulphide powder over a train of low-alloy steel gears. They mixed the powder with ferromagnetic cobalt disulphide powder, and mounted a permanent magnet beneath the gear case, which was made of a non-magnetic stainless steel. When the mixed powder was poured into the gearcase, it distributed itself in a fairly uniform film over the magnetised gears and the bottom of the gear case.

They found that the gear train ran successfully up to 400°C, with efficiencies as high as 97%. On conclusion of the tests the gear teeth were found to be coated with a thin film of molybdenum disulphide. Drozdov and Egorov[214] found that the technique was suitable for heavily-loaded gears in vacuum, and Pavlov[215] published calculations of the film thicknesses and contact stresses arising. Some further developments were carried out by Vaisfeld[216], who used nickel powder instead of cobalt disulphide as the ferromagnetic component in the mixture. He found that one part of nickel powder to three of molybdenum disulphide was effective in tests in air, vacuum and carbon dioxide.

Table 9.2 Lubrication by Molybdenum Disulphide in a Gas Stream

Gas	Bearing type	Load	Speed (rpm)	Temperature	Life	Ref
Air	25.4mm bore ball bearing	0.45kg thrust	1725	to 540°C	1hr each at 200° 315°,430° and 540°C	217
Air	75mm bore roller	168kg radial	12000	230°C	9.1hrs	217
Air	20mm bore ball bearing	50kg thrust	2500	to 540°C	20hrs	218
Nitrogen	20mm bore ball bearing	50kg thrust	2500	540°C	2hrs	219
Nitrogen	20mm bore ball bearing	45.4kg thrust	10000	660°C	10hrs	221
Nitrogen	Four-ball machine	5.3kg thrust	700	540°C	30mins	220

A more versatile way of applying free powder is in a gas stream. This has the potential advantages of more uniform distribution, penetration into less accessible positions, and re-supply during operation. Several applications of this technique have been described, and some are summarised in Table 9.2. These applications have apparently been restricted to the lubrication of rolling contacts, probably because of the inherent difficulty of applying satisfactory bonded or burnished films to rolling elements, and the problem of re-supply.

According to the published reports the tests were generally satisfactory, but there appear to have been no operational uses of the technique, and little or no further testing in the intervening forty years. This may be partly due to the complication of arranging a suitable feed system, and partly to the satisfactory development of alternative techniques such as composite retainers and transfer films, and lead lubrication for rolling bearings in vacuum[222]

K Müller[223] studied the effectiveness of various lubricants, including free molybdenum disulphide powder, in preventing fretting. His assessment of effectiveness was based on the friction energy per cycle, which is not in fact a satisfactory criterion for prevention of fretting. He found that initially molybdenum disulphide powder was very effective in reducing friction, but subsequently there was a steady deterioration until the performance was almost identical with that in unlubricated tests.

Table 9.3 Weight Loss under Fretting Conditions (Ref.223)

Surface Treatment	Wt.Loss(mg)
Unlubricated	44.6
Molybdenum disulphide powder	22.2
Tricresyl phosphate	9.5
Molybdenum disulphide + tricresyl phosphate	5.6,7.8

His results are presented in Table 9.3 and show that overall the best performance was obtained by rubbing molybdenum disulphide into the surfaces before testing, and then lubricating with tricresyl phosphate. However, comparison of those results with those for tricresyl phosphate alone or molybdenum disulphide alone gives no clear indication of any synergistic effect from using the two lubricants together, or of any benefit from rubbing the molybdenum disulphide onto the surface instead of using free powder.

Loose molybdenum disulphide powder does not have a strong tendency to adhere to surfaces, and can be easily scraped off unless the conditions are such that the powder is burnished onto the surface. This probably occurs in rolling or sliding/rolling contacts such as in rolling bearings or many types of gear, but is ineffective in pure sliding contacts.

9.3 DISPERSIONS

Some of the difficulties in applying free powder to a surface can be overcome by dispersing the molybdenum disulphide in a volatile liquid. This enables it to be applied uniformly, in controlled quantity, and with much less tendency to escape from the desired working space. Evaporation of the liquid carrier then leaves a film of powder of remarkably uniform thickness on the surface.

The usual commercial forms of molybdenum disulphide powder are not readily wetted by water or by some organic liquids. They tend either to float on the surface of the liquid or to sink to the bottom, and do not form satisfactory dispersions. The technique of floating powder on the surface of water, and lifting it off onto a surface, has been used experimentally to produce uniform films by Matsunaga and others, as described in Chapter 6. Such films can be very uniform, but are usually too thin for practical applications.

In order to produce satisfactory dispersions for lubricant use, the wettability of the powder by the carrier liquid must be increased, and there are basically two ways of achieving this. The conventional way is to use a stabilising or wetting agent. It seems possible that most commercial dispersions are prepared in this way, although some published accounts imply that satisfactory dispersions have been produced in acetone, ethyl alcohol or polyglycol without any use of wetting or stabilizing agents.

An alternative method of improving wettability was first described by Groszek[175,224], and has been described in Section 7.6. This is the production of oleophilic molybdenum disulphide by ball milling in oil. The product is readily wetted by organic liquids, and can in fact be used as a grease thickener. Groszek described dispersions in several different organic liquids[225] and it is clear that the improved wettability leads to the formation of much more stable dispersions.

Dispersions can be applied to a bearing surface by brushing, dipping or spraying. The most suitable process depends on the size and shape of the component, and the viscosity and concentration of the dispersion. The liquid then evaporates, with or without the use of heat or enhanced airflow, leaving a thin but uniform and often adherent film of molybdenum disulphide powder. The adherence of the film is adequate in many applications to retain the molybdenum disulphide in position during subsequent burnishing or during the initial stages of operation while a run-in cohesive film is being formed. Multiple applications of a dispersion can be used to produce a thicker film, and the subject of optimum film thickness has been discussed previously in Chapter 6. Multiple applications seem to be effective in thickening the film only if only mild or no burnishing or running-in takes place between applications, but repeated applications can be used with some degree of success to repair a worn film.

One practical application of dispersions was the use of a brushed-on 50 : 50 mixture of molybdenum disulphide powder and petroleum solvent for the lubrication of freight car centre plates, which was at one time specified by the Association of American Railroads.

Dispersions have been very widely used in many metalworking applications, especially where high forming pressures are used, such as in hot or cold drawing of rod, tube, or wire, extruding, deep drawing, ironing or broaching, but there have been relatively few detailed published reports. Nittel[226] reviewed the current status of lubrication in cold extrusion of steel in 1992. He described the use of molybdenum disulphide dispersions for high reductions, high extrusion pressures and high temperatures. Compared with the alternative of reactive soaps they gave higher pressing and ejection pressures, and the extruded component surfaces were smoother. However, he also mentioned the innovation of "black soaps" in which aqueous dispersions of molybdenum disulphide are added to reactive soaps.

Kuznetsov and Simonov[227] reported the use of a dispersion of molybdenum disulphide and graphite powders in ethyl alcohol applied to the surface of a bore to provide lubrication for a broach burnishing operation. They found that a satisfactory finish was achieved provided that all or most of the alcohol had evaporated before burnishing began, and the results were better than with conventional boundary lubrication.

A Japanese study assessed several different lubricants in a test designed to simulate a cold forming operation[228]. A tool steel flat was heavily loaded radially against the apex of a trapezoidal thread cut into a cylindrical rod of annealed aluminium. The load was high enough to deform the thread apex plastically and thus generate freshly-exposed surface. The ability of a lubricant to lubricate satisfactorily in relation to the amount of fresh surface generated was defined as its coverage or coverability. The lubricant powders were applied by spraying in the form of a dispersion in acetone. Molybdenum disulphide gave a much greater coverability than graphite or boron nitride, and maintained a constant coefficient of friction of 0.25 to 0.30 regardless of the extent of coverage.

The use of dispersions in polyglycol is a special case, because the purpose is not simply to deposit a uniform film of molybdenum disulphide by evaporating an inert carrier. The technique is in fact more comparable with the incorporation of molybdenum disulphide in an oil or grease, and will be described in Chapter 13.

One disadvantage of dispersions in water or light organic solvents is that they are generally not very stable in long-term storage. In order to achieve long-term stability it is necessary either to use larger quantities of stabiliser, or to use a very fine powder. The finest powders tend to have higher friction, lower load-carrying capacity and shorter life than coarser particles.

A number of molybdenum disulphide dispersions are listed in Table 13.4.

9.4 COMPACTS

In theory an alternative to the production of a lubricating film of molybdenum disulphide on the surface of a component is to manufacture a self-lubricating component out of solid molybdenum disulphide. This can readily be done with composites, in which the structural integrity is provided by a polymeric, ceramic or metallic matrix, but the potential advantage of using molybdenum disulphide alone has encouraged a number of investigators to study the manufacture and properties of compacts made from molybdenum disulphide powder alone. Compacts can be readily made by compressing molybdenum disulphide powder in a suitable die, usually under vacuum. Their formation and properties have been described in Chapter 8 in connection with their use for transfer lubrication.

At one time it was hoped that they might be used for the manufacture of self-lubricating machine components, but compacts of molybdenum disulphide alone had inadequate structural strength. Because the crystallite orientation is random, their coefficients of friction tend to be high, for example Matsunaga et al[229,230] reported 0.16 to 0.29 for compacts formed at 800 MPa_n and Brendle and Colin[231-233] 0.11 to 0.21 for a compact formed at 400 MPa.

Matsunaga prepared his compacts from dried molybdenum disulphide powder, compressed in a die without a binder, and studied their friction and wear behaviour in sliding tests against a low-carbon steel. Friction and wear both decreased with increasing sliding speed, and he suggested that this was caused by desorption of adsorbed contaminants due to frictional heating at the higher sliding speeds. This suggestion was supported to some extent by plotting all the friction results and all the wear results against a frictional heating parameter fPV. In each case the results all fell approximately on a single curve. Matsunaga also found that within the experimental repeatibility the friction decreased with increasing contact pressure, although the high friction values show that the crystal orientation in the compacts was random.

Tyler and Ku[201] carried out a useful study of the mechanical properties of compacts produced from well-characterised molybdenum disulphide powder. Compacts formed under mechanical pressure in a die with a single plunger showed variations in physical and mechanical properties throughout their length. This was considered to be due to the frictional resistance of the powder during pressing, which caused more effective compression near to the moving plunger. As a result the specific gravity, and especially the hardness, varied along the length of a compact, the hardness varying from 70 (on an arbitrary scale) at the plunger end to 34 at the other end. An alternative press system, in which two plungers moved simultaneously from opposite ends of an open die, gave more uniform compacts.

Some compacts obtained by an explosive forming technique were more uniform, with better structural integrity, and were more highly compressed, with a specific gravity of 4.62, compared with 4.26 for the pressed compacts and about 4.9 for fully dense molybdenum disulphide crystals. The maximum ultimate compressive strength for press-formed material was 20 MPa for a compact formed at 4000 MPa, but the ultimate tensile strength of similar material was only 1.7 MPa. This is generally inadequate for practical use, and in fact when a simple cylindrical die was used, many compacts fractured on removal from the die due to tensile strain. Although well-formed compacts could be machined satisfactorily, their low tensile strength made them difficult to use. Some practical applications of pure molybdenum disulphide compacts seem to have been made but their performance was considerably improved by the use of binders, and the resulting compacts are described in Chapter 12.

The use of antimony tetrasulphide (Sb_2S_4), also known as antimony thio-antimonate, is in a rather different category. Centers[192] showed that when 36 vol%

of the tetrasulphide was added to molybdenum disulphide prior to pressing, the resulting compacts gave lower friction and considerably lower wear loss. He suggested that the effect was caused by the soft additive facilitating the orientation of the lubricant crystallites during sliding. If this explanation is correct, then it would appear possible that a similar effect would occur during pressing of the compacts, thus improving consolidation and orientation, and this would also presumably reduce friction and wear. He also described similar but less powerful effects for other antimony sulphides and for antimony trioxide, and believed that the improved orientation was the cause of the improved life of films containing antimony trioxide, rather than the more generally-accepted anti-oxidant mechanism. Unfortunately he made no measurements of the physical and mechanical properties of the compacts produced, so that their suitability for practical applications was not clear.

9.5 IN-SITU FORMATION

Molybdenum metal will react on heating in hydrogen sulphide gas to form a surface film of molybdenum disulphide. The technique appears to have been first used for the production of low-friction surfaces in 1950 by F P Bowden[234], who obtained coefficients of friction as low as 0.06 at 300°C with sintered molybdenum having a surface coating of molybdenum disulphide formed in this way. He also found that porosity of the substrate enabled the conversion to penetrate to a considerable depth below the surface. In later work[235] he extended the procedure to include heating vacuum-cleaned molybdenum metal in hydrogen sulphide at 850°C, and heating oxidised molybdenum in carbon disulphide vapour. The coefficients of friction obtained by using several different processes are compared in Table 9.4.

There has been some discussion about whether pure molybdenum will in fact react with hydrogen sulphide to produce a surface film of molybdenum disulphide. The vacuum used by Bowden and Rowe to clean the surface of their molybdenum[235] was relatively soft, and it seems probable that some oxide remained on the surface. Later work with cleaner surfaces[236] showed that it is very difficult or even impossible to produce a molybdenum disulphide film on a really clean molybdenum surface by reaction with hydrogen sulphide.

Probably the most widely-studied and widely-used procedure for *in situ* formation of molybdenum disulphide films is based on conversion of an electroplated molybdenum coating by reaction with hydrogen sulphide at 204°C (400°F), and was first described in a U S Patent by Brophy and Ingraham in 1956[237]. In a later

Table 9.4 Friction of In Situ Molybdenum Disulphide (Ref.235)

		Coefficients of Friction								
Temperature (°C)		18	600	700	750	810	870	940	1020	1130
Denuded Mo	Vacuum	2.0	-	2.1	2.3	2.5	-	2.8	3.6	Seiz -ure
Denuded Mo	H_2S	0.9	.45	0.3	0.2	0.2	0.2	0.3	0.45	- 0.5
Oxidised Mo	CS_2	0.9	0.6	0.5	0.4	0.4	-	0.3	0.4	1.1 0.5
Rubbed MoS_2	Vacuum	0.1	0.2	0.3	-	0.3	.35	0.4	0.7	
In situ MoS_2	Vacuum	0.2	-	0.2	-	0.2	-	0.3	0.45	

publication the authors described the reaction conditions[238] as 100° to 200°C and 0.69 to 2.76 MPa (approximately 7 to 38 bars) gas pressure. The resulting coatings were said to be less than 2.5μm thick, but to have quite good lubricating properties and lives.

The validity of the description of this patented procedure was also challenged[239] on the grounds that molybdic oxide needs to be present in the plated film for the conversion to take place. It is interesting, therefore, that C E Vest[240] introduced an important modification to the process in order to electrodeposit molybdenum trioxide instead of molybdenum metal when studying the application of the technique to the lubrication of spacecraft components. Vest's modification was to use a mixed plating bath of molybdenum trioxide in an alkali salt solution such as ammonium formate. The component being coated formed the cathode, and the molybdenum was probably present in the bath in the form of ammonium molybdate.

A typical process would include some or all of the following stages:-

(a) Solvent degrease in an ultrasonic bath.

(b) Vapour blast to give the required surface finish (about 0.5 μm).

(c) Acid or caustic pickle to reduce surface oxide and activate the surface.

(d) Rinse with hot distilled water.

(e) Electroplate with a metallic base film such as copper.

(f) Electrodeposit molybdenum trioxide from the mixed bath with a current density typically about 18 - 20 A/m^2, and deposition time about 3 to 10 mins., depending on the thickness required.

(g) Place the coated component in a pressure vessel, evacuate to 10^{-3} Torr and fill with hydrogen sulphide to 25 - 30 bars, raise the temperature to 195° ± 5°C and maintain for between 4 and 12 hours.

(h) Bleed off surplus hydrogen sulphide either for re-use or into absorber bottles containing potassium hydroxide solution.

The film thickness should be between 1 μm and 6 μm, and can be controlled to within 1 μm. Figure 9.1 shows the variation of final molybdenum disulphide thickness with electrodeposition time for a series of tests at 18.6A/m^2 (12 mA/in^2).

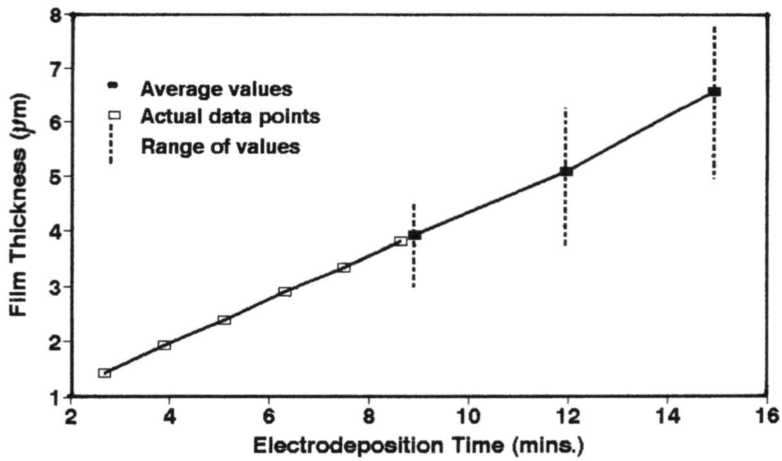

Figure 9.1 Effect of Deposition Time on In Situ Film Thickness

The coefficient of friction was found in Vest's work to be between 0.025 and 0.05, which suggested that the coating was well-oriented with a hexagonal crystal structure, but this was not definitely proved. The wear life was claimed to be better than that of burnished or inorganic-bonded films, but not as good as for organic-bonded films. However, this will obviously depend greatly on coating type and test conditions.

The basic Brophy and Ingraham technique[237] was studied by several other authors. Bayer and Trivedi[241] found that the effectiveness of the technique depended more on the nature of the coating than on its thickness, and that retained moisture in the electroplate was essential for effective conversion. They recommended a current density of 21.5 A/m^2 for 5 minutes to produce a coating thickness of 1.25 to 2.5 μm. Nishimura and co-workers[242,243] found that the presence of air or water or both in the conversion gas improved the wear life. Table 9.5 compares the properties of the *in situ* films with those of burnished and sputtered films, and shows superior wear life for the *in situ* films. Their friction results were curious, in that they found that the initial films which were formed gave low friction in air or nitrogen but not in vacuum. Low friction in vacuum was obtained when the initial product was heated in vacuum to 400°C.

Table 9.5 Performance of Different Molybdenum Disulphide Films (Based on Refs.242-244)

Film Type	Film Thickness μm	Friction	Wear Life, (revs to failure)	Wear Life/ Film Thickness
In Situ	4	0.04	1,577,000	39revs/Å
	4	0.02	1,865,000	47revs/Å
	4	0.03	1,391,000	35revs/Å
	4	0.03	1,526,000	38revs/Å
	4	0.03	1,392,000	35revs/Å
Burnished	0.57	0.035	11,700	2.1revs/Å
	0.42	0.05	12,400	3.0revs/Å
Sputtered	0.56	0.02	376,000	67revs/Å
Test conditions:440C ball/440C disc, 1kgf, 1.5m/s, dry air				

X-Ray diffraction examination and electron probe microanalysis showed[244] that the initial film was amorphous and contained an excess of sulphur. When the film was heated in vacuum at 400°C for one hour, the excess of sulphur decreased, and the X-ray diffraction pattern showed peaks possibly indicating the presence of molybdenum disulphide crystals. They inferred that the excess of sulphur initially present reacted with more molybdenum to produce additional molybdenum disulphide, while at the same time the amorphous molybdenum disulphide became converted to hexagonal crystals.

Several Russian papers described an alternative conversion process in which the initial molybdenum plating was converted to molybdenum disulphide by heating with elemental sulphur. Dukhovskoi[245] and co-workers heated molybdenum in contact with sulphur at various temperatures between 300° and 950°C. At temperatures up to 600°C they found the crystallographic structure to be a mixture of hexagonal and rhombohedral. Annealing at 800°C in a vacuum of 10^{-4} to 10^{-5} Torr converted the structure to hexagonal, but only if the surface of the film was covered for example by a second specimen. The authors suggested that residual sulphur was left in the film after the original conversion, and that the presence of this free sulphur was essential for the re-arrangement of the crystals. They assumed that if the film was not covered during the annealing process the free sulphur was lost and re-arrangement did not take place.

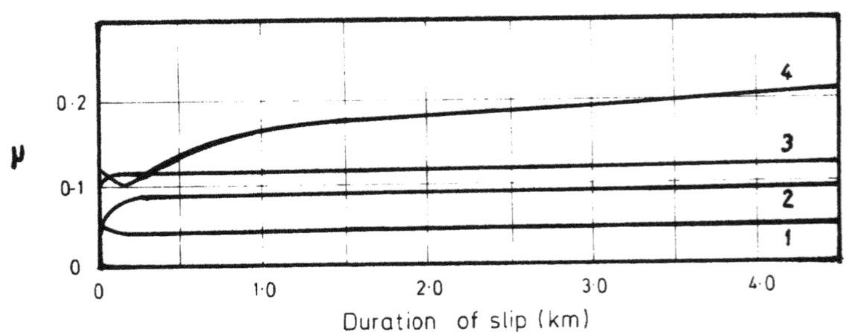

1 - 25°C   2 - 300°C
3 - 600°C   4 - 700°C

Figure 9.2 Variation of Friction with Life for an In Situ Film at Different Temperatures (Ref.246)

Figure 9.2 shows some results of friction testing at 5 x 10^{-6} Torr on coatings produced by Ermakov[246] by heating molybdenum in contact with elemental sulphur at 500° - 600°C. It is interesting that he found an increase in friction as the test temperature was raised, but that there was no film failure even at 700°C.

Table 9.6 Effect of Pretreatment on Wear Resistance of In Situ Coatings (Ref.247)

Molybdenum Pretreatment	$R_a(\mu m)$	Failure Load (kgf)
Abrasion	0.8-1.0	1.25-2.25
Abrasion + etching in nitric and sulphuric acid	3.3-5.05	2.25-2.5
Abrasion + etching in potassium hydroxide and potassium ferricyanide	3.0-3.4	1.75-2.0
Abrasion and sandblasting	3.3-3.6	3.5-4.5
Grinding with 50μm abrasive powder	3.8-4.2	8.5-10.0

The effect of surface texture of the molybdenum on the wear resistance of *in situ* coatings produced by a similar process was studied by Aparin[247] and co-workers. The molybdenum samples were annealed in vacuum at 800°C for one hour, and then sulphided at 575°C for 5 hours, giving a 35 - 40 μm thick film of molybdenum disulphide. They assessed the wear resistance in a pin-on-disc machine in which the contact load was increased progressively until failure occurred, and their results are summarised in Table 9.6. The results indicate wear resistance rather than seizure load, because wear took place progressively in all load stages, and the highest load attained is therefore a function of the wear life. Surprisingly, the best wear resistance was obtained with a ground surface of 3.8 to 4.2 μm, whereas most other comparable studies have recommended a grit-blasted surface finish of 0.5 to 1.0 μm. It is possible that in this work some change in surface texture took place during the conversion process.

Several other techniques for producing *in situ* coatings have been described. A French paper[248] described direct electrolytic deposition of molybdenum disulphide and molybdenum sesquisulphide from a mixed aqueous bath of ammonium molybdate and ammonium sulphide. G W Rowe and his co-workers produced a molybdenum disulphide coating on molybdenum wire[249] by electrochemical deposition from a molten bath of lithium chloride/potassium chloride eutectic containing dissolved sodium sulphide at 400°C. Yajima[250] and co-workers obtained molybdenum disulphide by heating molybdic oxide with carbon in a stream of sulphur dioxide at temperatures above 1000°C, although at lower temperatures the product also contained molybdic oxide.

It seems clear that Mizutani et al[251] produced something approaching an *in situ* molybdenum disulphide, although this was not the objective of their work. They studied the self-lubricating behaviour of a series of ternary iron-molybdenum-sulphur alloys containing 3% by weight of sulphur and between 15% and 20% by weight of molybdenum. By X-ray diffraction they detected the presence of a molybdenum sulphide $MoS_{1.8}$ with a structure similar to that of molybdenum disulphide. This is interesting in view of the occurrence of similar products in sputtering of molybdenum disulphide (see Chapter 10). They measured coefficients of friction as low as 0.02 in vacuum, but in air the low-friction properties were rapidly lost due to oxidation.

More recently there has been some interest in producing molybdenum disulphide by electrochemical deposition. Much of this work has been aimed at producing large crystals for use as semi-conductors, but Ponomarev et al[252] reported the production of highly-textured films with basal plane orientation by cathodic electrochemical deposition from tetrathiomolybdate solutions, followed by annealing at 550°C in argon.

The published results of studies on *in situ* molybdenum disulphide films have almost always been very promising, and there have also been many reports of successful practical applications[240,253]. In view of this it seems surprising that since about 1982 there has been relatively little information published about them, whereas sputtering has become much more widely used. This is in spite of the fact that sputtering requires more sophisticated equipment and techniques, and in general produces less highly oriented films with higher friction than has been reported for many of the *in situ* films. One important cause of this situation is probably that with experience a high level of quality control and consistent performance can be achieved with sputtering. It can only be a matter for speculation whether a similar high level of quality control could have been obtained with *in situ* films if the same amount of effort had been devoted to their development. However, it is certain that sputtering cannot be applied to some components because of their size or shape, and a better understanding of *in situ* film use might be of great value for such applications.

Considerable doubt and controversy exist about the mechanism of action of a number of other molybdenum compounds or mixtures in lubrication, and in particular about whether their action involves *in situ* formation of molybdenum disulphide. Braithwaite[254] has stated that one of the reasons for the development of soluble molybdenum compounds was that it "was thought by some (erroneously) that Organo-molybdenum sulphur compounds would decompose to MoS_2 which would then provide lubrication." Nevertheless Braithwaite goes on to refer to several other authors, such as Feng et al[255], Isoyama and Sakurai[256], and Yamamoto et al[257], all of whom maintain that they have shown the formation of molybdenum disulphide from organo-molybdenum sulphur compounds. Similar results have been reported by Farr[258], Kasrai et al[259], and Nagakari et al[260].

Many different organo-molybdenum sulphur compounds have been studied in this context, and some are listed in Table 9.7. All those listed have been found to have friction-reducing, anti-wear or extreme-pressure properties, and it is significant that several authors have reported that the presence of sulphur is essential.

Detailed investigations of the mechanism of action of these oil-soluble molybdenum compounds have not yet been performed, but there is some evidence to suggest that the mechanism is similar to that of molybdenum disulphide.

(i) All the effective compositions apparently contain both molybdenum and sulphur.

(ii) Surfaces of iron-containing components showed the presence of molybdenum after the tests[54].

(iv) With or without the presence of iron, DTA curves for the oil-soluble compounds were similar to those for molybdenum disulphide[54].

It was suggested that "as in the case of MoS_2" the oil-soluble compounds react chemically with the metal surface, giving a system including molybdenum, iron, sulphur and oxygen. As has been shown earlier, this comment is inaccurate. The oxidised condition in molybdenum disulphide lubrication represents the end of its useful life, so that this mechanism provides no explanation for the lubricating action of the oil-soluble molybdenum compounds.

Isoyama and Sakurai[256] investigated the mechanism of pyrolitic breakdown of some molybdenum-containing dithiocarbamates. They found that the products contained molybdenum disulphide and inferred that this at least partly explains the EP activity of such additives in lubricants.

Table 9.7 Some Organo-Molybdenum Compounds Studied for Lubricant Performance

Molybdenum dithiocarbonates
Molybdenum dithiophosphates
Molybdenum dithiolates
Molybdenum dialkyldithiocarbamates
Molybdenum diallyldithiocarbamates
Molybdenum naphthenate + sulphur compound
Alkylammonium molybdate + sulphur compound
Tris(N-p-methoxyphenylthiosemicarbazido)molybdenum

The various types of compound and their mechanisms of action have been reviewed by Mitchell[261]. He concluded that the question of the extent to which the molybdenum additives decompose to produce molybdenum disulphide, if at all, was not yet resolved and certainly depended on the type of compound and the operating conditions. There was little doubt that molybdenum dithiocarbamates form molybdenum disulphide, probably by decomposition at hot spots caused by asperity interactions. It was less certain that molybdenum dithiophosphates form molybdenum disulphide, although mixtures of zinc dialkyldithiophosphate and molybdenum complexes had been shown to do so[255].

Similar materials were used[262] by Skeldon et al to produce a self-lubricating surface film on aluminium. They first produced a porous alumina film on the surface by anodizing, and then re-anodized in an electrolyte containing 0.01M ammonium tetrathiomolybdate. They then found that the pores of the primary anodized film contained mainly amorphous molybdenum trisulphide, and this was converted by vacuum annealing to hexagonal molybdenum disulphide. The film gave a marked improvement in wear resistance.

Jin and Zhou[263] found that molybdenum disulphide was formed by a different approach which did not depend on the use of a soluble molybdenum compound. They were studying the performance of a mixed ether/ester oil containing 3% of a zinc dialkyldithiophosphate at 450°C, in contact with a molybdenum alloy. They found that an *in situ* film of graphite and molybdenum disulphide was formed, and the presence of the graphite produced a reducing atmosphere which helped to delay oxidation of the molybdenum disulphide. The films were said to give excellent lubrication at 450°C with friction between 0.04 and 0.07.

The effect of operating conditions in the formation of these in situ films is likely to be complex. In 1972 Forbes showed[264] that even under fairly mild conditions with an anti-wear additive such as tricresyl phosphate surface films are formed which contain both phosphorus and organic fragments. Similarly Cann[265] showed that a cross-linked thick surface film is formed from zinc dialkyldithiophosphate under mild rubbing conditions. It seems probable that at least some of the soluble molybdenum compounds would also experience partial breakdown under mild rubbing conditions, and that these would encourage the formation of protective surface films.

Under more severe rubbing conditions, further breakdown occurs, and the products are likely to be mainly inorganic, such as phosphates and sulphides. Finally, where contacts are severe enough to remove the protective oxide and leave exposed fresh unoxidised metal, Morecroft[266] found that organic molecules, even saturated hydrocarbons, will break down to give fragments such as carbon, hydrogen and carbon monoxide.

If such a progression occurred with the organo-molybdenum sulphur compounds, a very likely final product would be a molybdenum sulphide. The Russian investigations[54-57] showed the breakdown of the complex oil-soluble compounds to simpler products and the presence of molybdenum in the steel surfaces. The system also contained sulphur and almost certainly nascent hydrogen, whose presence in the boundary layer has been demonstrated in other situations. Under those circumstances the production of some molybdenum disulphide would be not only possible but probable, even if only as a transient phase and as a monomolecular film at the asperity contacts. However, the formation of both molybdenum disulphide and its hexagonal crystalline form seem to be favoured at higher temperatures, so that there is nothing inherently unlikely about hexagonal molybdenum disulphide being an end-product from the organo-molybdenum sulphur compounds.

At one time it seemed less likely that molybdenum disulphide could be produced in sufficient quantity to provide effective lubrication, even of asperity tips, by any mechanism resembling the conventional engineering applications of molybdenum disulphide. However, recent studies of sputtering have given strong indications that thin residual films of sputtered molybdenum disulphide only a few hundred Å thick can provide useful lubrication in practical engineering components, and this will be discussed in the next chapter.

It is therefore difficult to avoid the conclusion that where molybdenum and sulphur are present together in conditions of temperature and shear which can encourage molecular rearrangements, there is a probability of molybdenum sulphides being formed. If then high enough temperatures are present, either superimposed or from frictional heating, the formation of a hexagonal crystalline sulphide similar or identical to molybdenum disulphide will be formed, and this may be in sufficient quantity to provide effective lubrication.

The behaviour of these oil-soluble compounds is of great practical importance, because of the potential value of effective oil-soluble additives, especially for friction reduction. Whether they act by *in situ* formation of molybdenum disulphide is of more academic interest, since an understanding of their mechanism is important mainly in indicating the best lines of future development.

8.6 BURNISHED FILMS

The fundamental aspects of the formation and properties of burnished films have been discussed in detail in Chapters 6 and 7. At this point it is proposed only to describe some practical factors in the preparation and use of burnished films produced directly from powder.

The process of preparing burnished films of molybdenum disulphide from powder can be summarized as follows. The substrate material is selected at a suitable hardness to meet the design requirements of the component to be lubricated, but keeping in mind the fact that very high substrate hardnesses (probably greater than 800 VPN) will adversely affect film life. The surface is degreased and a surface finish of 0.5 to 1.0μm is obtained, preferably by grit-blasting. If desired, a surface coating such as phosphating can be applied, and this may be particularly useful on a very hard substrate. (For substrates other than steel, other conversion coatings such as those listed in Table 11.6 may be useful.)

A small quantity of molybdenum disulphide powder, preferably between 1 μm and 10 μm particle size, is then placed on the surface and rubbed smoothly into it with a soft cloth, using a firm orbital polishing action. Alternatively the powder can be placed on a cloth or impregnated into a piece of synthetic sponge and applied in the same way. The technique for applying the powder to the surface is not critical, and satisfactory films can be obtained by many different techniques. Additional powder can be added during the burnishing process to ensure complete coverage and to increase the film thickness. The application pressure should be progressively increased during the burnishing, and the lowest friction and highest load capacity will be obtained with a high final burnishing pressure.

Figure 9.3 Device Used to Apply Burnished Coatings to Flat Surfaces (Ref.122)

A high level of repeatibility of the burnishing process requires careful control of the process, including control of the rate and total quantity of powder applied, and preferably the use of some form of mechanical burnishing system. The following are some examples of equipment which has been used.

(i) Johnston and Moore[122] used the device shown in Figure 9.3 to apply burnished coatings to flat surfaces. The burnishing pad was a steel cylinder covered with six layers of terylene cloth, and was oscillated at 40 cycles/min. along a 6 cm. straight-line path under a load of 100 to 1000 g. through a heap of loose molybdenum powder on the flat surface. This gave satisfactory repeatible coatings for experimental work, but later information showed that reversing the direction of burnishing had an adverse effect on friction and film life.

(ii) Giltrow[267] used a modified polishing machine to coat steel discs. The discs were rotated for five seconds under a load of 1 kg. against powder spread on a nylon cloth stretched tightly over a rotating brass table.

(iii) De Gee and Salomon[121] used the specially-constructed machine shown in Figure 6.1 in which the outer surface of a ring or cylinder was coated with a burnished film by means of a neoprene sponge impregnated with molybdenum disulphide powder. The sponge was pressed against the cylindrical surface under a load of 1.4 kg while the cylinder rotated at 100 rpm. The sponge moved in a reciprocating path about 12 mm long parallel to the axis of the cylinder in order to ensure more uniform coverage. In this case there was no actual reversal of the contact movement so that the potential problem of film life impairment did not arise.

(iv) Brewe, Scibbe and Anderson[268] burnished ball bearings by running the assembled bearing under a 5lb load at 1725 rpm on a burnishing fixture. Molybdenum disulphide was worked into the races by means of a wire brush before assembly, and was fed into the bearing during rotation from a plastic squeeze bottle every 10 minutes for 3 hours.

Burnishing of more complex small shapes can be done by tumbling them in molybdenum disulphide powder, and pieces of wood, cork, walnut shells or pine cones have been added to help the burnishing process. Larger components of complex shape can be burnished by a similar process, but the component should be held between suitable chucks and rotated at low rpm, with the molybdenum disulphide powder and the pieces of burnishing material held inside a shroud around the component. Quality control in such processes can be difficult because of the problem of achieving uniform film deposition.

The thickness of the burnished film tends to increase with continued burnishing provided a supply of powder is assured, and the film life increases with film thickness. The rate of film build-up increases at higher relative humidity, and it is difficult to obtain burnished films thicker than about 10 μm except in high humidity. It also follows that changes in relative humidity during film preparation will affect the repeatibility of the process. The endurance of a burnished film increases with the thickness of the film, but decreases with high relative humidity during operation. The greatest endurance will therefore be obtained with films produced at high humidity and operated at low humidity. Where the film is to be operated under high load, better performance will be obtained if a running-in stage at intermediate loads is introduced after the burnishing. The coefficient of friction is reduced after operation at high load, and values as low as 0.05[269] can be achieved.

The optimum conditions for producing burnished films with good endurance and load-carrying capacity and low friction are:-

1. Substrate surface finish 0.5 to 1.0μm
2. Medium grade of powder, possibly about 2 μm, high purity

3. Burnish at a load of several kilograms
4. Burnish with smooth polishing action
5. Burnish at higher humidity (> 40% R H)
6. Run-in at intermediate and then at high load for a short period

The application technique is straightforward, although more complex than for bonded films, but there is good control over film thickness, and the performance may be as good[269] as for bonded films.

If a burnished film is prepared under mild rubbing pressure it will be relatively non-reflective and the friction will be high ($\mu > 0.08$). Running-in during the early stages will then convert it to a highly reflective film with lower friction. The course of the running-in process can be followed by monitoring the friction, and Figure 6.3 shows a friction trace for a burnished film from the commencement of sliding to the eventual failure. There are some benefits in using the running-in process to achieve the final consolidation of a burnished film, since the pre-treatment effort is minimised, the final film conforms well to the counterface, and the degree of consolidation is suitable for the operating conditions.

Since commercial dispersions and bonded coatings became widely available, the use of powder-burnishing has probably declined, but applications to ball-bearings have been reported[270,271], and a 1981 report described a cold-extrusion machine for steel components in which a burnished film of molybdenum disulphide was applied to the steel billets automatically prior to extrusion[272]. This resulted in a reduction in extrusion loads and improved quality.

Other industrial applications still exist, but the use of dispersions is cleaner and more convenient. Burnishing of molybdenum disulphide films applied by means of dispersions can be carried out in exactly the same way as for free powder, and the resulting burnished coatings have similar properties, but there are no detailed reports about them other than those of Matsunaga[118] described in Chapter 6. Films from dispersions will also be burnished in use by the effects of sliding under contact load, and their eventual form and behaviour are likely to be similar in all respects to those produced from loose powder. Similar burnished films are likely to be the end-product of many of the softer bonded coatings, and these will be discussed further in Chapter 11.

~ ~ ~ ~ ~ ~ ~ ~ ~ ~ ~ ~ ~ ~ ~ ~

CHAPTER 10.

SPUTTERING AND OTHER PHYSICAL DEPOSITION PROCESSES

10.1 THE SPUTTERING PROCESS

There are many physical deposition (PD) processes which can be used to deposit lubricating films on surfaces, and several of them have been used, either separately or in combination, for depositing molybdenum disulphide. They include Ion Beam Enhanced (or Assisted) Deposition (IBED or IBAD), and Pulsed Laser Deposition (PLD), but the most important so far is sputtering, or more precisely sputter-coating.

Sputtering, or cathodic bombardment, has been intensively investigated and is now widely used for producing molybdenum disulphide coatings. It was first described for that purpose by Spalvins and Przybyszewski[273] in 1967, and for the past fifteen years it has probably been the one single aspect of molybdenum disulphide lubrication about which most papers and reports have been published. There are several reviews of the subject[274-276], although none gives a really comprehensive picture of this rapidly-developing field.

The basic principle of sputtering is most easily explained in relation to DC (direct current) sputtering, and is shown diagrammatically in Figure 10.1. The chamber contains a gas, almost invariably argon, at low pressure. A potential of the order of several hundred volts is established between the anode and the thermionic filament cathode, and the latter is heated to emission temperature. The resulting flow of electrons through the low-pressure argon creates a glow discharge and a plasma of ionised argon atoms.

A negative potential of the order of 2 to 4 kV is then applied to the specimen which is to be coated. Positive argon ions from the plasma bombard the specimen substrate, and this high-energy bombardment generates a clean etched surface. The high negative potential is then switched to the target, which consists of the material which is to be used for coating. In the case of molybdenum disulphide sputtering, the target will usually be a pressed compact of high purity molybdenum disulphide. The ion bombardment of the target causes sputtering, with the emission of particles of atomic or molecular dimensions from the target surface. Sputtered particles re-deposit on the substrate surface, and the sputtering process is continued until a coating of the required thickness is obtained.

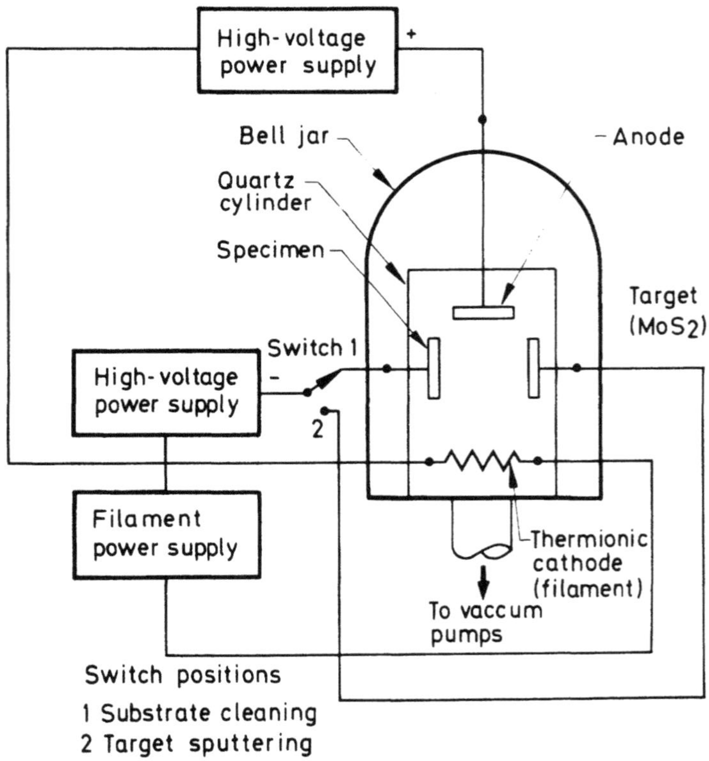

Figure 10.1 Schematic Diagram of a Typical D-C Sputtering System (Ref.559)

DC sputtering cannot be used with targets which are electrically non-conducting, because impingement of the positive argon ions creates a positive charge on the surface of the target, which then ceases to attract the ions. It can be used for semi-conductors such as molybdenum disulphide, and some of the earliest work on molybdenum disulphide sputtering was by the DC process[19,277], but there can still be some tendency for polarisation, which limits the flexibility of the procedure.

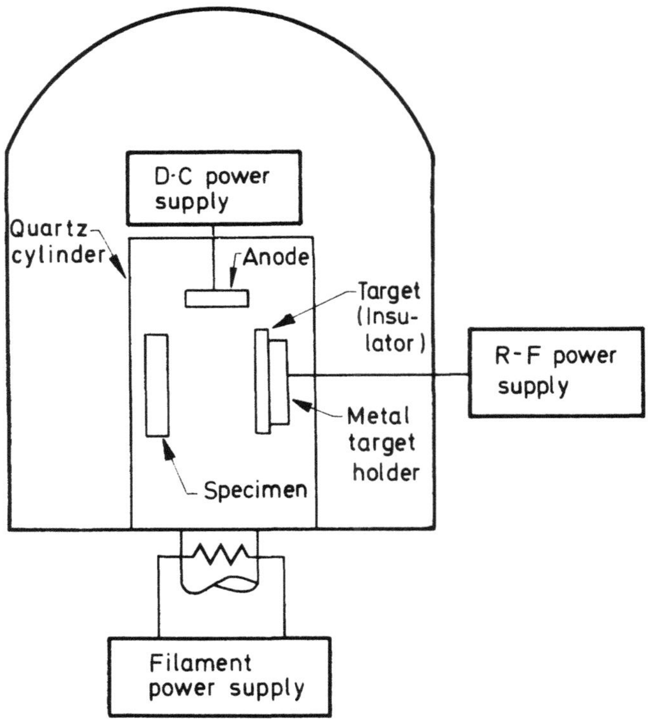

Figure 10.2 Schematic Diagram of a Typical R-F Sputtering System (Ref.559)

The polarisation problem is overcome by applying a radio-frequency potential to a metal support behind the target. The frequency is in the HF band, typically between 5 and 15 MHz, and the effect is to generate an alternating positive and negative charge. While a negative charge is present, the argon ions are attracted, and sputtering takes place. When a positive charge is present, electrons are attracted from the plasma, and any positive charge built up on the target is neutralised. One effect of the alternating power is to reduce the effective sputtering time, so that deposition rates are slower than for DC sputtering. The system is called R F sputtering and is shown in Figure 10.2.

Improved control of the sputtering process is obtained by using magnets to control the position and shape of the plasma. Permanent magnets placed behind the target have the effect of confining the plasma to an area close to the target, and thus increasing the sputtering and deposition rates[278]. A second effect is to constrain the paths of secondary electrons emitted from the target, practically eliminating their impact on the substrate and thus reducing substrate heating. Alternatively magnets can be placed in an annular ring around the plasma[279]. This presumably has the usual effect of producing a more focussed and linear plasma. The sputtering process with the use of permanent magnets is called RF magnetron sputtering.

Further control of the ion paths can be obtained by introducing a positively charged screen or ring either between the target and the substrate[280] or below the substrate[274]. This can be used to apply a positive DC bias in the region of the substrate, and provides further control of the plasma and sputtering conditions.

The introduction of unbalanced magnetron sputtering, in which part of the argon plasma can impinge on the substrate, allows ion bombardment of the coating as it forms. The use of ion-beam assisted deposition can be used concurrently or sequentially. This technique is described separately later in the chapter.

10.2 EFFECTS OF SPUTTERING VARIABLES ON FILM STRUCTURE

Ten of the significant variables in RF sputtering are listed in Table 10.1, together with some of the ranges of values which have been reported. Many of these variables are inter-related, and are also influenced by the size and shape of the vacuum chamber. For example, the overall power consumption is influenced by the voltage, the argon pressure and the distance from the thermionic cathode to the target, as well as, to a lesser extent, the screen voltage, the RF power, any substrate bias, and the position and strength of permanent magnets.

Because of the complex interaction of many of the variables, and the influence of the shape and size of the vacuum chamber and the sputtering module, no attempt will be made here to analyse or define all the effects of the variables. In practice it seems that different operators have established their own optimum operating conditions, and that these differ considerably between different operators, even after thirty years of experience.

Table 10.1 Sputtering Variables

Variable	Range of Values
Argon pressure	0.1 to 2.67Pa
Sputtering voltage	2 to 5kV
Sputtering power	0.1 to 0.9kW
RF frequency	7 to 14MHz
RF power	0.1 to 2kW
Target-substrate distance	20 to 40mm
Substrate bias	-100 to +10V
Screen voltage	0.2 to 1kV
Sputter-cleaning time	10 to 30mins
Sputtering time	

It has, however, become clear that certain specific coating characteristics are important to ensure low friction and long life, and the sputtering conditions to achieve those characteristics are also understood. The required characteristics are good adherence to the substrate, a high level of crystallinity, a high degree of purity, something approaching the stoichiometric composition of MoS_2, and high film density.

High purity is achieved by several steps. The system is first operated with the specimen substrate shielded or clear of any plasma or sputter contact to warm up the target and other components. This removes any condensation or absorbed water, and prevents condensation from taking on a cold substrates. The vacuum chamber is evacuated to high vacuum of about 10 mPa (0.07μ; 7 x 10^{-5} Torr), and backfilled to between 0.1 Pa and 5.0 Pa with high-purity argon, and this process may be repeated once or twice to remove contaminants, especially water and oxygen. The specimens themselves are cleaned before being introduced into the vacuum chamber, for example by ultrasonic cleaning in a degreasing solvent such as trichlorotrifluoroethane. After system start-up the specimens are sputter-cleaned, typically for 15 to 30 minutes, before film sputtering begins.

The influence of the substrate will be discussed later, but in general substrates for sputtered molybdenum disulphide films are very smooth, with surface textures of the order of 0.1 μm. The gross keying effect which has been found important for adhesion of bonded or burnished films is therefore not normally available. It has been suggested[274] that the sputtered particles impinge on the substrate surface with high kinetic and thermal energy, so that they are capable of penetrating and mingling with the surface atoms to create strong bonding between sputtered particles and the substrate, although this concept is not universally accepted. However, it is important that the surface atoms of the substrate are themselves firmly attached, and the purpose of the preliminary sputter-cleaning is to remove loose surface oxide or other contaminants, as well as to etch the surface to ensure strong adhesion.

The most important factor in ensuring crystallinity of the sputtered coating is the substrate temperature during sputtering. Spalvins[281] studied the effect of temperatures from -195°C to 320°C, and found that at temperatures below about 7°C the film became increasingly amorphous. At cryogenic temperatures it consists entirely of small amorphous particles, the friction is high ($\mu = 0.4$), and the film is brittle and even abrasive. At temperatures above 25°C the film is largely crystalline, with a coefficient of friction about 0.04, and as the temperature increases further the average particle size rises from 50Å to 110Å.

Lavik and Campbell[277] studied coatings sputtered at temperatures between 150°C and 427°C, and found a high degree of crystallinity. They estimated that about 30 per cent of the crystals were oriented with their basal planes parallel to the substrate surface, and 70 per cent perpendicular to the surface. Most workers now appear to assume that above 70°C an even higher proportion of the crystals have the perpendicular orientation and this is discussed later.

In considering the composition of the sputtered coating, it must be remembered that the sputtered particles are generally believed to consist at least partly of atoms, or of ions which consist of a single charged atom. In other words, the molybdenum disulphide is sputtered partly as separate molybdenum and sulphur atoms, and not as molybdenum disulphide moleculesor crystals. The composition of the sputtered film therefore depends on the proportions of molybdenum and sulphur atoms retained. In fact it is possible to produce similar coatings by using a molybdenum target and introducing hydrogen sulphide gas into the plasma[282,283].

By varying the sputtering parameters, coatings have been produced varying in composition from $MoS_{1.1}$ to $MoS_{2.26}$[276] Two factors have been shown to exert a strong influence on the stoichiometry of the films. The presence of small amounts of moisture has a major effect on stoichiometry as well as on crystallinity. The effect may be due to the incorporation of oxygen from the water into the films, and it has been suggested that oxygen directly displaces sulphur, so that MoS_{2-x} is in fact $MoS_{2-x}O_x$[284]. It has also been found that a negative bias applied to the substrate during sputtering leads to sulphur deficiency[285]. The effect of negative substrate bias on the friction of the coating is shown in Figure 10.3.

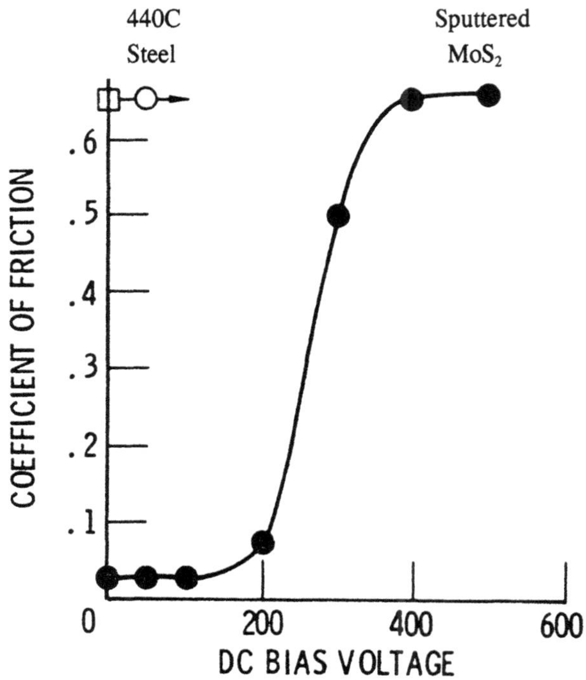

Figure 10.3 Effect of Negative DC Bias on Coefficient of Sliding Friction for Sputtered Molybdenum Disulphide (Ref.274)

The deposition rate affects film stoichiometry[278,286] as shown in Figure 10.4. Highly sulphur-deficient coatings are formed at deposition rates below about 150Å/min. This is an important reason for the use of high deposition rates, and therefore of magnetron sputtering. However, there is a corresponding disadvantage in that the magnetic fields concentrate the argon plasma on a reduced area of the target surface. As a result, the substrate size over which uniform film formation can be obtained is also reduced.

One final factor which can influence stoichiometry has been described by Stupp[287]. He reported that fresh molybdenum disulphide compacts initially emit sulphur faster than molybdenum. As a result the target surface becomes molybdenum-rich. As sputtering continues, sulphur diffuses to the surface at a rate which balances the removal of sputtered atoms, so that an equilibrium is established and the composition of the sputtered material remains constant. The problem of the initial variation in composition is overcome by a preliminary sputtering of the target before the specimen substrate is exposed to the sputtered particles.

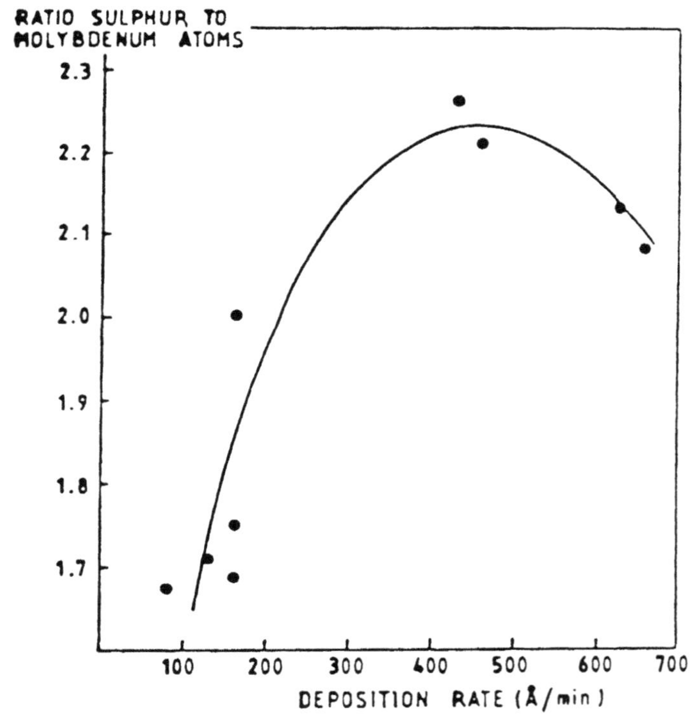

Figure 10.4 Sulphur Content of Sputtered Molybdenum Disulphide as a Function of Deposition Rate (Ref.278)

It is interesting and curious that even with significant levels of sulphur deficiency, the crystals in the sputter-deposited coating retain their normal hexagonal crystal structure. This is one reason for considering the probability that sulphur atoms missing from the lattice are replaced by other atoms such as oxygen.

In most publications there has been a lack of clear indication of the factors determining the optimum operating pressure of argon. Rochat et al[279] reported that good adhesion was obtained with a gas pressure chosen such that the sputtered particles experience no more than two collisions during their trajectory, because with more than two collisions they lose too much kinetic energy and adhesion is diminished. This implies that in principle a low gas pressure is desirable, but on the other hand too low a gas pressure will reduce the availability of argon ions and reduce the sputtering rate.

Roberts[278] studied the effect of argon pressure and sputtering power on friction, specific wear rate and endurance of films. He found that the optimum conditions were an argon pressure of 2.15 Pa (16μ) and 0.75 kW RF power. However, the performance was not very sensitive to argon pressure, and varied very little in the pressure range 1.7 Pa to 2.7 Pa (12.5μ to 20μ), although at the even lower pressure of 0.67 Pa, the friction, wear rate and endurance all deteriorated. Spalvins[280] used virtually the same argon pressure of 15 - 17μ for RF sputtering.

10.3 EFFECT OF SUBSTRATE

Sputter-coating can be performed on a wide variety of substrate materials, and molybdenum disulphide has been successfully sputtered on steels, nickel, Inconel, cobalt, molybdenum, tungsten, titanium, aluminium, gold, glass, ceramics, mica, sodium chloride crystal and plastics. There are several exceptions, including copper, bronze and silver, where adherence is poor due to reaction between the substrate metal and the sputtered sulphur. The problem could be overcome by oxidising the surfaces before sputter-coating. If the sputtered atoms impinge on a substrate with high energy and have a significant tendency to react, then for effective film formation it is obviously important that any reaction product does not adversely affect adhesion.

Most of the successful applications have been on bearing steels, where there is no difficulty in obtaining strong adhesion. There is of course another important factor in coating substrates such as bearing steels, and that is that the sputtering process should not degrade the properties of the substrate. This is another reason for ensuring that the substrate temperature does not rise excessively during sputtering, since this could cause annealing and loss of hardness of the bearing materials.

Apart from the chemical composition of the substrate, the surface finish is also important. It has been found that surface defects such as scratches and inclusions act as nuclei for accelerated film growth. The resulting film defects extend above the normal film surface, for example in the form of nodules, which may be far larger than the original substrate defect. These defects represent weak points, and may be detached in service leaving holes in the film coating. It is therefore important to ensure that the substrate surface is smooth and uniform in texture and composition.

However, although surface defects have adverse effects on coating quality, Roberts[288] et al showed that the smoothest surfaces do not give the lowest friction or maximum endurance. They studied roughnesses between 0.04 μm and 0.40 μm CLA. The coefficient of friction decreased with increasing surface roughness for all three substrates, and this is in accordance with friction theory, which predicts a decrease in friction with increasing Herzian contact pressure. The endurance of the coatings on 52100 steel reached a maximum at a substrate roughness of 0.2μm C L A, while with the softer IMI 318 titanium alloy the endurance increased with surface roughness over the whole range tested. With a silicon nitride substrate the endurance also showed a maximum at about 0.2μm CLA, but the situation was complicated by the fact that there were several incidences of partial failure followed by recovery.

Gardos[289] showed that there was some degree of correlation between coating thickness and the optimum substrate surface roughness. Coatings thicker than about 8,000Å had a longer life on surfaces of 0.28 μm CLA, while thinner films performed better on surfaces polished to 0.063 to 0.11μm CLA.

The best coating performance is often obtained with hard substrates. This is consistent with the discussion of friction of thin films in Chapter 5, but with sputtered coatings there may be an additional factor involved. The endurance of sputtered films has been shown to depend critically on the strength of adhesion to the substrate, and if adhesion involves intermingling of sputtered atoms with those of the substrate, it seems likely that the strength of adhesion will be inherently greater with hard substrates.

Spalvins[274] showed the advantage of depositing a hard coating on a substrate to form an underlayer for a molybdenum disulphide coating. He deposited a 1000Å thick underlayer of chromium silicide on 440C angular contact ball bearings, followed by a 6000Å thick sputtered coating of molybdenum disulphide. The endurance of the bearings was over a thousand hours, compared with only 187 hours for an identical system without the chromium silicide underlayer. Similar results have been obtained with titanium carbide, titanium boride and boron nitride, while multilayer coatings of molybdenum disulphide and such hard interlayers appear to have the additional benefit of encouraging basal plane orientation of the molybdenum disulphide film.

Coating life in moist atmospheres is also influenced by the effects of moisture on the substrate-coating interface, and marked improvements in life have been claimed by the use of moisture-protective pre-treatments of the substrate. Niederhäuser et al[290] studied a wide range of metals and titanium nitride, titanium carbide and chromium carbide as pre-treatments. The material was sputter-deposited on a steel substrate, and then sulphided by introducing hydrogen sulphide into the sputtering chamber in order to improve molybdenum disulphide adhesion. They found a marked improvement in life, particularly with a rhodium or palladium interlayer, but the actual degree of improvement is confused because they also used co-sputtered PTFE, and this is discussed further in Section 10.6.

Fleischauer[291] investigated the effect of sputtering molybdenum disulphide onto a substrate of molybdenum disulphide single-crystal oriented so that the surface consisted of basal plane, and found that the sputtered coating also adopted a basal plane orientation.

10.4 STRUCTURE OF THE SPUTTERED COATING

Once the sputtering conditions have been standardised, the rate of film deposition can be controlled quite accurately, so that a required film thickness can be obtained simply by controlling the deposition time. The most uniform coating thickness is obtained on a flat substrate mounted parallel to, and facing, the target. Even in this configuration a decrease incoating thickness is likely towards the edges of large substrates, and this imposes a limit on the size of substrate which can be effectively coated. Spalvins found[280] that in RF sputtering a specimen at 7.6 cms from the target, exposed on all sides to the sputtering beam, was completely coated on all surfaces. However, the film thicknesses were not reported, and Roberts[292] found that in RF magnetron sputtering there were considerable (±80%) variations in film thickness over the faces of steel cubes depending on their location and orientation relative to the target. Gardos[289] also found significant thickness variations.

Practical lubricating coatings are usually prepared in thicknesses between about 2000Å and 10,000Å, and Spalvins[280] reported that stress-induced peeling became significant at thicknesses greater than 15,000Å.

Figure 10.5 Structure of a Type I Sputtered Film (Ref.278, Courtesy of E W Roberts)

Until the late nineteen-eighties it was generally accepted that the desirable sputtered films of molybdenum disulphide had the type of structure shown in Figure 10.5. The plate-like or rod-like crystals are oriented with their basal planes perpendicular to the substrate surface, and are superimposed on an amorphous, or perhaps partly micro-crystalline sub-layer. Such films are often referred to as Type I or Type A films, and their production and properties are now well understood.

The perpendicular orientation of the main crystal structure in Type I films is not the desirable orientation for effective lubrication, but coatings with this structure are consistently found to give excellent lubrication in vacuum, with friction coefficients better than 0.05. It is clear that some mechanism must exist by which the crystals

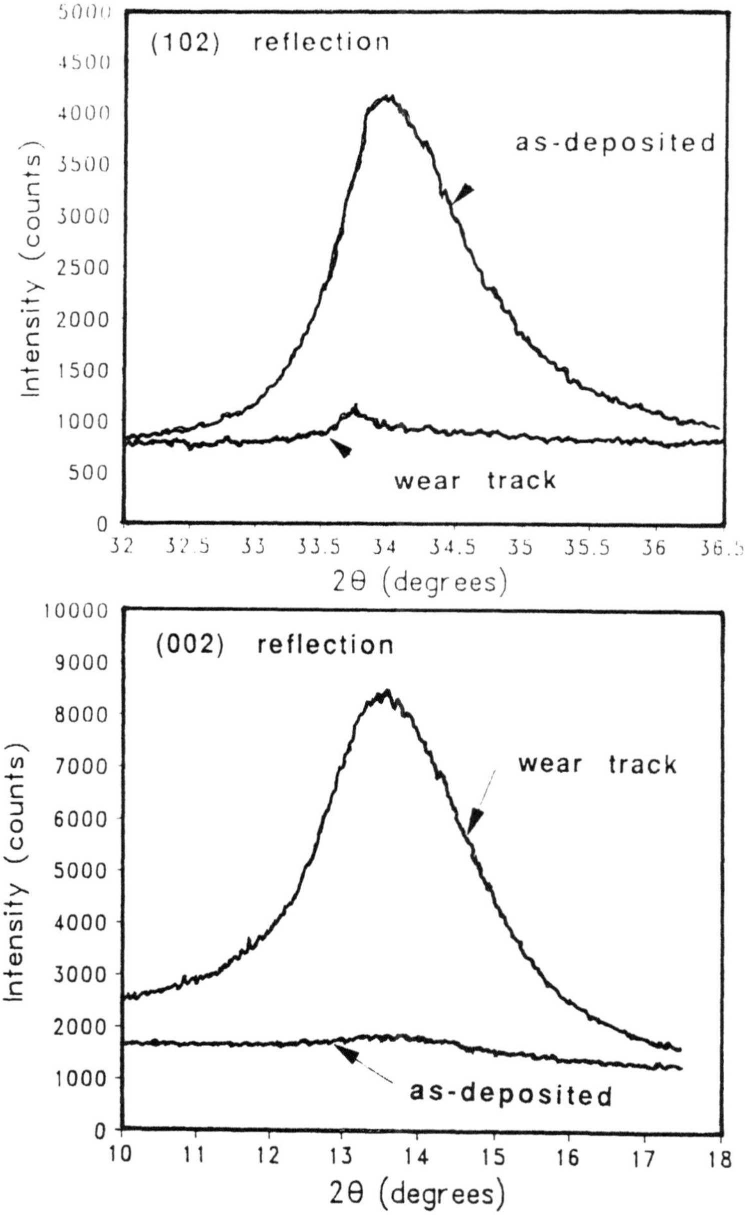

Figure 10.6 X-Ray Diffraction Intensities of an As-Sputtered Molybdenum Disulphide Film and a Wear Track Showing Re-Orientation of the Crystal Structure (Ref.294)

become re-oriented parallel to the substrate. This is confirmed by X-ray diffraction, as in Figure 10.6, which shows that the Type I (102) reflection of an as-deposited coating has changed to a basal plane (002) orientation on the wear track[294].

Spalvins[293] showed that under a sliding load crystals fractured within the columnar zone just above the isotropic sub-layer, as shown in Figure 10.7. The greater part of the lubricating life was then considered to be provided by the remaining thin film 2000Å thick, although the mechanism as portrayed in Figure 10.7 would indicate that the immediate surface layer still consisted of crystallites wrongly oriented to give low friction. The alternative possibility would be that the fractured crystal fragments might re-attach with a basal plane orientation. Buck[295] in fact suggested that the columnar crystals are capable of bending plastically without fracture to provide a full basal orientation. He also suggested[296] that if high levels of water contamination are present during sputtering, the crystal platelets will be brittle and will fracture more readily.

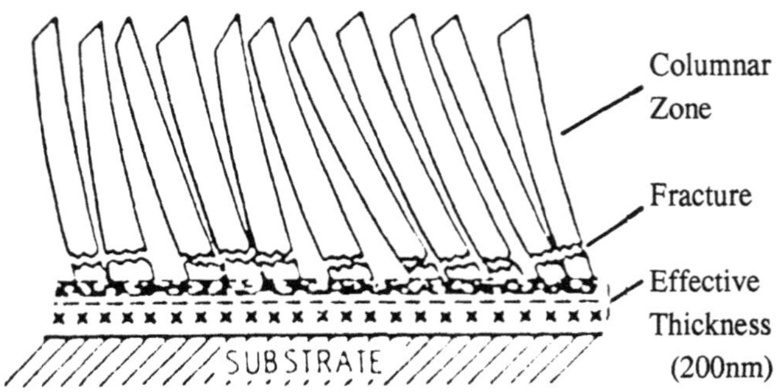

Figure 10.7 Fracture of Columnar Sputtered Film in Sliding (Ref.293)

All the highly crystalline Type I sputtered films are more or less porous, with densities of the order of 3.8 gm cm^{-3}, compared with a typical value of 4.92 gm cm^{-3} quoted for fully dense single crystals. Fleischauer et al[297] pointed out that the ease with which the crystals can bend to a basal plane orientation will depend on the crystal size and the density of packing. They also suggested that an increased degree of pre-etching of the substrate would increase the number of reactive sites available for crystal formation, and therefore the density of crystal packing, as well as the quality of film adhesion. Other factors which probably affect the packing density of the crystals are the rate of deposition and the total film thickness.

In recent years there has been considerable interest in the production and properties of sputtered films in which the dominant crystal orientation is with the basal planes parallel to the substrate surface. These are often loosely described as Type II films but their technology has been developing rapidly, and they cannot yet be considered as being at all precisely defined. Fleischauer reported[298] in 1983 the preparation of both Type I and Type II films, but the different conditions which produced the two different films were not known. The films were produced with different targets, and the deposition rate for the Type II films was slower. He speculated that conditions which favour the closest approach to MoS_2 stoichiometry would favour the formation of Type II films, and later[200] that different substrate temperatures might be responsible.

Buck[296] studied the effect of the partial pressure of water during sputtering, and found that very low moisture content encouraged the formation of Type II films, which occurred if there was an excess of sulphur in the sputtered coating. More recently, a number of research groups have produced Type II coatings, but with a variety of sputtering conditions, so that no clear picture of the conditions favouring basal orientation has yet emerged. Grosseau-Poussard et al[299] prepared highly sulphur-deficient films which were described as "quasi-amorphous", with small layered domains having a parallel orientation, embedded in an amorphous structure.

Aubert et al[282] produced parallel-oriented crystalline films by dc sputtering with a molybdenum target in an argon/ hydrogen sulphide atmosphere. Their films were also sulphur-deficient, with formulae varying from $MoS_{1.5}$ to $MoS_{1.7}$. Obeng and Schrader[283] used a similar technique, but with magnetron sputtering, and also obtained good crystalline sulphur-deficient films with parallel orientation. These Type II films therefore vary from highly sulphur-deficient to sulphur-rich.

Lince et al[284,300] studied a range of sputtered coatings containing between 9 and 40 atoms % of oxygen in place of sulphur, and showed by extended X-ray absorption fine structure that they consisted predominantly of a crystalline material with the formula $MoS_{2-x}O_x$, together with much smaller quantities of pure crystalline MoS_2. Increasing the oxygen content of the films had the effects of increasing the value of x in $MoS_{2-x}O_x$ and of reducing the relative quantity of the MoS_2 phase. Even at a 40 atoms % concentration of oxygen the $MoS_{2-x}O_x$ phase retained the same crystal structure as molybdenum disulphide, although with some measurable changes in specific bond lengths. However, high oxygen contents caused increasing disorder in the crystallites, causing reductions in crystallite size and increased film density.

The work of Fleischauer[140] showing the completely non-bonding nature of the molybdenum disulphide basal plane has been discussed earlier in Chapters 5 and 6. However, it must be remembered that this work was carried out in the specific context of sputtering, and has two implications for sputtered films. Fleischauer concluded that the undisturbed (0001) basal surface of molybdenum disulphide has no capability of forming bonds or reacting unless its molecular orbital scheme is altered by physical or chemical manipulation. For all types of sputtered coating, if atomic mixing takes place between the substrate and coating materials, then this would be a significant form of physical manipulation which could provide effective adhesion in the absence of chemical bonding. For oxygen-containing sulphur-deficient films, the lattice disorder resulting from the incorporation of oxygen in the lattice is a significant form of chemical alteration which may increase the reactivity of basal planes and thus improve adhesion, as well as encouraging basal plane adhesion in place of the edge-site adhesion of Type I films.

Type II films appear to be more reliably produced by certain co-sputtered materials, and by ion beam bombardment, and these are discussed in Section 10.7.

10.5 PERFORMANCE OF SPUTTERED COATINGS

The performance of sputtered molybdenum disulphide coatings in vacuum is outstandingly good. Good quality films under high contact pressures give coefficients of friction as low as 0.01, with specific wear rates of the order of $10^{-18}m^3/Nm$. It is more difficult to give meaningful life data, because the test conditions used by different groups have varied considerably, but on ball bearings, for example, the lives obtained have extended to millions of revolutions. There have been many reports of sliding tests showing performance in vacuum comparable with the lives of fully-burnished bonded coatings under similar test conditions in air.

The friction of Type I films is highly dependent on the purity or cleanliness of the coatings and on their stoichiometry. This was shown dramatically by Donnet and co-workers[301,302], who reported a coefficient of friction as low as 0.002. They deposited a coating 1200Å thick on a cleaned AISI 52100 bearing steel surface by RF magnetron sputtering at room temperature using a previously degassed molybdenum disulphide target. The coating was shown by analysis *in situ* to have the formula $MoS_{2.04}$, and was therefore very nearly stoichiometric. Auger electron spectroscopy showed no trace of contamination by elements such as carbon or oxygen. They found that the crystal structure initially consisted of nanometre-scale domains, mainly edge-site (i.e. perpendicular) oriented, which became re-oriented to a basal plane orientation by rubbing friction.

Table 10.2 Friction of Uncontaminated Coatings (Ref.302)

Atmosphere	Friction Coefficient	Calculated Shear Strength (MPa)
Ultra-high vacuum (5×10^{-8}Pa)	0.002	0.7
High vacuum (10^{-3}Pa)	0.013	4.9
Dry nitrogen (1bar)	0.003	1.1
Ambient air (1 bar)	0.150	56.0

The friction behaviour was studied in a high-vacuum reciprocating pin-on-flat tester coupled directly to the sputtering chamber, so that it was possible to avoid all contact with any contaminating atmosphere. Tests were carried out in four different environments, and the results are shown in Table 10.2. Structural examination showed no abnormal reason for the exceptionally low friction in ultra-high vacuum or in dry nitrogen . They concluded that the results were due simply to an inter-lamellar shear strength as low as 0.7 MPa, far lower than the level of 15 to 20 MPa normally assumed. The exceptionally low shear stress is apparently caused by the complete exclusion of contaminants. This implies that the normal friction values of 0.01 for optimum sputtered films in high vacuum and 0.04 or more for burnished films result from an increase in inter-lamellar shear stress caused by contaminants. This is consistent with Fleischauer's conclusions regarding the complete inertness of uncontaminated and unmodified basal planes.

It is interesting that the authors referred to the need for a transfer film to form on the counterface as a condition of very low friction, although the necessity for the formation of an effective transfer film has been ignored by many of those working with sputtered films. Fayeulle et al[187] had in fact made a very informative study of the formation of transfer films from Type I coatings. They showed that although the primary orientation of the Type I coatings was perpendicular to the surface, and the density low, the transfer films which they produced on steel counterfaces were basally-oriented and their density was close to that of bulk molybdenum disulphide.

In contrast with their exceptionally good performance in vacuum, the performance of Type I coatings in air is generally poor. The friction level in air shown in Table 10.2 is fairly typical, and life in air is also poor. In this context it is interesting that Fayeulle et al found that the transfer films formed from Type I coatings in air were highly oxidised, although still basally-oriented.

Spalvins[303] attributed the poor performance in air to the presence of water, and Roberts[278,292] further showed that the effect on friction occurs at two different levels and is reversible when testing alternates between moist air and high vacuum. It is significant that the coefficient of friction in moist air of the ultra-pure films produced by Donnet et al[301], 0.15, is almost identical with that of the less-rigidly contaminant-free films of Roberts, 0.16. This suggests that the high friction in moist air is almost entirely due to the water present, and is not significantly affected by any other contaminants which may have been present.

It should be noted that the friction and wear life of Type I films in air are both very much worse than those of good quality burnished films. The poor endurance of the sputtered films is generally attributed to oxidation, and there are two features of sputtered films which might reasonably be expected to render them susceptible to oxidation. The first is the porosity of Type I sputtered films, which would allow easy penetration of oxygen into the bulk of the film, and the second is the very small quantity of material present, which would not provide any reservoir for replacement of oxidised material. Fleischauer, Hilton and Bauer[297] reported that very low friction can be obtained with improved resistance to oxidation by using sputtering conditions (mainly higher substrate temperatures) which encourage larger crystallite and grain formation. However, this effect is not great enough to provide coatings with satisfactory lives in air.

It is generally accepted that coating endurance, especially in moist air, is very dependent on the strength of adhesion of coating to substrate. Fleischauer and Bauer[304] came to the conclusion that, just as slight oxidation of the lower layers of molybdenum disulphide and an excess of sulphur were beneficial in improving the adhesion of transfer films to substrates, similar control of the interface chemistry was likely to be beneficial for the endurance of sputtered coatings. The problems posed by poor performance in air are not confined to components intended to be used in air. Even for components intended to be used in high vacuum it is generally necessary, or at least highly convenient, for them to be stored, assembled or tested in air. The oxidation problem restricts the performance of sputtered coatings even in dry nitrogen at elevated temperatures. Anderson and Roberts[305] tested a sputtered molybdenum disulphide coating in nitrogen containing less than 15 ppm of oxygen, and found a marked deterioration in both friction and endurance at 400°C, which they ascribed to oxidation.

In contrast with the beneficial effects of MoS_2 stoichiometry and chemical purity on the performance of Type I films, good performance has been generally reported both in vacuum and in air for sulphur-deficient Type II coatings. Typical coefficients of friction are in the range 0.02 to 0.05 in vacuum, and only a little higher in air, and endurances in moist air are also better than for Type I films. The improved life in air is probably due to the parallel orientation of the Type II films, which exposes unreactive basal planes to oxidation instead of reactive edge-sites. The densification effect reported by Lince et al[300] would also probably assist by reducing oxygen or water penetration. The generally low friction values may be partly explained by the initial parallel orientation, while the small changes in bond lengths reported by Lince were also such as to reduce inter-lamellar shear forces.

Improvements in the performance in air have been reported from the use of co-sputtered substances and from ion bombardment, and these effects will be described in the next two sections.

10.6 EFFECTS OF CO-SPUTTERING

The incorporation of other elements or compounds during the production of sputtered coatings ("co-sputtering") has been intensively studied in the last few years, mainly to overcome the defects in the performance of sputtered coatings in moist air. Some of the earliest studies were performed by Laboratoire Suisse de Recherches Horlogères (LSRH) in Switzerland and Hohman Plating and Manufacturing (HPM) in the United States.

LSRH found that co-sputtering of PTFE and molybdenum disulphide[290] from a mixed target gave a very considerable improvement in life in humid air. Several other materials similarly tested, including lead, lead oxide, silver, antimony trisulphide, antimony trioxide and a silane, were largely ineffective. In conjunction with a pre-coating of sulphided rhodium or palladium the co-sputtered PTFE and molybdenum disulphide gave a thousand-fold improvement in sliding wear life in air of 50% relative humidity. Incidentally, the sputtering of PTFE would appear to cast some doubt on any assumption that sputtering always involves the transfer of atoms or monatomic ions, and not larger groups.

Stupp at HPM[306] described the effect of co-sputtering any one of several different metals with molybdenum disulphide in a DC triode sputtering system. The best results were obtained with between 5% and 7% of nickel or titanium, which gave improved life and much more uniform friction in sliding friction tests in air than were obtained with conventional DC-sputtered molybdenum disulphide films. The co-sputtering could be performed satisfactorily using either a composite nickel/molybdenum disulphide target or two separate targets.

Table 10.3 Effects of Co-Sputtered Nickel in Different Atmospheres (Ref.287)

Environment	Sputtered MoS_2		Co-Sputtered MoS_2 and Ni	
	Friction	Cycles to failure	Friction	Cycles to failure
Argon	0.03	40×10^4	0.03	96×10^4
Dry air	0.035	20×10^4	0.03	70×10^4
Ambient air	0.06	12×10^4	0.045	60×10^4
Moist air	0.12	0.8×10^4	0.10	1.2×10^4

Further work was carried out[287] with 3% - 5% of co-sputtered nickel to evaluate the effects of film thickness and test speed and load. He found that the life of a co-sputtered film increased almost linearly with coating thickness, while that of a conventional film reached a maximum at a thickness of about 4000 nm (40,000Å). These are far thicker than coatings described by other authors. A comparison of life results in different atmospheres is presented in Table 10.3. The coefficient of friction of the co-sputtered coating was generally about 20% lower than that of the conventional molybdenum disulphide coating in all the tests in air.

Roberts and Price[307] investigated the co-sputtering of gold in a molybdenum disulphide coating by RF magnetron sputtering, and found a three-fold or four-fold life improvement in air compared with a similar coating without the co-sputtered gold. The optimum concentration of gold was about 14 atoms %, and the variation of film life in air with gold concentration is shown in Figure 10.8. They found that a multi-layered coating in which alternate layers of gold and molybdenum disulphide were deposited showed no improvement in performance in air compared with a simple molybdenum disulphide film, although the film density was high. This result is probably consistent with other published work, since in general it is the hard substrates and interlayers which give the best performance.

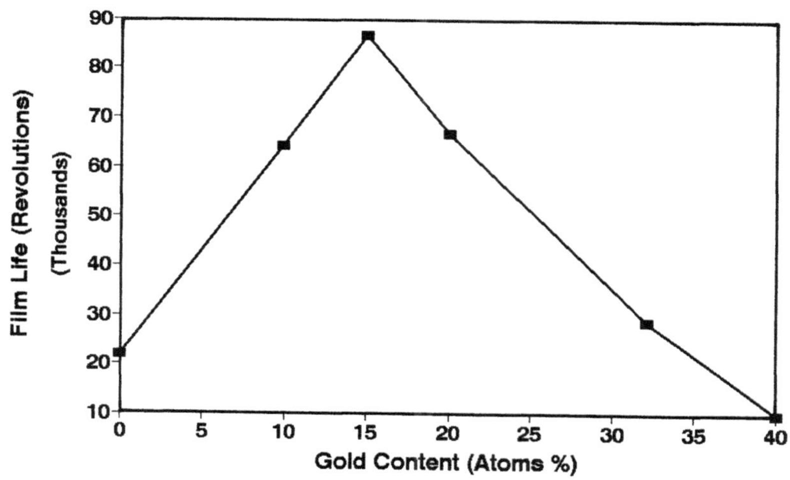

Figure 10.8 Variation of Sputtered Molybdenum Disulphide Film Life with Gold Content (Ref.307)

In general the effect of co-sputtered metals[308] seems to be to increase film density and hardness, as well as modifying molybdenum disulphide crystal size. The increase in density may have some beneficial effect by reducing penetration of oxygen or water vapour into the bulk of the coating, but otherwise there is no obvious reason for the beneficial effect of co-sputtered metals on life in air. The co-sputtering of antimony trioxide has also been shown to improve life in air, and this effect is presumably similar to the effect of antimony trioxide in bonded coatings.

10.7 EFFECTS OF ION BOMBARDMENT

Ion bombardment can be used in several different ways to alter the properties of sputter-deposited coatings. The simplest of these is unbalanced magnetron sputtering.

In normal magnetron sputtering the magnetic fields are arranged so as to confine the plasma to an area near the target. In unbalanced magnetron sputtering the magnetic fields are arranged so that part of the plasma impinges on the substrate. As a result the sputter-deposited coating is subjected to some degree of argon ion bombardment as it grows. This results in compaction of the coating, and some early results[309] showed sulphur-deficiency. This is believed to be caused by preferential re-sputtering of sulphur from the coating, since the degree of sulphur-deficiency increased with the bombarding beam current density, as shown in Figure 10.9. It can presumably be corrected by varying the deposition conditions.

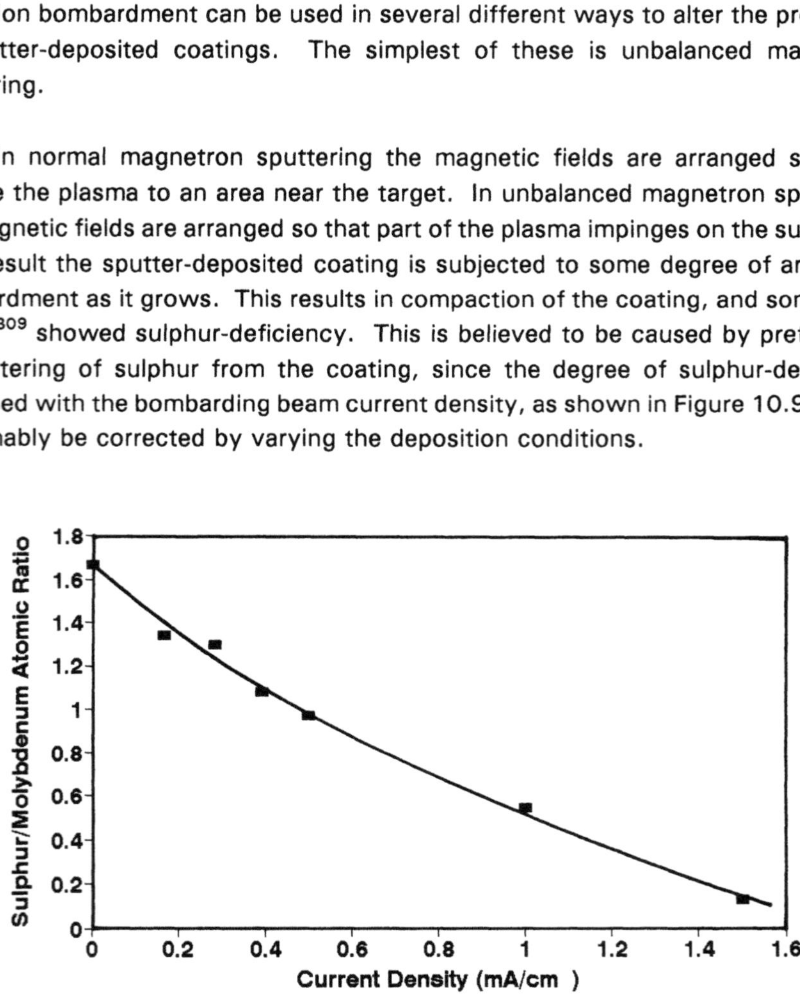

Figure 10.9 Effect of Bombarding Current Density on Sulphur/Molybdenum Ratio in a Sputtered Film (Ref.309)

Fox et al[310] reported low friction (0.02) in 45% RH for molybdenum disulphide coatings deposited in an unbalanced DC magnetron sputtering system, and this implies a dense coating with basal plane orientation. They also used a separate titanium target to pre-coat the substrate with titanium and to produce a multi-layer coating with alternate thin layers of titanium and molybdenum disulphide. This type of coating gave low friction and extended life in air of 40 - 50% R H, and is presumed to possess basal plane (Type II) orientation of the molybdenum disulphide.

An alternative is to carry out post-deposition bombardment of a sputter-deposited coating. This provides great flexibility by the selection of ion beam power and ion composition. A general result of the process is increased coating density and improved performance in humid air.

At low bombardment energy with neon, argon or xenon ions Roberts and co-workers[309] reported compaction of the coating and postulated that it was caused by sputtering and re-distribution between the columnar Type I crystals. With high-energy bombardment the bombarding ions are capable of completely penetrating the coating. Improved adhesion has been reported with silver or yttrium ions[311], and this suggests that the bombarding ions have penetrated to the substrate surface, improving mixing of substrate and coating materials.

Hirvonen and co-workers reported[294] that high-energy neon ion bombardment resulted in amorphization of the crystal structure and increased film hardness. However, the friction remained low, and this suggests that the amorphous coating readily recrystallised with basal plane orientation under a sliding stress.

Bhattacharya et al[312] studied the effect of irradiation of DC magnetron sputtered films with high energy (2MeV) silver $Ag+$ ions. They found a marked increase in film life for the irradiated films in 1% humidity, and attributed this to improved adhesion to the substrate and increased densification. Tests in argon gave low friction coefficients of 0.02 for the irradiated coatings and the as-sputtered coatings, but the life of the irradiated coatings was again much longer. They reported that all their films were amorphous, but the low friction levels suggest that some re-orientation under sliding load had occurred.

It is clear that the variety of processes and conditions available in conjunction with sputtering is currently leading to an enormous empirical increase in the available range of sputtered coatings, which is rapidly improving the performance obtainable both in vacuum and in atmospheric applications. A clear understanding of the chemistry and physics involved may take longer to establish.

10.8 PULSED LASER DEPOSITION

Like sputtering, pulsed laser deposition (PLD) is a high-energy vacuum process in which a lubricant is transferred from a target reservoir to the surface of a specimen substrate in a vacuum chamber. It differs from sputtering in that the energy is supplied in the form of a pulsed laser beam. The principle of the technique is shown diagrammatically in Figure 10.10.

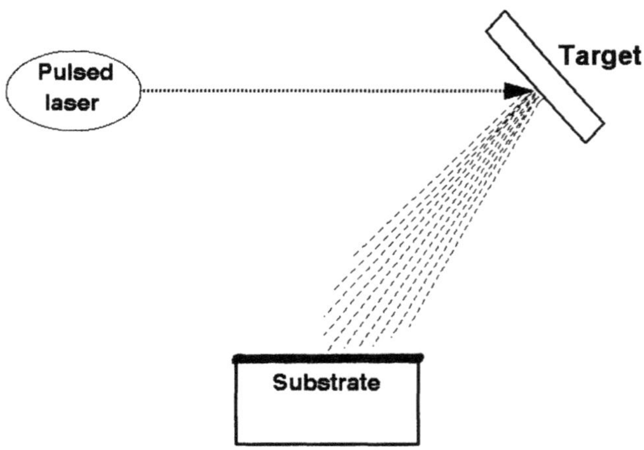

Figure 10.10 Schematic Layout of a Pulsed Laser Deposition System

Typical operating details[313,314] are a vacuum of about 10^{-6} Pa, with an argon fluoride excimer laser at 193 nm wavelength or a krypton fluoride laser at 248 nm wavelength, pulsed at 10 Hz with a pulse width of 17 ns. A sophisticated computer system is used to monitor and control the process. Because the process is essentially optical, roughening of the target reduces the deposition rate. The hot pressed molybdenum disulphide target is therefore polished before each run, and the laser beam is rastered over the whole target surface. Film uniformity is improved by slowly rotating the target and the substrate. Calculations have indicated[313] that transitional temperatures are as high as 110,000°C, and the transfer mechanism is ablative.

Films deposited with the substrate at room temperature primarily contain molybdenum disulphide with some excess elemental sulphur near the surface, but also contain other chemical states. Higher temperatures have the usual effect of improving crystallinity, and the excess of free sulphur is reduced, but in this respect post-depositional laser annealing, as shown in Figure 10.11, is more effective than simply increasing the substrate temperature. It has been shown that the PLD coating surface consists of crystal basal planes, and friction in dry nitrogen is less than 0.05, but otherwise tribological evaluation is at an early stage.

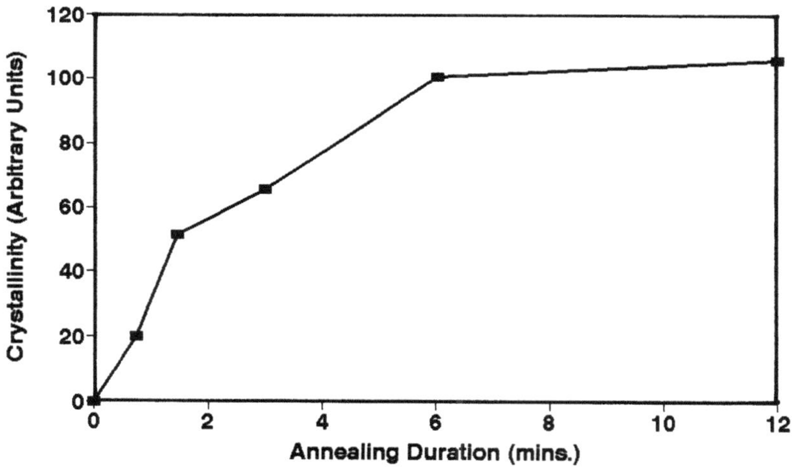

Figure 10.11 Effect of Post-Deposition Laser Annealing on the Crystallinity of a PLD Molybdenum Disulphide Film (Ref.313)

PLD has also been used to produce so-called adaptive lubricant coatings containing molybdenum disulphide and lead oxide. The theory of adaptive lubricants is that a lubricant which is effective at low temperature changes on exposure to higher temperature into a substance which lubricates effectively at the higher temperature. In the case of a molybdenum disulphide coating which contains lead monoxide PbO, the molybdenum disulphide provides effective lubrication to about 400°C in air, and the lead monoxide to about 650°C. The two react together in air to form lead molybdate $PbMoO_4$ which can lubricate between about 700°C and 1000°C.

Zabinski[315] used PLD to incorporate lead monoxide into molybdenum disulphide films. With annealing, or increasing the substrate temperature during deposition, crystalline lead molybdate and molybdenum disulphide could be identified in the coatings, together with molybdic oxide. An optimised film had a coefficient of friction which decreased from 0.2 to 0.04, and a life ten times as long as films of molybdenum disulphide produced at room temperature. The technique therefore gives promise of effective lubrication over a wide temperature range in air.

~ ~ ~ ~ ~ ~ ~ ~ ~ ~ ~ ~ ~ ~ ~ ~ ~

CHAPTER 11

BONDED FILMS

11.1 TYPES OF BONDED FILM

The most straightforward technique for producing a thin uniform adherent coating is by the use of an adhesive binder, and bonded coatings were introduced very early in the modern development of molybdenum disulphide technology. They are probably now the most widely used form of molybdenum disulphide lubricant, and as long ago as 1968 Gresham[316] estimated that over 95% of the solid lubricants commercially used were resin-bonded.

The earliest binders were those which were already well known, either in adhesives or paints. They included[317] corn syrup, glycerol, ethylene glycol and dextrose (all dispersed in water), asphalt-based varnish (in naphtha), and silicone varnish (in xylene). The range of binders which has been studied has since been extended enormously, to include lacquers, polymers, soluble salts, fused salts, fusible oxides and fluorides, ceramics and metals, and a number are listed in Table 11.1.

Much of the early research and development of bonded films was carried out by governmental organisations such as NASA and various military research establishments. The results were usually published and often used as a basis for military specifications, so that information about compositions and performance was readily available. It is ironic that because of the general success and popularity of bonded coatings, much of the development in the succeeding thirty years has been commercial, so that information about compositions has been less readily available. Although a certain amount of information is still published concerning performance, it is less meaningful when it cannot be related to composition.

Table 11.1 Bonded Film Components

Binders	Lubricants	Other Components
Organic Air-Drying	**Dichalcogenide**	**Solvents**
Cellulosics	Molybdenum disulphide	Water
Acrylics	Tungsten diselenide	Isopropyl alcohol
		Toluene
Organic Heat-Cured	**Inorganic**	Amyl acetate
Alkyd	Graphite	Ethyl acetate
Epoxy	Lead oxide	Naphtha
Phenolic	Lead sulphide	
Silicone	Boric oxide	**Oxidation Inhibitor**
Polyimide		Antimony trioxide
Polyvinylbutyral	**Organic**	Antimony tetrasulphide
	PTFE	
Inorganic Salt	Phthalocyanine	**Corrosion Inhibitor**
Sodium silicate		Dibasic lead phosphate
Aluminium phosphate	**Metal**	Sodium phosphate
Sodium phosphate	Gold	Stannous chloride
Potassium silicate	Silver	Potassium aromate
Sodium borate	Indium/lead	
Titanates		
Inorganic ceramic		
Boric oxide		
Silica		
Alumina		
Calcium fluoride		
Metal		
Silver		
Nickel		
Gold		
Tantalum		

However, there can be little doubt that many of the commercial formulations are still based on the research work carried out between 1950 and 1980, and the work published in that period will therefore be still relevant.

The basic technique involved in the production and application of a bonded coating is simple. A binder is dissolved in a suitable solvent, and molybdenum disulphide powder is dispersed in the solution. The proportions of binder and lubricant are selected to give a suitable consistency for application, usually by spraying, but alternatively by dipping or brushing. The bearing surface is prepared, the liquid is applied to give a uniform coating, and the solvent is removed by evaporation. Other materials may be added to stabilise the dispersion, to increase coating life, to improve oxidation resistance or to reduce corrosion. After evaporation of the solvent, a final treatment may then be given to the coating by heating to cure the binder, and by running-in or burnishing.

Table 11.2 lists several classes of formulation with some of their general properties. The simplest class of binders in formulation and application are probably the air-drying organic binders, usually cellulosic or acrylic. They have the advantage that no heat treatment is necessary, so that, apart from reducing effort, they can be used on components or substrate materials which would be adversely affected by heat. The solvent evaporates in air at room temperature, leaving a cohesive film of binder incorporating the molybdenum disulphide powder. In general they contain a relatively low concentration of molybdenum disulphide, less than about 40% of the dried film, so the initial friction may be high. They also have low resistance to organic solvents, and their thermal stability is limited, but they are relatively inexpensive, and can be used for a wide variety of lubrication situations.

The heat-cured organics include phenolics, epoxies, alkyds, silicones, polyimides and polybenzimidazoles. The alkyds have the advantage that they will provide a useful film at room temperature and a better one on curing at 100°C, but their films are rather soft. Epoxies will also cure at room temperature but give much stronger films on heating to 150° to 200°C. They are harder than alkyds and have very good solvent resistance. Phenolics give very hard adhesive films, but also need a curing temperature between 150°C and 200°C, so their use is not advisable on certain aluminium alloys. Silicones have better thermal stability, but their films are relatively soft and have poor adhesion, and their use is limited. They normally require curing temperatures of 250°C - 300°C, but Benzing et al described an air-drying version[318]. Polyimides and polybenzimidazoles give strong adhesive films with good solvent resistance and excellent thermal stability but they require curing temperatures between 200° and 300°C. They have been used successfully in a number of spacecraft applications. Lubricant concentration in heat-cured films may be as high as 90%.

Hopkins and Campbell[319] described a film based on polyphenylenesulphide (PPS). This is a very strong and inert polymer with good thermal stability to over 300°C after curing. It can be slush or melt-coated onto metal surfaces, but gives a hard and rather thick coating. It is one of many binders which have been described as satisfactory in the research literature, but have not subsequently achieved any practical prominence. These include polybutyltitanate, vinyl resins, ethylene-acrylic acid copolymer, oil-modified polyurethane, pyrrones, and polybenzothiazoles.

Table 11.2 Some Bonded Film Formulations

Binder	Conc. MoS_2 (%)	Other Typical Components	Cure Temp. °C	Max.Temp. in air/ vacuum °C	Wear Life	Example
Cellulosic or acrylic	30-40	Graphite	Room	65	Good	MIL-L-23398
Alkyd	30-70		Room-120	200	Very good	
Epoxy or phenolic	40-90	Graphite, antimony trioxide	150-200	350-500	Very good	MIL-L-8937
Silicone			Room-200	300	Fair	DCI3943
Polyimide	33-70	Antimony trioxide 33%	200-300	350/500	Very good	MLR-2
Sodium silicate	71	Graphite 7%	150-200	350/500	Very good	NAML
Sodium silicate	43	Graphite 4% gold 22%	150-200	350/500	Good	MLF-5
Aluminium phosphate	29	Graphite 3% bismuth 40%	200	350/500	Good	MLF-9
Boric oxide	70		400-600	350/500	Fair	
Ceramic	46		800	350/600	Fair	

To overcome the temperature limitation on organic binders, especially in air, many inorganic binders have been tested and developed. They fall into two categories, the salts which can be applied in solution and the oxides and ceramics which are normally applied by fusion, sintering, plasma spraying, and so on. Whereas bonded coatings with organic binders are usually limited by the thermal or oxidative stability of the binder, those with inorganic binders are usually limited by the stability of the molybdenum disulphide. Thus in Table 11.2 the temperature limit of most of the formulations in air is the oxidation limit for extended use of the molybdenum disulphide component.

At one time, especially during the nineteen-sixties and early nineteen-seventies, the most widely-used inorganic binder was the salt sodium silicate, and several commercial lubricants, such as Molykote X-15, were available with a sodium silicate binder. Most of them were based on the US Naval Aircraft Materials Laboratory[88] or MLF-5[320] formulations, details of which are given in Table 11.2. The carrier liquid for these formulations was water, and because the molybdenum disulphide and other insoluble components tended to agglomerate or sediment out on standing it was necessary to keep the dispersions agitated during application. After application the coatings were first dried at 50° to 90°C and then cured at 150°C to 200°C. They had good adhesion and cohesion, but gave no corrosion protection, and it was eventually realised that the graphite content could actively cause corrosion. They were therefore most suitable for use at high temperatures and in vacuum, but they were eventually superseded by other formulations.

Other salts which have been used successfully as water-soluble binders are potassium silicate, sodium phosphate, sodium borate and aluminium phosphate[321], the last two being the only ones apart from sodium silicate which have apparently been widely used. Their application techniques are similar to those described for sodium silicate. The concentration of molybdenum disulphide in inorganic salt binders is usually between about 20 and 75%.

Hartley and Wainwright[322] gave detailed compositions of two bonded film formulations based on an aluminium phosphate binder which were developed for coating metalworking tools. The compositions are shown in Table 11.3. The chromium trioxide was also described as a binder, but in such small quantities it presumably acted mainly to improve the performance of the aluminium phosphate. The alkyl phenol polyglycol ether is a non-ionic wetting agent. Coating A was designed to be used in the presence of fluid lubricants, and Coating B in dry situations.

Table 11.3 Compositions of Two Coatings with Aluminium Phosphate Binders (Ref.322)

Component	Concentration (%)	
	Film 1	Film 2
52% aluminium phosphate solution	24.7	12.35
Molybdenum disulphide	20.0	20.0
Chromium trioxide	1.3	0.65
Alkylphenol-polyglycol ether	0.4	0.4
Water	53.6	66.6

The oxide and ceramic binders have been studied by several research groups, but do not seem to have been much used in practice or made commercially available. They are a rather heterogeneous group with considerable theoretical potential, being capable of withstanding temperatures of the order of 1000° compared with 400°C for molybdenum disulphide. There has therefore been a tendency to use the ceramics and oxides themselves as lubricants to retain the best high-temperature capabilities.

In vacuum or very inert atmospheres it would seem to be possible to achieve operating temperatures to 1000°C with low friction and wear by combining ceramic binders with molybdenum disulphide, but for such applications the use of bonded films has been largely overtaken by sputtering techniques. One possible approach to high-temperature lubrication is to use molybdenum disulphide as an expendible component to provide low friction at temperatures below those at which the oxides or ceramics become effective lubricants. This technique has been studied in recent years in the search for suitable lubricants for low-heat-rejection ("adiabatic") automotive engines. The most promising formulations for this have not been based on molybdenum disulphide, and the friction levels have been generally rather high, at around 0.2.

Boric oxide (B_2O_3) has been more intensively studied than any other oxide binder. It was originally tested as a binder for lead sulphide, but to improve the low temperature performance[323] molybdenum disulphide was added to give a MoS_2: PbS ratio of 2:1, and later studies[324] simply used molybdenum disulphide in boric oxide. The coating was applied in the form of an aqueous slurry by dipping, brushing or spraying, dried at 50° to 90°C and cured at 400° to 600°C, at which temperatures the boric oxide softens to a molten glass-like consistency. When properly applied and cured at the higher temperatures the film was tough, but otherwise could be liable to cracking or chipping. Clow[324] also studied the importance of wetting of molybdenum disulphide particles by molten boric oxide, and showed that improved wettability had a beneficial effect on film wear life.

Silica (SiO_2) has also been studied as a binder for solid lubricants, including molybdenum disulphide. One technique is to use lead oxide and silica together in a eutectic and to heat the film to fusion temperature. An alternative is to apply a film of molybdenum disulphide dispersed in ethyl orthosilicate solution and to hydrolyse the silicate to silica[325]. This process has the advantage that it can be carried out at low temperatures, but the integrity of the film is better with the high temperature method. In general silica films are brittle and chip easily, but they have good thermal stability to 600°C. Colloidal silica was the bonding agent in some commercial films[326] which could be cured at 190°C or higher temperatures, the higher cure temperatures giving better wear life but slightly higher friction.

Useful properties have been claimed for molybdenum disulphide coatings bonded with alumina (Al_2O_3) formed from a colloidal fibrous hydrated alumina called Boehmite[327]. By controlling the pH of a dispersion to about 10 a gel was formed, and the viscosity could be controlled by means of the pH and solids content. An aqueous dispersion containing 3 to 30% of Boehmite and 30 to 60% of molybdenum disulphide at pH 10 could be spray coated to give a dry film thickness of 5 to 10 μm. After spraying the coating was fixed by dehydrating at 230° to 300°C to render it water-resistant.

More complex ceramic binders have been used, but in general the detailed composition of the binder has not been described. One very complex one[328] consisting basically of molybdenum disulphide in silica, had the composition in Table 11.4. Molybdenum disulphide has also been incorporated in fused-fluoride lubricant coatings[329] to improve their properties at temperatures below 500°C. In tests over 450°C in air the molybdenum disulphide was effective for one test, but was then no longer available because of oxidation. However, such coatings would presumably have useful lives at high temperature in vacuum or inert atmosphere.

The use of metal binders has been generally disappointing. Silver and tantalum have been successfully plasma sprayed with molybdenum disulphide[330] by means of

Table 11.4 Composition of a Ceramic-Bonded Film (Ref.328)

Component	Concentration (%)
Molybdenum disulphide	45.6
Silica	10.5
Alumina	1.9
Cadmium oxide	0.3
Lithium nitrate	1.1
Calcium nitrate	3.6
Lead sulphide	21.4
Sodium carbonate	5.8
Cobalt trioxide	0.3
Calcium oxide	0.7
Potassium perborate	8.3

a dual-port-entry plasma gun to give films with useful friction and wear properties. Nickel-coated powders have also been plasma-sprayed[330] and electro-codeposited[331]. Silver has been applied as an organometallic solution[332], fired to reduce the compound to metallic silver and then burnished to give a cohesive film. Although silver-bonded molybdenum disulphide has been reported[333] to give room-temperature friction and wear characteristics comparable to those of phenolic-bonded films, there seems to have been little practical application of metal-bonded films.

11.2 OTHER COMPONENTS OF BONDED FILMS

Apart from the three major components of the bonded film, the binder, the molybdenum disulphide, and the solvent or dispersant, many other substances may be incorporated, and several have already been mentioned.

At one time graphite was probably the most common additive incorporated. It was originally studied as an additional dry lubricant, to establish whether there would be any synergistic effect in reducing friction. Under certain conditions some reduction in friction was obtained[334], but the more important effect of graphite was to increase wear-life. The optimum ratio of graphite to molybdenum disulphide was shown[88] to be 1:10, and this proportion was used in several widely-used formulations. Gardos[335] suggested that the beneficial effect of graphite is a result of it reducing both the rate of oxidation and the catastrophic blistering which follows oxidation and leads to film failure. This view has been challenged by Centers[336]. It is interesting to note that after the use of graphite was discontinued in many formulations because it was found to encourage galvanic corrosion[337], there was an increasing trend to use antimony trioxide, which had also been shown to inhibit oxidation, and which also improved the wear-life of coatings.

The mechanism of action of antimony trioxide has been discussed in some detail in Chapter 7. It has been widely used in bonded film formulations since the mid-nineteen-sixties to improve the wear life of coatings. The greatest improvement is obtained at higher temperatures, and Kinner[132] found a small deterioration in performance at low temperatures when antimony trioxide was used. He also found that the use of lead oxide, PbO, together with antimony trioxide gave markedly improved performance at both high and low temperatures.

Lavik et al[190] reported up to 400% improvement in wear life, and lower friction at high or low temperature, when antimony trioxide was used in polyimide or polybenzothiazole-bonded films. Lower improvements, up to 270%, were obtained with antimony trisulphide, while Bartz[183] reported that the addition of both graphite and antimony tetrasulphide (or antimony thio-antimonate $Sb(SbS_4)$) gave a much greater improvement in the life of a polybutyltitanate-bonded molybdenum disulphide film than either of the additives separately.

A number of different metals have been used in powder form as additives to bonded molybdenum disulphide films. They include gold, silver, nickel and tantalum, and it has been suggested[179] that they improve film life by helping to facilitate re-adhesion of lubricant debris to the substrate. Many other components have been added to bonded films as corrosion inhibitors, anti-oxidants, dispersion stabilisers and biocides, and many different solvents have been used. There is therefore an almost infinite variety of possible formulations, and several hundred have been produced commercially.

11.3 SUBSTRATE PREPARATION AND PRE-TREATMENT

The object of surface preparation is to obtain the best performance and life of a bonded film for a reasonable expenditure of effort. In general bonded films will adhere to almost any surface which is clean and free of grease, oil and other contaminants, but various types of pre-treatment can be used to improve the adhesion, corrosion resistance and life.

For critical components, such as in aircraft or spacecraft, a full cleaning process might involve a warm aqueous detergent solution, followed by a water rinse, acetone rinse, and final cleaning and degreasing with a chlorinated solvent. For less critical components, or where the only possible contaminants are cutting or grinding oils or oil-soluble temporary protectives, it may be enough to use only the chlorinated solvent. For the degreasing stage vapour degreasing is usually the most effective, but where a shape is complex or there is very adherent grease it may be more effective to use a similar solvent in an ultrasonic bath. There have been reports of corrosion caused by moisture on a surface which has been cleaned with certain chlorinated solvents, so that a further cleaning with a petroleum solvent may be desirable if a component is to be stored for a long period or handled with bare hands. Once cleaning is completed, the parts should in any case be protected as far as possible from corrosion by moist air and should be handled with gloves. It follows that it may be desirable to carry out the cleaning immediately before the next stage of preparation.

The most important aspect of pre-treatment is probably to obtain the optimum surface roughness to give the maximum degree of mechanical "keying" of the coating to the surface. It is generally accepted that the optimum surface finish is between 0.5 μm and 2.0 μm cla, and peak performances have been variously reported at 0.5 μm[338] 0.75 μm[339], and 1.5 μm[209] rms or cla, but the weight of evidence suggests that the lower figure of 0.5 μm is to be preferred.

The way in which the surface finish is produced is also important, and the objective is to obtain a surface in which the depressions are distributed as uniformly as possible. Table 11.5 shows some wear lives measured with a Falex machine with surface textures obtained by grit-blasting and vapour-blasting, while Figure 11.1 shows the effect of various pre-treatments on wear life of a bonded coating. It can be seen that the wear life on a ground surface is far shorter than on an equivalent sand-blasted surface. In general it is therefore recommended that the bearing surfaces are roughened to a finish of 0.5 to 1.0μm cla by grit-blasting, and the results may be marginally better with wet grit-blasting using an alumina grit of between 220 and 100 mesh size.

Table 11.5 Effect of Grit-Blasting Pretreatment on Wear Life of a Silicate-Bonded Molybdenum Disulphide Film (Ref.209)

Grit Material	Mesh Size	Surface Finish (μm rms)	Wear life (minutes)
Sand Blasting			
Silicon carbide	60	1.5-1.7	281
	120	1.0-1.1	137
	240	0.6-0.8	11
Aluminium oxide	60	1.5-1.9	294
	120	0.9-1.1	175
	240	0.6-0.7	12
Iron shot	50	2.9-3.5	157
	120	1.1-1.4	218
Quartz	80	1.4-1.6	82
Vapour-Blasting			
Quartz VB	60		187
	80		150
	100		35
Aluminium oxide VB	100		277
	150		177
Novaculite VB	100		15

The other important aspect of pre-treatment is the use of a chemical conversion treatment. Conversion treatments such as phosphating[148] and sulphiding[150] can be used to give an improvement in load-carrying capacity and wear life, and suitable treatments for a number of metallic substrates are listed in Table 11.6. The improvement with phosphating of steel seems to be relatively small if the surface has previously been given an optimum surface finish by grit-blasting, but otherwise the improvement may be considerable[136]. Phosphate coatings will break down at high temperatures, and should probably not be used if the service temperature is much higher than 200°C. A secondary advantage of phosphating and some of the other conversion coatings is that they provide some degree of corrosion protection which helps the generally mediocre corrosion protection provided by the bonded films themselves.

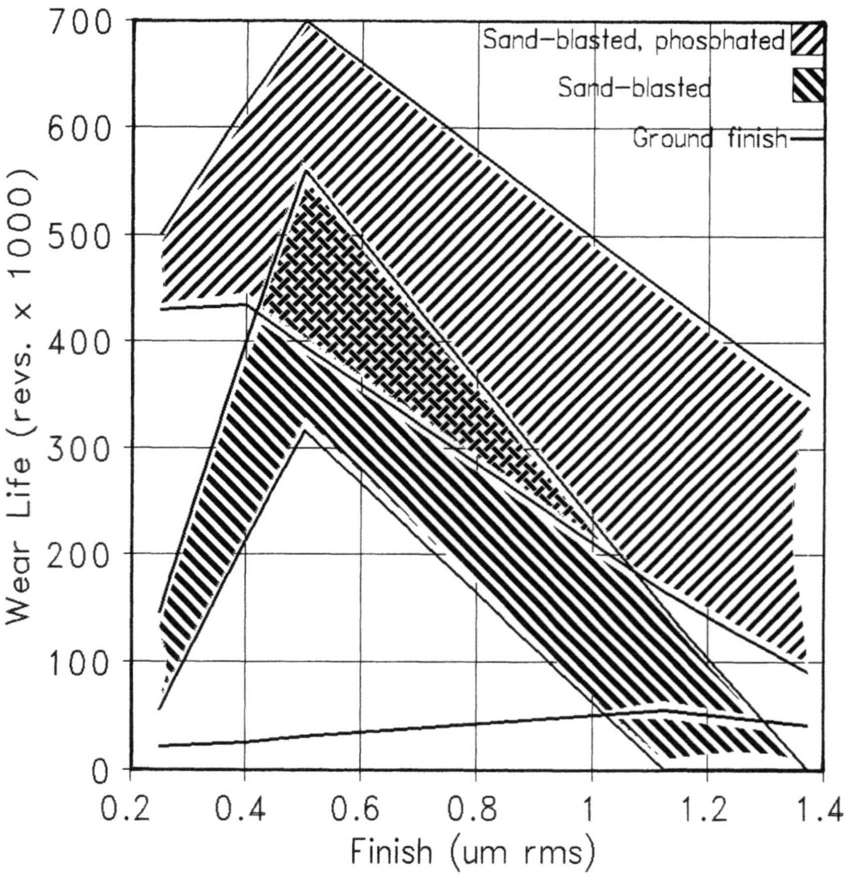

Figure 11.1 Effect of Various Surface Finishes on Wear Life of a Bonded Molybdenum Disulphide Coating (Data from Ref.338)

Nitriding of steels is not in itself beneficial to bonded film performance, but Kawamura et al[340] showed an interesting synergistic effect of soft nitriding and manganese phosphating on the wear life of a bonded molybdenum disulphide coating on an annealed steel.

Table 11.6 Chemical Conversion Coatings

Substrate	Conversion Coating
Steel (carbon or low alloy)	Phosphate; sulphide
Aluminium	Anodize; Fluoride-phosphate (Alodine)
Magnesium	Dichromate
Titanium	Phosphate-fluoride
Copper	Bright dip; oxalate
Cadmium	Phosphate (excellent combination for corrosion resistance)
Zinc	Phosphate

For consistent performance the bath concentrations, temperature and immersion time should be carefully controlled, and the component should be cleaned to ensure complete removal of contaminants such as cutting oils, grinding fluids and residues from grit-blasting before the conversion treatment is carried out. For certain substrates, such as stainless steels and nickel, there are no generally recommended conversion coatings, although a special phosphate treatment exists. The use of an optimum surface roughness is therefore of particular importance for these materials. An alternative to chemical conversion coatings is the use of a porous sintered or plasma-sprayed layer to provide a keying surface for the bonded film.

The reason for the effectiveness of conversion coatings has not been established. It has been suggested[341] that they provide small pits or reservoirs of lubricant for film replenishment. Alternatively they may provide a surface of optimum hardness for the re-embedding of molybdenum disulphide crystallites when a bonded film is approaching the end of its wear life. It was mentioned in Chapter 8 that Lancaster[134] in studying transfer found that molybdenum disulphide on phosphated steel had the very low coefficient of friction of 0.04. This suggests that a benefit of phosphating in that case was to facilitate the orientation of the molybdenum disulphide to a full basal orientation. One final possibility is that conversion coatings increase wear life by preventing transient seizure when asperity contact begins to occur towards the end of film life.

11.4 APPLICATION OF THE BONDED FILM

The objective in applying a bonded film is to obtain a smooth, uniform, adherent coating of a predetermined thickness. Coatings can be applied by dipping or brushing, but the best control of coating thickness and uniformity is usually obtained by spraying. Whichever method of application is used, the consistency of the dispersion is quite critical, but, as with painting, the optimum consistency for spraying is usually thinner than for dipping or brushing.

The subject of coating thickness has been discussed in detail in Chapter 7. Most of the evidence quoted in that discussion is particularly relevant to bonded coatings. The consensus is that the optimum coating thickness is generally between 4μm and 10 μm. Where the operating stresses will be low in relation to the hardness of the film, the thickness is much less critical and greater thicknesses may lead to longer wear lives. In other situations it may be important to achieve the recommended thickness, and this may require some experimentation. The main difficulty in developing the application technique to produce the required thickness is the problem of measuring the coating thickness.

Some guidance can be obtained simply from the appearance of the dried coating. The coating should be semi-transparent, and the colour and texture of the substrate should be visible through the coating. This is in marked contrast to paint films, which are applied to obscure the colour and texture of the substrate, and will usually be from 50 μm to 70 μm thick. For this reason it is often found that expert spray-painters are not suitable for applying bonded coatings, because they find it hard to maintain the very thin coating thickness required.

A rough idea of the coating thickness can be obtained by spraying across the edge of a piece of adhesive tape, masking tape or adhesive paper. After curing the paper or tape is stripped off, leaving a sharp edge to the coating, whose height can be measured by means of a profilometer, especially a non-contacting profilometer.

There are several thickness measuring devices which can be used for more accurate measurement. Disapio[342] used an automatic bench micrometer with indicator readings to 0.5 μm. Hopkins and Campbell[179] used a dial indicator or a four-place micrometer, and also used a Magne Gauge, as did Sauer et al[176]. The Magne Gauge works on the basis that the attractive force between a magnet and a ferromagnetic substrate material is inversely proportional to the thickness of the non-magnetic film. All these techniques have some disadvantages. The Magne Gauge can only be used on ferromagnetic substrates, and this excludes stainless steels and many non-ferrous metals. All contact devices require care in positioning because the modulus of most coatings is low, and they may deform under contact.

In a programme at the Swansea Tribology Centre in 1987, an Elcometer Minitector FN was found to give satisfactory results. This instrument uses electro-magnetic induction and eddy-current techniques, and can be used for non-magnetic coatings on a ferromagnetic substrate or for non-conducting coatings on a conducting substrate. Where the actual test specimens were not suitable for direct use of the Minitector, a blank mild steel specimen was mounted between the test specimens and coated alongside them, so that the blank and the test specimens were coated under identical conditions. Regular checks of coating thickness on specimens and blanks were carried out by sectioning the coatings and measuring the thickness with a scanning electron microscope, using the Link Energy Dispersive X-Ray Analyser to scan for sulphur in the film. The correlation between the two techniques was generally better than 1 μm.

The coating thickness with brushing or dipping can be adjusted by varying the consistency of the dispersion, preferably using a thinner which is the same as the dispersant or solvent in the dispersion. For spray application, possible suitable conditions are listed in Table 11.7. The molybdenum disulphide and other lubricating constituents will be insoluble in the liquid medium and will tend to agglomerate or settle out, especially when the mixture is thinned to spraying consistency. The mixture should therefore be well stirred before application, and a swirling motion given to the spray-gun reservoir during or between spraying operations to keep the constituents uniformly dispersed. It may also be useful to put glass or steel balls in the reservoir to help mixing.

The following procedure will usually give a uniform, adherent film of the required thickness, and the operator should be prepared to experiment in order to achieve a satisfactory performance.

1. The nozzle should be held at a distance from the component such that the spray reaches the surface wet, but dries quickly.

2. The spray density should be tested by placing a sheet of paper beside the component. The paper should be uniformly darkened in a single pass.
3. The gun should be continually swirled to keep the constituents dispersed.

Table 11.7 Spraying Conditions for Bonded Films

Variable	Requirements
Spray gun and nozzle	Not critical, but should be kept the same
Gas pressure	2-3.5 bars (0.2-0.35MPa)
Gas	Dry nitrogen or clean dry air
Spray distance	0.20-0.25m
Spray condition	Should reach surface wet but dry quickly
Thickness per coat	2μm
Number of coats	2 to 6
Time between coats	Several minutes
Total thickness	4-10μm

There are many different spray guns available and the type used does not seem to be critical[343], but the small type known as an artist's air brush is suitable for small components, and standard paint spray guns for larger components. The operator should standardise on one type of gun and nozzle for the sake of consistency. Where a large number of similar components are to be sprayed it is desirable to set up a semi-automatic or automatic system to ensure uniformity.

Because spray-guns are designed to supply the thicker films used for applying paint, it will often be found that the duration of spraying for a bonded coating is very short. The droplet distribution in the spray is usually found to be less satisfactory at the beginnning and end of the spraying, and this problem is accentuated when the spraying time is short. A more uniform coverage may be obtained if a travelling or rotating mask or screen is placed between the spray-gun and the specimen, and the period of exposure to the spray is controlled by the speed of movement and dimensions of a window in the screen which moves past the specimen and through the spray. For satisfactory bonded film application it is desirable to standardise all the coating conditions and to select one operator, who is trained particularly to understand the importance of applying thin coatings.

For coatings dispersed in water it is recommended that component surfaces are pre-heated to a uniform temperature of about 60°C to improve the drying characteristics of the film, but where organic solvents are used they are selected to give satisfactory drying at ordinary room temperature. However, film application will often be unsatisfactory if the atmosphere is too cold or damp.

Where only a few components are to be coated it may be more convenient to use a commercial film in an aerosol, but it is then not possible to check the quality of the material before application, and the subsequent inspection of the coated component will be particularly important. It is often found that the performance of coatings applied by aerosol is inferior because they are considered as a low-technology process and the pre-treatment has not been carefully done.

11.5 CURING THE FILM

Before curing at elevated temperature the coated items should be allowed to dry thoroughly at low temperature (room temperature for organic solvents or 50° - 80°C for water dispersions) to remove any residual solvent. They should then be brought smoothly up to the recommended curing temperature. Any residual solvent will otherwise tend to bubble or "ciss".

Typical curing temperatures are listed in Table 11.2. In general a manufacturer's recommendations for the cure schedule should be followed, but where these are difficult because of either excessive temperature or excessive time it may be possible to reduce the temperature and increase the time or vice versa. However, for every type of binder there will be a temperature below which satisfactory curing will not take place.

11.6 PLASMA SPRAYING

An alternative to conventional spraying and curing of bonded films is plasma spraying[344], in which a plasma jet is formed by passing a compressed gas through an electric arc struck between two electrodes. Typical electrodes are a rear cathode of thoriated tungsten and a hollow front anode of copper, both being water cooled. The plasma-forming gas is usually argon or nitrogen, and the operating temperature is probably between 2500° and 8000°C, but the total heat capacity of the plasma is relatively low and the plasma is chemically inert, so that substrates which are sprayed will not be subjected in depth to very great temperatures.

A plasma spray can therefore be used to apply sintered coatings to substrates which would normally be adversely affected by the sintering temperature of the coating. However, the powders carried by the plasma are exposed to greater thermal stress than the substrate, and some experimenting with specific powders is usually necessary to define the best application conditions. Where molybdenum disulphide in a metal, resin or ceramic binder is applied by this technique, the optimum conditions will usually be different for the two materials. This difficulty has been overcome by the use of two separate entry ports into the nozzle for the two components. The use of this technique has even been applied to molybdenum disulphide in a polyethylene binder.

Pretreatment of the substrate for plasma spraying consists only of cleaning and mild grit-blasting. If the bulk substrate is not capable of withstanding the temperatures involved, it may be necessary to apply a film of a metal such as nickel or molybdenum to provide a stable substrate. Plasma spraying has not been widely used, and for many applications has probably been superseded by sputtering.

11.7 FRICTION AND WEAR PROPERTIES OF BONDED FILMS

The initial properties of bonded films as applied vary considerably depending on the type of binder, the ratio of binder to molybdenum disulphide and other components, the nature of the drying or curing process, and the way in which the film was applied. In particular there is a wide variation in hardness from very soft (silicones and alkyds) to very hard (metals and ceramics), and this has important implications for the way in which a film is used and even the purpose for which it is used.

In a bonded coating as applied, the molybdenum disulphide crystallites are randomly oriented[345]. The static coefficient of friction is therefore quite high, typically between 0.1 and 0.3, and the films can be quite abrasive. The actual degree of abrasivity depends on the hardness of the coating, the particle size of the coating, and the way in which the coated surface is loaded against a counterface.

As soon as sliding begins, there is an immediate tendency for crystals to become basal-plane oriented, at least at the surface[345,346]. This is similar to the process in burnishing of films, as described in Chapter 6, and can occur either by complete rotation of a particle, or by fracture of particles and re-orientation of smaller crystallites, or possibly by plastic bending of crystals. The extent of re-orientation will be higher with high contact loads and speeds, with soft films, and to a lesser extent with the duration of sliding. The depth to which re-orientation occurs will depend very much on the hardness of the film and the extent of rubbing.

As a result of re-orientation, the film becomes less abrasive, and the coefficient of friction decreases. Because re-orientation begins rapidly with sliding contact, measurements of kinetic friction will often not show the decrease in friction unless friction monitoring is continuous from the commencement of sliding, when a trace such as that shown in Figure 11.2 may be obtained[132,347]. This is very similar to the trace shown in Figure 6.3 for a rubbed film of molybdenum disulphide powder.

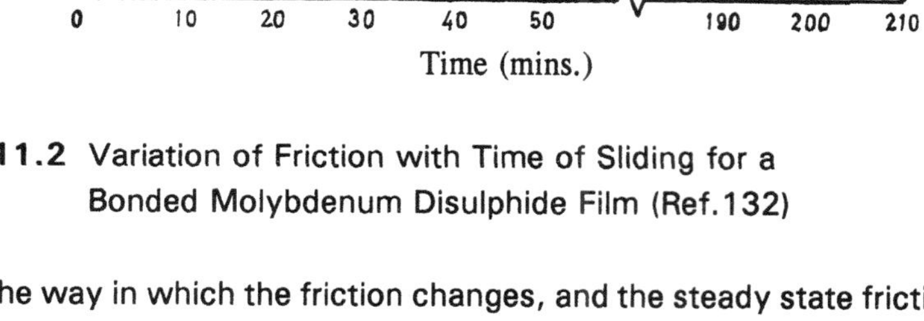

Figure 11.2 Variation of Friction with Time of Sliding for a Bonded Molybdenum Disulphide Film (Ref.132)

The way in which the friction changes, and the steady state friction in use, will depend on the hardness of the film, the concentration of molybdenum disulphide and certain other components, and the nature of the consolidation process to which it is exposed. The subject of consolidation has been described in Chapters 6 and 7 in connection with burnishing of films. If the contact geometry and operating conditions are suitable, consolidation and orientation of the film will take place satisfactorily during running-in of the film during operation. Suitable contact geometry requires conformal contact with any edges well radiused and preferably a gentle lead-in angle or curve on the counterface. Suitable operating conditions are low initial loads and speeds, but if necessary these can be ensured by a mild running-in procedure before full-scale operation.

Better control and more repeatable consolidation can be achieved by a separate burnishing process before the coated surface is put into normal operation. Burnishing is carried out in much the same way as in the burnishing of loose powder. It is in fact easier because a uniform layer of molybdenum disulphide is already in position, thus avoiding the problems of controlling loose powder.

Films can be burnished with almost any inert smooth soft material such as leather or a low-modulus polymer, but a simple method is with a hand-held soft cloth, or a similar cloth mounted on a harder backing device. As with spraying, better control and repeatability are achieved with some form of mechanical system in which the contact pressure, speed and duration can be accurately determined and maintained. These requirements are similar to those of a wear test machine, and in fact a simple wear test rig can often be used as a burnishing device. The contact pressure should initially be low, and should be increased progressively to a pressure approaching that which will be applied to the film in service.

A fully-burnished film will have a coefficient of friction between 0.04 and 0.08 at high contact pressure, depending on the final burnishing pressure, and figures in this range indicate that the surface is well oriented. There is a considerable loss of material in heavy burnishing of a soft coating. An initial cured coating 6 μm thick burnished at 0.1 MN/m^2 and 5 cm/s was reduced to 4 μm thickness in 30 seconds with one type of soft coating. Part of the reduction in thickness is accounted for by the consolidation of the film, but probably about 25% of the coating material was removed in the burnishing process.

If a soft film is put into service without pre-burnishing, a similar consolidation process will take place during the early stages of operation, but the degree of consolidation will depend on the operating conditions, and the eventual friction and wear life will be less predictable. The surplus coating material removed during the consolidation process will remain in the system in the form of powdered debris, and this may be a problem in some systems. A soft film, finally burnished at high pressure, is likely to be fully consolidated, with low friction. Its wear rate can be very low, and ultimate failure of the film is likely to be by oxidation and blistering, as described in Chapter 7, rather than by steady wear.

A hard coating will behave quite differently. The effect of sliding against a counterface will be to cause some re-orientation at the surface of the film, but this may be no more than 50Å or 100Å thick. With such a thin oriented film, contact loads will be carried partly by the surface film and partly by the randomly-oriented bulk of the coatings. Surface asperities and anomalies on the counterface will lead to distortion of the thin surface film, and defects in the orientation. As a result, crystallographic analysis often fails to show strong indications of clear basal orientation. In one such situation[345] a strong indication of basal plane orientation was only obtained after severe burnishing with a wire brush at about 30 m/s.

As a result of the very thin and defective oriented film, the friction tends to be higher, perhaps between 0.09 and 0.15 at high contact pressures. Judged by the criterion of friction reduction, this is therefore not highly efficient lubrication, and its benefits must be considered partly as the prevention of galling or seizure rather than simply reduction of friction. The wear behaviour of such a coating will depend on the load applied to it. Under light loading the surface film may survive for a lengthy period, so that wear may be negligible. Under heavy loading, although the oriented surface film will be thicker, it is more likely to be disrupted due to deformation of the bulk of the coating. There will therefore be a continuing cycle of surface film disruption and loss followed by renewal from the bulk of the coating. The effect will be of steady wear of the coating, and failure will occur when the coating is completely worn away.

In practice there is a continuous range of coating hardnesses varying from very soft to very hard, and the hardness is determined not only by the binder but by the concentrations of molybdenum disulphide and other solid components. The definition of hardness is also relative to the service contact loads. A coating of intermediate hardness may behave like a hard film under low contact pressures or like a soft film under high contact pressures. As a result, it is very difficult to compare the quoted load-carrying capacity and wear lives of different films tested under different conditions by different research groups.

It is unfortunate that with very few exceptions the effect of film hardness on the consolidation behaviour has not been taken into account in reporting coating performance. Even where the authors of a report have obviously been aware of the problems of running in, their test results may show enormous differences between similar films. For example, in a paper by Hopkins and Campbell[159] the authors made the comment "All friction coefficients were measured after a brief but complete run-in", but the static friction measurements for four phenolic-bonded coatings of a similar type varied from 0.14 to 0.33. The static friction for four sodium silicate-bonded coatings under the same conditions varied from 0.20 to 0.39. This variation shows how important it is to ensure effective film consolidation or burnishing either for consistent research results or for consistent service operation.

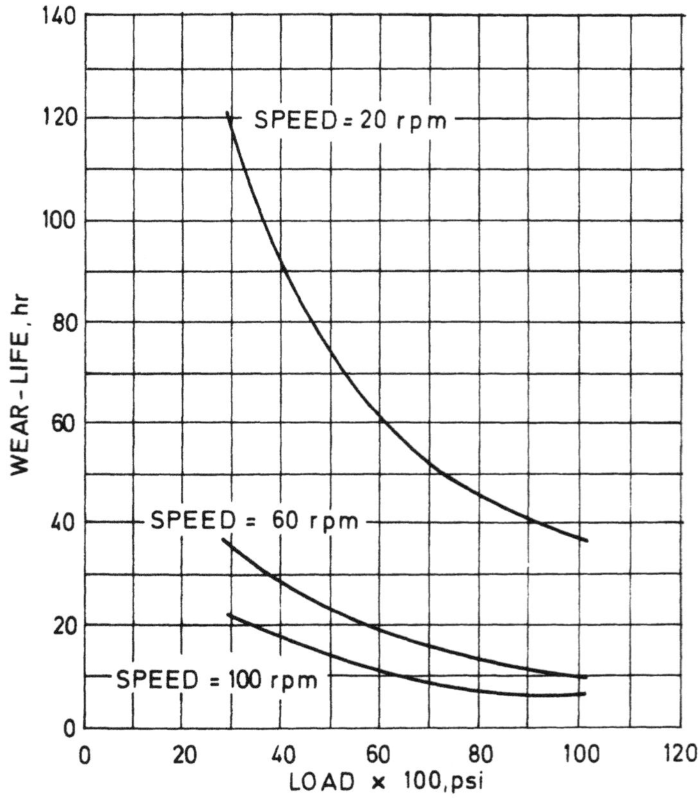

Figure 11.3 Effect of Load on Wear Life of a Phenolic-Bonded Film (Ref.560)

Some useful generalisations can be made about friction and wear behaviour. The coefficient of friction decreases with increasing contact pressure although the relationship will vary with the type of coating and other test conditions. The effects of contact load and speed on wear rate or wear life are less easily defined, although the general patterns are as shown in Figures 11.3 and 11.4. However, as has been

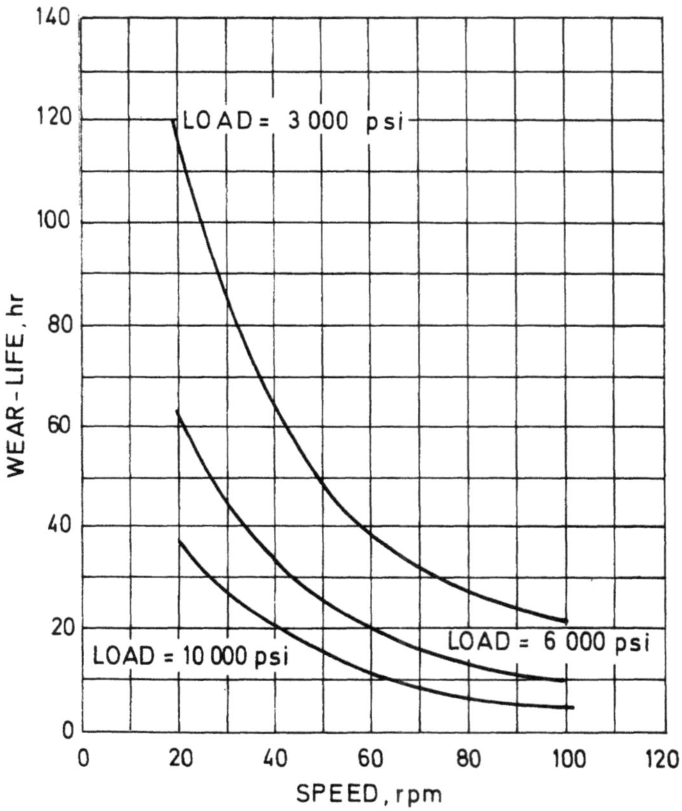

Figure 11.4 Effect of Speed on Wear Life of a Phenolic-Bonded Film (Ref.560)

pointed out previously, the friction of a highly-burnished film may be influenced more by the burnishing pressure than by the subsequent operating pressure.

The effect of humidity on the friction of a bonded molybdenum disulphide film has been shown to be broadly similar to the effect on a burnished film,[156] as described in Chapter 7. The coefficient of friction increased slowly as the relative humidity increased to about 60%, and then progressively more rapidly as it was further increased to 90%.

However, one interesting observation was that when phosphor-bronze was used as a substrate, there was a falling-off from the peak friction at the highest humidities, which was accompanied by the appearance of a dark red-brown film. This was presumed to show the occurrence of some reaction with components of the phosphor-bronze to produce metallic sulphides which acted synergistically with the molybdenum disulphide to improve lubrication, as described by Haltner and Oliver[188]. It is interesting that in this study the presence of the modified phenolic resin binder and graphite in the coating made no major difference to the effect of humidity, although, as the authors pointed out, this would not necessarily be true of other binders or additives.

There are indications that polymeric binders may protect films to some extent against oxidation by moist air in storage. A few tests on a fully-burnished soft film with an inorganic binder also showed no deterioration during extended storage in air at 70% relative humidity and 60°C. This gives some confirmation to the concept that in a fully-burnished film the exposed surface consists almost entirely of inert basal planes, and is therefore more resistant to chemical attack.

The effects of temperature on bonded films have been mentioned earlier in this chapter in connection with specific types of binder, and temperature limits are shown in Table 11.2. As a generalisation it is clear that up to 350°C in air the temperature limits are determined by the type of binder used, while at higher temperatures the thermal stability of the molybdenum disulphide is likely to be the limiting factor. However, where a soft coating has been heavily burnished, the resulting film will consist largely of molybdenum disulphide, and the thermal stability of the binder used will be less important.

Even below the temperature at which thermal breakdown becomes significant, there may be large variations in friction and wear with temperature. The nature and extent of the variations are generally inconsistent and unpredictable, depending on the type of binder, concentration of lubricant, other components present in the coating, completeness of film curing, degree of consolidation and the test method used.

For example, Tsuya and Kitamura[348] studied the change in wear rate of two bonded coatings with temperature. They found that the wear rate increased with increase in temperature for an epoxy-bonded coating, and decreased with increase in temperature for a silicate-bonded coating. On the other hand Hopkins and Campbell[319] found that coatings bonded with polyimide, silicate or aluminium phosphate all had decreased wear life with increasing temperature. The relationships are probably dependent on the composition of the film, the actual temperature range studied, and the possible contribution of frictional heating, but Figure 11.5 shows the relationships obtained by Lancaster[149], which appear to confirm some of the results of both Tsuya and Kitamura and Hopkins and Campbell. The complexity of the situation is shown by the different behaviour of the two silicate-bonded films. Variation of friction with temperature is even more variable, but this may often be due mainly to variations in consolidation.

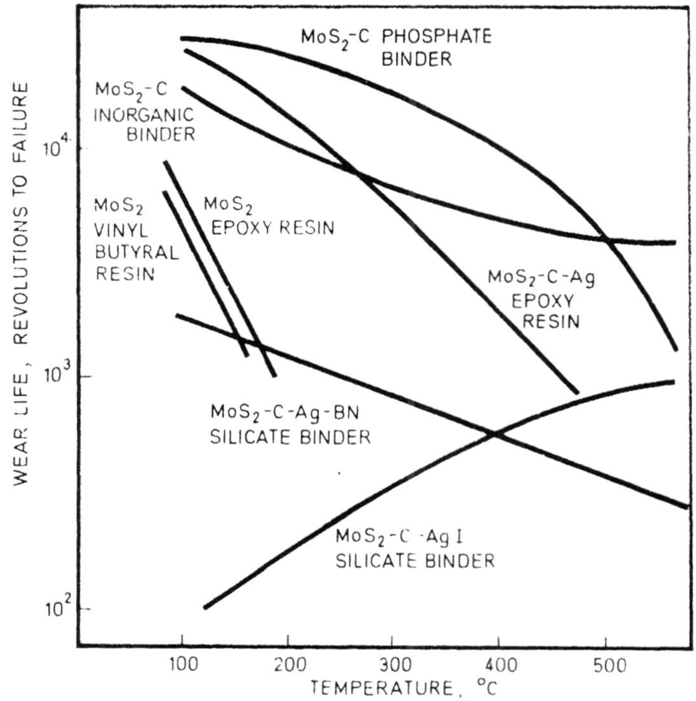

Figure 11.5 Effect of Temperature on Wear Life of Bonded Coatings (Ref.149)

As a broad generalisation, inorganic compounds are relatively unaffected by gamma-ray or neutron irradiation, while organic compounds are more seriously affected, as was pointed out in Chapter 7. The same generalisations are true in respect of bonded molybdenum disulphide films, although there are several reports of specific exceptions to the rule.

McDaniel[162,349] described a number of experiments in which inorganic-bonded films were subjected to combined gamma-ray (2.2 x 10^{11} ergs/gm) and neutron (5.2 × 10^{16}n/cm^2, E>2.9MeV) doses in a nuclear reactor. The films were then tested for friction and wear in a Hohman Rub-Shoe Test Machine at 26.7°C, 316°C and 538° or 649°C. There was no significant effect of radiation on a boric oxide-bonded film containing lead sulphide and molybdenum disulphide. With a sodium-silicate-bonded film containing graphite and molybdenum disulphide there was a 50% reduction in wear life and a slight increase in friction at 26.7°C. At 316°C, however, there was a 51% increase in wear life and a reduction in friction from 0.14 to 0.08, while at 649°C there was a 143% increase in wear life and a reduction in friction from 0.24 to 0.20. The reasons for this behaviour are not at all clear.

Taguchi et al[350] studied the effect of α-radiation from a cobalt-60 source on films bonded with a phenolic resin or sodium silicate. With the phenolic-bonded film there was an increase in wear life and a reduction in friction. The friction fell from 0.145 for non-irradiated specimens to the remarkably low figure of 0.004 after a radiation dose of 2 x 10^5R. With higher doses of 2.10^6R, 2 x 10^7R and 2 x 10^8R there was a progressive increase to 0.065, 0.069 and 0.075. It is difficult to explain the anomalous behaviour of these organic-bonded films, although analysis of the friction figures suggests that curing of the films at 105°C for 15 minutes may have not been complete, and that further curing occurred during irradiation.

Like McDaniel they found that irradiation increased the friction and decreased the wear life of the film bonded with sodium silicate in tests at temperatures up to 214°C. They reported rusting of these specimens during irradiation, whereas McDaniel had reported formation of a white powder, so clearly this type of film has limited stability under irradiation.

11.8 REPAIR AND RENEWAL OF FILMS

One practical difficulty with the use of bonded films in general engineering is in repairing or renewing them when they have become worn or failed. Such films are usually highly polished in places and worn through or blistered in others. In the early years of use of bonded films it was found impossible to add fresh coating on top of a highly burnished area, and very difficult to remove the old film. It was even suggested that a film be removed by grinding to below the original metal surface and then rebuilding the metal surface by electroplating or plasma deposition before renewing the bonded coating. This was obviously a serious disadvantage which would discourage many operators from using bonded molybdenum disulphide coatings.

Table 11.8 Results of Immersion Cleaning Tests of Molybdenum Disulphide Films (Ref.351)

Cleaner Type	Coating Remaining			
	1	2	3	4
Solvent emulsion	Light but loosened	None	Medium but loosened	None
High strength alkaline	None	None	Heavy but loosened	None
Alkaline permanganate	None	None	Medium but loosened	None
High strength alkaline solvent emulsion	Medium but loosened	None	Medium but loosened	None
Alkaline permanganate solvent emulsion	None	None	Light, some clean spots	None
High strength alkaline permanganate	Light trace	None	Light, some clean areas	None

As a result Bertrand and Vukasovitch of Climax Molybdenum carried out[351] an intensive study of different cleaning techniques applied to various forms of molybdenum disulphide film. The types of film used were a burnished film from a sub-micron sized powder, a burnished film from a soap-containing molybdenum disulphide powder, an air-cured inorganic-bonded coating, and a film formed from a dispersion in oil. The films were formed on steel pins which were then pressed through an undersized bush. This ensured highly-loaded contact on the film, and consolidation of the film. The burnished powder and the resin-bonded films could be assumed to achieve good adhesion under these conditions. The films from the soap and oil dispersions would not necessarily achieve a high level of adhesion by this technique but both the lubricant dispersions and the application technique are typical of industrial metalworking practice.

The pins were then cleaned by commercial cleaning companies using six standard industrial cleaning procedures, and the results are summarised in Table 11.8. The only film which proved difficult to remove was the inorganic-bonded film (Soil 3). This result was not thought to be typical of all bonded coatings, as many organic binders would be expected to be easier to remove, especially with solvent-containing cleaners. Furthermore, the cleaning companies used only standard cleaning techniques, and the use of techniques such as brushing and ultrasonic agitation would presumably be even more effective.

Their conclusion was that with the use of the proper choice of cleaning system, molybdenum disulphide does not pose a cleaning problem and could be used in commercial metalforming operations without difficulty. This was confirmed by Davison and Gilbert[352], who found that after using bonded coatings in a severe ironing operation with thin-gauge steel sheet, the lubricant could be satisfactorily removed.

Certainly, since the publication of those reports in 1976 there has been a vast expansion in the range of uses of bonded molybdenum disulphide coatings, and the problem of coating removal has apparently not been found to be serious, since it seems not to have been highlighted in recent years.

~ ~ ~ ~ ~ ~ ~ ~ ~ ~ ~ ~ ~ ~ ~ ~

CHAPTER 12

COMPOSITES

12.1 LUBRICATING COMPOSITES

In some publications bonded coatings are described as composites, and technically the description is correct, but it is convenient to use the term composite only for materials in three-dimensional or bulk form in order to differentiate clearly between them and bonded coatings. Composites can generally be described in terms of one or more dispersed phases in a continuous solid matrix, although there are some, such as resin-impregnated fabrics, which are not readily described in this way.

Incorporation of a dispersed lubricant phase in a solid matrix is not new. Porous metal bearings impregnated with a lubricating oil are composites. According to Morgan[353], they were probably first devised over seventy years ago, and even then they followed an earlier composite consisting of oil-soaked wooden bearings. The earliest use of a solid lubricant dispersed in a polymeric matrix was that of graphite in a fabric-reinforced phenolic resin, and this was first reported in 1937[354,355].

These examples illustrate the two primary purposes of lubricant-containing composites. The oil-impregnated porous metal bush is designed to supply lubricant to a bearing. The purpose of the graphite-containing phenolic bearing is to reduce the friction of the material itself, so that it is in fact a self-lubricating composite. In the latter case, the phenolic composite was intended to be used under marginally-lubricated conditions with water or oil lubrication. The benefit of the graphite was to reduce frictional heating and wear when lubrication became inadequate. Later composites achieved much lower friction and better wear-resistance in the complete absence of liquid lubricants, and they then became known as dry bearing materials.

The two purposes, of lubricant supply or friction reduction, are of course not entirely distinct, since continuing friction reduction requires a continuing supply of lubricant. The difference lies in the fact that the amount of lubricant required to maintain low friction of the composite itself is relatively small, probably restricted to a film which may be only a few molecules thick. The more general lubrication of a bearing system will require an additional supply of lubricant which is at least sufficient to provide an equivalent film on all the bearing surfaces involved. With the inevitable wastage inherent in movement of lubricant to other surfaces, the lubricant demand will be much greater than for self-lubrication alone.

Lubricant is made available by wear of the composite. The greater quantity required for supplying a bearing system must therefore be provided by a higher wear rate, or by a higher concentration of lubricant in the composite, or both. It follows that in practice the formulation of the composite and the design of the composite component are significantly different when lubricant supply is the objective, and not merely self-lubrication. The former case will therefore be considered separately in connection with Transfer Lubrication of Rolling Bearings in Section 12.5.

The existence of the earlier lubricating composites meant that the concept of incorporating a lubricant in a bearing material was well understood when the expansion of interest in molybdenum disulphide arose in the late nineteen-forties, and the first molybdenum disulphide-containing composites were reported in 1950, with either polymeric[356] or metal[357,358] matrices. Since then hundreds of different composites containing molybdenum disulphide have been described in the technical literature, and many of these have become available commercially. The technology of composites is now fairly well understood as a result of the enormous scientific effort which has gone into their development. It is still complex because of the wide variety of fillers and reinforcing materials available, and because many of the fillers have several different effects on the properties of the composites. The effects of fillers are generally quite different for polymeric and metal matrices.

12.2 POLYMER COMPOSITES

From the early days of polymer development, many polymers have been used in bearing applications, and there have been several useful reviews of the subject[179,202,359-363] as well as the comprehensive McGraw-Hill Encyclopaedia[364] and the excellent ESDU Design Data Item No 87007[365].

Compared with metals or ceramics, polymers are lighter, softer, weaker, less thermally stable and less wear-resistant. They are also poor conductors of heat and electricity. However, their properties can be enormously modified by the incorporation of fillers, reinforcements, and other components such as plasticizers. For most purposes it is useful to consider polymers in three separate groups, namely thermosetting, thermoplastic and PTFE.

Table 12.1 Common Thermosetting Polymers

Polymer Type	Characteristics
Alkyd	Excellent dimensional stability and heat resistance, surfaces tough, good susceptibility to fillers and fibres.
Epoxy	High physical strength and dimensional stability. Some types cold curing. Good filler susceptibility.
Phenolic	Low cost. Good physical strength and high temperature resistance. Good toughness with fillers and fibres.
Polyester	Good temperature and chemical stability. High physical strength with glass fibre reinforcement.
Polyurethane	High abrasion resistance, toughness and chemical resistance.
Silicone	Properties maintained from -75 to 260°C. Can be cold-cured. Fillers for increased strength.
Urea-formaldehyde, melamine	Good dimensional stability and abrasion resistance, surfaces tough. Good susceptibility to fillers and fibres.

A thermosetting polymer is one in which heat applied to a liquid or semi-solid resin causes a chemical change by which it becomes solid (sets). In fact several of them, such as epoxies, are commonly hardened by the use of a catalyst, called a curing agent or hardener, rather than by heat. It cannot be melted or softened significantly by heat, and the effect of excess heat is to cause thermal breakdown. A list of the commoner thermosetting plastics (thermosets) with their significant properties is in Table 12.1. Of these only alkyds, epoxies, phenolics and polyurethanes are important in molybdenum disulphide composites.

A thermoplastic polymer is one which can be softened by heat, formed into a desired shape, and re-solidified on cooling without any significant degradation. The absolute upper temperature limit for engineering use is therefore the melting-point, but with many of them some thermal degradation takes place below the melting-point, so that there is a lower temperature limit above which they cannot be used indefinitely. The commoner thermoplastics are listed in Table 12.2. Those which are of interest for bearing use are acetal, nylon, polyetheretherketone (PEEK), polyimide and polyphenylene sulphide (PPS).

PTFE (polytetrafluoroethylene) is a unique polymer. It is often classified as a thermoplastic, and does soften with heating, but it does not melt and is formed by sintering the powder. Its maximum operating temperature for continuous use is normally quoted as about 290°C, but in a glass fibre reinforced form it has been tested satisfactorily for thousands of hours at 300°C[366]. It is also usable down to cryogenic temperatures. Three other fluorinated polymers which are similar in some respects to PTFE are polytrifluoro chloroethylene (PTFCE), fluorinated ethylene-propylene (FEP) and polyvinylidene fluoride (PVF_2). Their friction and thermal stability are inferior to those of PTFE, but they have some advantages in ease of processing and adhesion to substrates.

Molybdenum disulphide may be incorporated in a polymer as a friction-reducing additive, to reduce the friction of the composite in sliding contacts, or to provide a reservoir for transfer lubrication. Transfer lubrication has been described in detail in Chapter 8, and the effects of the two different requirements will be discussed briefly in this chapter.

Polymers are now widely used in bearings, gears, cams and other moving contacts. The main reason for their use, as always, is economic, but they have several advantages over metals to offset their generally lower structural strength. One major advantage is that they have a much lower tendency to adhere strongly either to metals or to other polymeric components. As a result they give lower friction either in dry sliding or particularly in marginally lubricated situations, and have much less tendency to seize. Some properties of the commoner bearing polymers are listed in Table 12.3.

Table 12.2 Common Thermoplastics

Polymer Type	Characteristics
Acetal	Good dimensional stability and abrasion resistance. High tensile strength. Used for water or oil-lubricated bearings.
Acrylic	Hard, transparent, fair chemical resistance.
Cellulosic	Very tough, but poor weather resistance and embrittle with age.
Polyamide (nylon)	Excellent toughness and good wear resistance. Low friction. Poor dimensional stability.
Polycarbonate	Tough, rigid, transparent. Good dimensional stability, high resistance to impact loads.
Polyetheretherketone	Stable to 260°C, chemically inert, easily processed and attached to metals.
Polyethylene	Good toughness and chemical resistance, poor thermal stability, limit about 75°C.
Polyimide	Very high thermal stability, to 350°C, high physical strength and wear resistance, low friction.
Polyphenylenesulphide	Very tough, good dimensional stability, wide temperature range -170 to+260°C.
Polysulphone	High heat resistance, to 260°C, and mechanical strength, good chemical stability.

They can be used in many conditions in which conventional liquid lubricants are not usable, such as at high temperatures or very low temperatures, or where the risk of contamination from liquid lubricants is unacceptable, as in the food and pharmaceutical industries. When they are used in dry or marginally-lubricated systems, although their friction is lower than that of metals in the same conditions, it is still higher than that of oil-lubricated bearings. This leads to higher frictional heat loss, and polymers generally have low thermal conductivity, so that high temperatures develop in the region of the sliding contacts. Particularly with thermoplastic polymers, this causes softening, with loss of structural strength and even higher friction. Unmodified polymers also have relatively low wear resistance, which is further reduced when they are softened by heating. It therefore becomes important to use reinforcing materials to improve structural strength, fillers to increase hardness, strength and wear resistance, and lubricating materials to reduce friction. Some of the more common materials used in polymer composites are listed in Table 12.4.

Table 12.3 Properties of Some Principal Bearing Polymers

Polymer Type	Coefficient of Friction	Limiting Temperature (°C)	
		Short Periods	Long Periods
PTFE	0.03-0.1	310	250
PTFCE	0.07-0.3	210	200
FEP	0.06-0.15	220	205
PVF_2	0.1 -0.25	175	150
Acetal	0.15-0.35	110	85
Nylon	0.05-0.28	150	120
PEEK	0.1 -0.3	250	150
Polyimide	0.1 -0.25	480	260
PPS	0.1 -0.3	260	230

The formulation of composites is complex, because a component introduced to modify one property may have significant effects on other properties, and because there may be interactive effects between two or more components. Lancaster[361] published a paper on the role of fillers and fibre reinforcement in polymer-based bearing materials in 1972 which is still an excellent summary of the technology.

Because of the complexity of the subject, no attempt will be made here to discuss the use of fillers and reinforcements in detail, but only to explain briefly some of the more important effects and to describe some typical composites and their properties.

Table 12.4 Some Components Used in Polymer Composites

To Improve Mechanical Properties	To Improve Thermal Properties and Electrical Conductivity	To Reduce Friction
Carbon powder or fibres	Bronze powder	Molybdenum disulphide
Glass fibres	Lead powder	Graphite
Textile fibres	Carbon/graphite	PTFE powder
Metal powders	Silver powder	PTFE fibres
Metal oxides	Sintered bronze matrix	
Sintered metal reinforcement		
Mica		
Asbestos (obsolete)		

The most effective reinforcing materials are fibres of various types. For the thermosets, woven fabrics are commonly used for reinforcement. It is a relatively simple procedure to impregnate a fabric with a liquid monomer or mixture of reagents and catalysts and then to carry out the necessarily polymerisation process. It would be much more difficult to produce a uniform distribution of a woven fabric in a thermoplastic polymer. Woven fabrics can provide a high degree of structural integrity to polymers, especially in very large structures such as rolling-mill bearings and pit-head gear brake linings. However, textile fibres have limited thermal stability and structural strength.

The fibres generally used for polymer reinforcement are glass, carbon and less often PTFE or metals. For use in thermosets, glass or carbon fibre may be used in the form of a pressed mat, or bundles of fibres may be laid in a preferred orientation. This enables anisotropic properties to be obtained where this is beneficial. The alternative is to disperse chopped fibres randomly in the polymer matrix. This gives more isotropic strength and modulus characteristics, although streamlining can occur due to flow in subsequent moulding, and this will result in some degree of anisotropy. Lancaster[361] pointed out that even with reinforcing fibres, the increase in stiffness of a composite may be greater than the increase in strength. For particulate fillers there is also generally a greater increase in stiffness (i.e. modulus) than in structural strength. This is particularly true of solid lubricant particles, where because of their easy shear properties, they will not generally increase the strength, and will sometimes reduce it.

The wear of polymeric composites by metals depends on the roughness of the metal surface and to some extent the speed of sliding. With a rough metal counterface, individual asperity contact stresses may exceed the elastic limit, and the composite will wear by plastic deformation and cutting. With a smooth counterface the composite usually wears by fatigue. One interesting aspect is that the wear rate of unfilled polymers is almost independent of contact pressure over a wide range of pressures[367,368]. The wear resistance of the composite is improved by the use of fillers against rough or smooth counterfaces, and in one example the addition of 5% of molybdenum disulphide to a polyamide was reported to improve the wear resistance by a factor of five[369].

Like other composite properties, the wear resistance is influenced by the orientation of reinforcing fibres. Sung and Suh[370] found that with biaxially-oriented glass fibre and molybdenum disulphide in a PTFE matrix (Duroid 5813), the wear resistance was greatest when the highest proportion of fibre was normal to the sliding surface. The same effect was found with a graphite fibre/epoxy composite and a Kevlar fibre/epoxy composite.

However, the friction of the Duroid 5813 (about 0.34) was practically unaffected by fibre orientation, unlike the other two composites which gave the lowest friction when sliding was normal to the fibre orientation. This may indicate that in the Duroid 5813 the friction was to a considerable extent determined by the molybdenum disulphide, which was randomly oriented, or by the PTFE, whereas in the epoxy composites the friction was strongly influenced by the fibres themselves. Table 12.5 shows the relationship between fibre orientation and specific wear rate for the Duroid 5813. It should be noted that although the effect of fibre orientation is significant, all three wear rates are of the same order of magnitude.

Table 12.5 Relation Between Glass Fibre Orientation and Specific Wear Rate for Duroid 5813 (Ref.370)

Axis	% of fibres oriented along axis	Specific wear rate perpendicular to axis
x	67%	$2.9 \times 10^{-9} cm^3/kg.m$
y	33%	$4.1 \times 10^{-9} cm^3/kg.m$
z	0%	$7.8 \times 10^{-9} cm^3/kg.m$
Composition: 69% PTFE, 15wt.% MoS_2, 16wt.% glass fibre		

The effect of speed on wear rate is probably due at least partly to the fact that polymers commonly exhibit visco-elasticity. As a result, increased contact speeds can lead to higher contact stresses. Higher speed also causes higher frictional heating, which will usually result in softening and increased wear. However, the use of polymer composites is best restricted to speeds below one or two metres/second because the combination of frictional heating and poor thermal conductivity leads to overheating and breakdown.

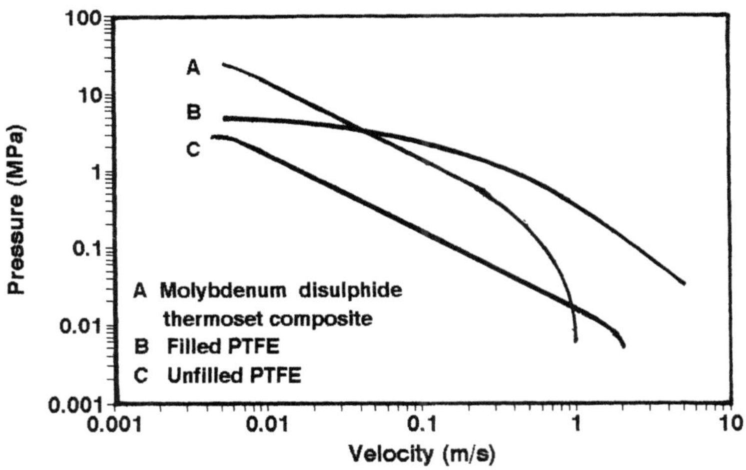

Figure 12.1 Pressure/Velocity Curves for Polymeric Bearing Materials (Ref.361)

The influence of speed on the performance of bearing materials can be expressed in terms of the PV factor, the limiting product of specific load and sliding speed, on the basis of either short-term survival or the acceptable wear rate. The concept is in fact an over-simplification because the limiting PV may vary with the actual pressure and velocity. This can be seen in Figure 12.1, where the limiting PV is almost constant for unfilled PTFE, constant for only part of the range for the thermoset containing molybdenum disulphide, and variable over the whole range for filled PTFE.

An important aspect of fillers in composites is their effect on the wear of metal counterfaces. Unfilled polymers do not cause abrasive wear of metal counterfaces but various fillers can cause significant counterface wear and thus modify the surface finish. The surface finish in turn affects the wear rate of the composite, as shown in Figure 12.2[371]. It follows that the selection of filler has important consequences for the wear of both composite and counterface, and Lancaster suggested that the incorporation of an abrasive filler in a composite could be an important contribution for optimum wear life. Some of the changes occurring to a counterface during sliding of a composite are shown in Figure 12.2.

Molybdenum disulphide is usually incorporated in polymeric composites in concentrations less than 30% unless the composite is specifically designed to provide transfer lubrication. At concentrations of 10% or more, molybdenum disulphide can reduce the structural strength of composites significantly except for the case of PTFE. In general molybdenum disulphide tends to be randomly-oriented in composites, although Griffin[372] has reported both uneven orientation and uneven distribution of molybdenum disulphide in composites depending on the nature of the moulding processes. In particular he found that there was a reduction in filler concentration near to a moulded surface. On the other hand, the usual tendency to form an oriented film on a surface, especially a machined surface, can lead to low sliding friction. Such factors as these can make the friction and wear performance of composites even less consistent than in other aspects of molybdenum disulphide use.

PTFE is a unique polymer in the formulation of composites, since it may be either the material of the matrix or a friction-reducing filler. It is a very soft polymer, which in the absence of reinforcement will wear rapidly, and it will cold flow under load. As a matrix material it must therefore be effectively reinforced. As a friction-reducing filler it may be used in the form of particles or fibres. Table 12.6 shows the effect of different fillers on the properties of PTFE. Spengler et al[375] reported an

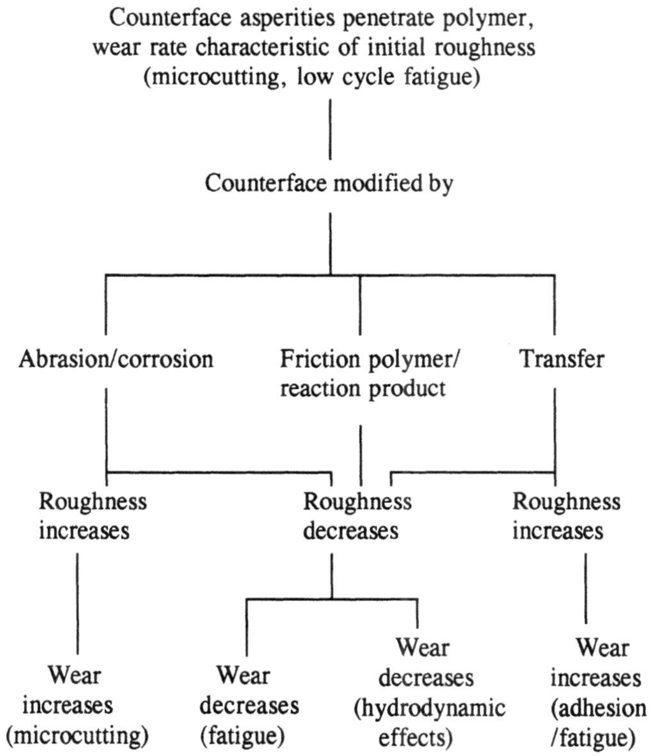

Figure 12.2 Changes to Counterface During Composite Sliding (Ref.361)

even more striking ten-fold improvement in wear life of PTFE from the incorporation of 12% of molybdenum disulphide, but clearly this must depend on the operating conditions. In general its beneficial effect on wear life is much less than that of other fillers such as bronze or glass fibre.

It has become the preferred solution for many applications to use molybdenum disulphide and glass fibre together in PTFE, and Table 12.7 shows the properties of two such ternary composites. This development reflects the results of many pioneering research projects such as those of Young et al[377], Scibbe et al[378] and Smith and Vest[379]. As a generalisation it can probably be said that PTFE containing 5 to 20 vol. % of molybdenum disulphide and 10 to 30 vol. % of a reinforcing filler

Table 12.6 Effect of Fillers on the Properties of PTFE (Data from Refs.361,373,374)

Filler	None	MoS_2		Bronze		Glass Fibre
Concentration (vol.%)		30	40	30	40	13,15
Coefficient of Friction in Air						
At 1.32cm/s						
Initial	0.05	0.03	0.04	0.07	0.07	
After 4hrs	0.20	0.16	0.14	0.17	0.17	
After 20hrs		0.17		0.20		
At 196.2cm/s	0.26	0.25		0.26		
On steel at 1cm/s	0.10					0.09
On st.st. at 6cm/s	0.12	0.13				0.22
On st.st. at 600cm/s	0.26	0.28				0.25
Wear Rate						
Specific wear rate	137	110	87			1.4
Physical Properties						
Elastic modulus (MPa)	558	404		307	479	
Compressive strength(MPa)	0.12	0.20		0.17	0.18	
Tensile strength (MPa)	9				6.2	
Thermal conductivity	0.57	1.63		1.08		
Specific gravity	2.2					2.19

such as glass fibre, carbon fibre or bronze powder has potential for many applications in which low friction is required with an unlubricated component where high structural strength is not needed. Where structural strength is important it is inferior to many filled thermoplastics.

When molybdenum disulphide is incorporated in nylon there is a general tendency to lower friction, lower wear rate, higher strength, and lower thermal

Table 12.7 Properties of Two Ternary PTFE Composites
(Data from Refs.361,376)

Filler Content: Molybdenum Disulphide	12.5%	10%
Glass fibre	12.5%	15%
Coefficient of Friction		
Static against steel, 27.6MPa		0.02
Dynamic against steel,0.01m/s	0.09	
0.6m/s		0.018
6.0m/s,0.1MPa		0.14
Wear Rate, mm^3/Nm	1.2×10^{-7}	
Physical Properties		
Tensile Strength (MPa)	13	28
Coefficient of Expansion (/°C)	11×10^{-5}	$2.9\text{-}27\times10^{-5}$
Thermal Conductivity (W/m.°C)	0.51	0.3
Specific Gravity	2.3	2.25

expansion. Table 12.8 shows data from several sources which tend to confirm these generalisations, but there are discrepancies with regard to friction, wear and tensile strength, where individual results (underlined) conflict with the general trend. Two of these discrepancies are for nylon containing only 1% or 5% by volume of molybdenum disulphide, and it seems that such low concentrations were ineffective in improving the friction and wear properties[381]. Concentrations from 5 to 35% by volume show progressive improvement of friction and wear resistance with increasing molybdenum disulphide concentration.

The effect of load on filled and unfilled nylon is shown in Figure 12.3, and it is clear that the presence of the molybdenum disulphide filler conveys a similar relative improvement at all the loads tested. One successful commercial range of filled nylons are the Nylatrons, such as Nylatron GSM, a composite of molybdenum disulphide in nylon which has been extensively used in heavily loaded bearings in heavy industrial plant such as steel mills. In one crane application Nylatron bearings were reported to last twenty times as long as the bronze bearings which they replaced[384].

Table 12.8 Properties of Nylon With or Without Molybdenum Disulphide

Property	Nylon without MoS_2	Vol % MoS_2	Nylon with MoS_2	Ref.
Coefficient of Friction				
Nylon 6,6	0.2-0.3		0.16-0.2	380
Nylon 6	0.19	1	0.20	381
Nylon at 30°C and 70°C	>0.6	40	0.17-0.2	382
Nylon (unspecified)	0.17	5	0.17	383
		20	0.15	383
		35	0.13	383
		50	0.15	383
Wear				
Wear volume in 400hrs.(mm^3)	6.0	1	7.5	381
Time to wear 0.254mm.(hrs)	6	5	30	383
		20	37	383
		35	47	383
		50	56	383
Physical Properties				
Tensile strength (MPa) at 23°C	76-81		83-103	380
Tensile strength (MPa)	162	5	148	383
		20	122	383
		35	93	383
		50	62	383
Elastic modulus (MPa) at 23°C	2760		3960-4140	380
Izod impact strength (J) 23°C	1.29		0.84	380
Coefficient of expansion (/°C)	10×10^{-5}		$4\text{-}6\times10^{-5}$	380

Risdon and Loban[385] studied the effect of different concentrations of molybdenum disulphide in Nylon 6.6. They found a steady decrease in wear rate as the concentration of molybdenum disulphide increased to 20%, the wear rate at 20% concentration being only 38% of that of the unfilled nylon. With a further increase in concentration to 40% there was a small increase in wear rate, which may have been caused by an increase in brittleness. They found improved wear resistance even at a 2.5% concentration, in contrast to the results in Table 12.6. However, in further

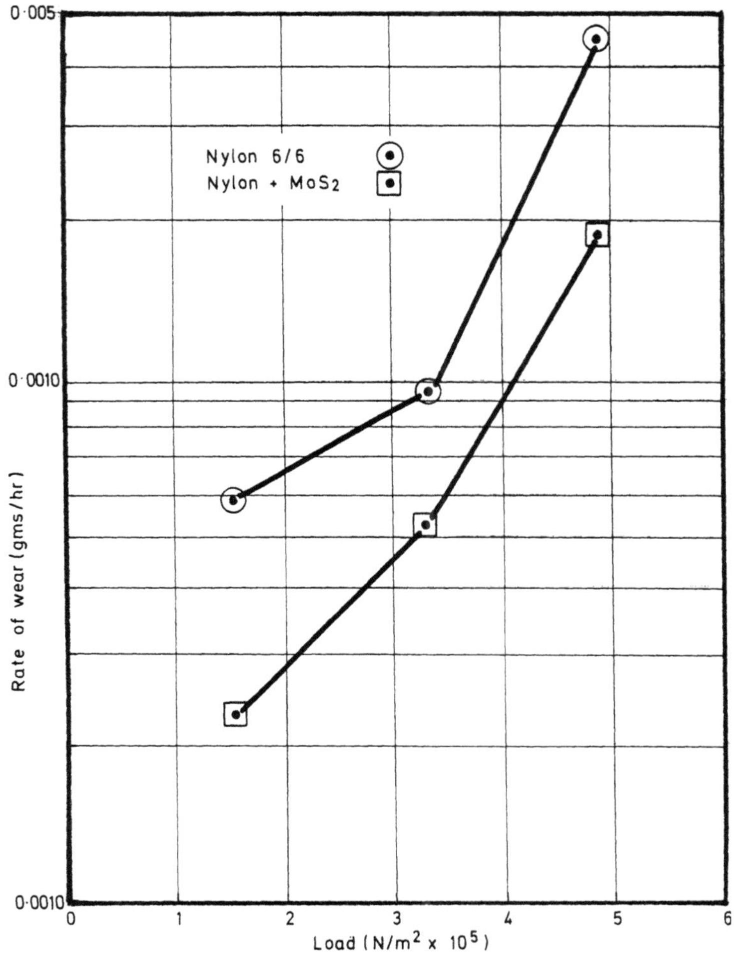

Figure 12.3 Effect of Load on Wear Rate of Nylon With and Without Molybdenum Disulphide (Ref.561)

tests with 4% w/w of molybdenum disulphide in an alternative supply of nylon 6.6 they reported a major influence of the particle size of the molybdenum disulphide. The linear wear rate varied from 96.5 mm/hr with a 0.3μm particle size to 15.7 mm/hr with a 4μm particle size. The friction was also lower with the large particle size, although all the friction values were higher than for the unfilled polymer. The interaction of these various influences on performance makes it very difficult to derive simple guidelines to composite formulation. They also reported[387] that molybdenum disulphide improved the wear resistance of a polyamide-imide.

Betts et al[388] reported satisfactory performance of polyimide containing molybdenum disulphide and either graphite powder or graphite fibres, with friction coefficients down to 0.05 at 28 MPa. The best composition contained 5% molybdenum disulphide and 25% graphite powder. The composites containing graphite fibres caused abrasion of the steel counterfaces. Similar composites studied by Sliney and Jacobson[363] with either high- or low-strength graphite fibres in a spherical bearing gave no significant abrasion of the counterface. They used both linear and cross-linked polyimides with 50% w/w of graphite fibre, and 10% w/w of either molybdenum disulphide, graphite fluoride, cadmium iodide or cadmium oxide added as a lubricant. They reported that all the composites gave satisfactory lubrication in dry air at temperatures of 20°C, 200°C or 315°C, with average coefficients of friction between 0.06 and 0.15.

The main advantage of polyimides is that they retain their structural strength to temperatures approaching 350°C, but they also have good outgassing characteristics in vacuum. For such applications the thermal stability of a PTFE filler is a limitation, and 5% of molybdenum disulphide may be added to a polyimide for friction reduction. The wear resistance of this material was inferior to some of the filled nylons, acetals and polycarbonates, and it would probably be chosen mainly for its high temperature stability.

Polyphenylenesulphide is unusual among the thermoplastics in that with prolonged heating it undergoes degradation and recombination to a crosslinked form which is no longer thermoplastic. Vinogradova et al[389] found that a composite of 85% molybdenum disulphide and 15% PPS behaved in a similar manner, producing a strong composite with high impact strength. The material had a low coefficient of friction varying from about 0.025 between 75°C and 180°C to 0.09 above 235°C. The wear resistance was high, with a linear wear rate quoted as $0.5 - 1.0 \times 10^{-10}$. This appears to represent a specific wear rate of about $2 - 5 \times 10^{-6}$ $mm^3/kg.m$ under the test conditions described.

Several of the thermoplastics benefit from lubrication with small quantities of liquids, including oils and water, and their friction and wear resistance may then be superior to those of conventionally lubricated bearing materials. The use of molybdenum disulphide fillers probably conveys little or no advantage in these conditions, but may be valuable if such a system runs dry at any time. Other thermoplastics used for bearing purposes are polycarbonates and acetals, and one of the earliest reports[356] was on perspex (polymethylmethacrylate), but there is a general tendency to use PTFE as a filler for friction reduction with these materials, rather than molybdenum disulphide.

Thermosetting resins are usually used lubricated, often with water. When they run in dry situations they benefit considerably from the addition of PTFE, graphite or molybdenum disulphide. Their wear resistance is still not as good as that of the better PTFE composites[139], but they can be used with higher loads because they are stronger and are not so catastrophically affected by heat. The effect of heat on the polymer is to cause surface charring, which does not seriously degrade performance.

As an example of the use of a phenolic composite at high temperatures, Lavik and Hopkins[390] described a composite of molybdenum disulphide, antimony trioxide and graphite fibres in a phenolic resin matrix carbonised at 400°C which operated satisfactorily up to 315°C. This material illustrated the importance of considering sliding speed and specific load independently of PV, since the highest successful operating speed was 3 m/s and specific load 44 MPa, but the limiting PV was 4.4 MPa. m/s. The coefficient of friction was in the range 0.04 - 0.15. The composite was said to compare favourably with some of the best current self-lubricating composites, but it had a tendency to brittleness which needed to be compensated for in equipment design. Other applications of molybdenum disulphide in thermosets include an epoxy composite for slideways with intermittent or contaminated grease lubrication[391], and more surprisingly, phenolic-bonded brake materials[392]. The importance of the molybdenum disulphide in brake applications is to provide a more consistent level of friction over a wide temperature range.

Brendle[393] described a process which he called polymer grafting, in which particles of molybdenum disulphide or other solid lubricants were coated with various polymers, including polystyrene, polymethyl methacrylate and poly-isobutylvinyl ether. The process was most conveniently carried out by grinding coarse molybdenum disulphide powder in a 20% - 30% solution of the appropriate monomer in a solvent. The quantity of polymer added to the molybdenum disulphide particles was very small, and could not be detected by scanning electron microscopy or infrared spectroscopy. The carbon content (1-4%) indicated a polymer content up to 6% maximum. Brendle considered that the polymer was preferentially grafted onto surface freshly exposed by grinding. This may be partly true, but in view of later evidence of the chemical inertness of the basal planes, it now seems probable that most if not all of the polymer attached to crystallite end-sites.

This would also explain the major effects of the grafting, which were reduced friction and easier plastic deformation. The marked reduction in friction is shown in Figure 12.4. In retrospect it seems probable that the grafted molybdenum disulphide would have been particularly useful in polymeric composites because of improved wettability, reduced friction and easier plastic deformation, but it is not clear that they were ever used in this way.

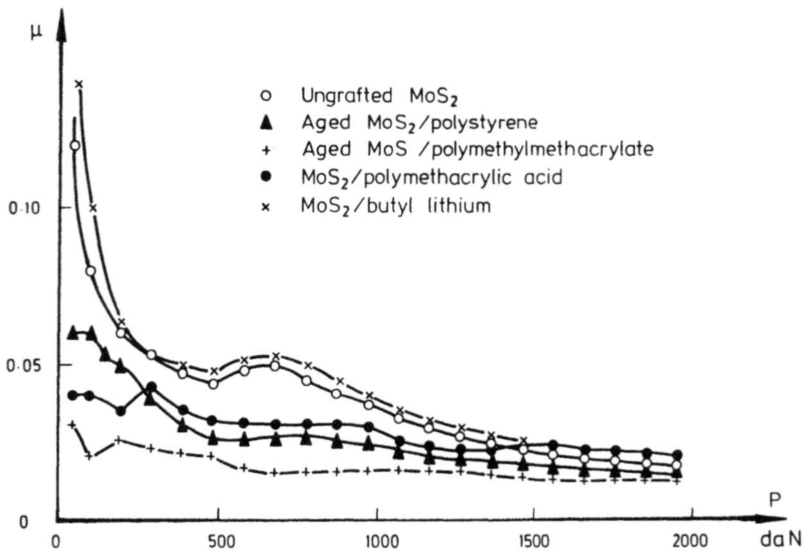

Figure 12.4 Variation of Friction with Applied Load for Various Polymers Grafted on Molybdenum Disulphide (Ref.393)

The use of polymers for bearing applications is limited to speeds below about 2 - 3 m/s. The limiting PV values for polymers may be increased by a factor of three with fillers. In general the best performance within the permissible PV range will be obtained at low speeds and high pressures within the strength limitation of the material.

An important factor in the selection of a polymer is its availability in a suitable form. In this respect the thermoplastics have an advantage in that they can be injection moulded, and nylon can also be cast or press-formed from powder by sintering.

The fabrication techniques for thermosetting resins are complicated because they are normally used with fabric reinforcement, but some filled epoxies containing molybdenum disulphide are available as two-component systems for casting or as solid bar for machining. Dough moulding compounds are used for rapid production of components.

Polymers containing molybdenum disulphide may be used as materials of construction for components subjected to rubbing, especially where fluid lubricant is marginal or absent. Braithwaite and Greene[52] described several examples in the automotive industry:-

(1) Timing gears of nylon 6.6 containing 30% of glass fibre and 1.5% molybdenum disulphide.
(2) Rod packing for water-pump gears made of nylon 6.6 with 15% of glass fibre and 5% molybdenum disulphide.
(3) Windscreen wiper motor bearings of polyacetal containing 30% of glass fibre with 1% of molybdenum disulphide.
(4) Steering rod sleeves made of polyurethane with 1.5% of molybdenum disulphide.

Reinforced nylon containing molybdenum disulphide is very commonly used for gears in low-stress situations such as windscreen wipers and car door window winders. It gives good wear and impact resistance and quiet operation, and is particularly useful where oil or grease lubrication is marginal or intermittent.

Filled PTFE is often used for piston rings in oil-free reciprocating gas compressors, and an alternative material consisting of PTFE, carbon and molybdenum disulphide with an epoxy binder has been reported[394] to show a considerable improvement. The applications were all at fairly high temperatures, between 110° and 200°C, and with the additional effect of frictional heating, the surface temperatures may have been too high for the PTFE matrix. In these circumstances the main design criteria are those for structural materials rather than for lubricants, but the following are a few guidelines.

(i) Applications are generally limited to sliding speeds below about 2 - 3 m/s.
(ii) The load limit is determined by the limiting PV, which will probably be specified by the supplier.
(iii) Initial wear is determined by the roughness of the counterface so that counterfaces should be as smooth as possible, down to about 0.1μm.
(iv) Soft counterface materials such as aluminium alloy should be avoided, but if they must be used the more abrasive fillers such as glass fibre should not be used.
(v) If fluid lubricants are present the preferred polymers would be nylon 6.6, acetals, phenolics or epoxies.
(vi) Initial clearances should be high, of the order of 50 μm/cm, with a minimum of 125 μm.
(vii) For increased structural strength the polymer can be bonded or mechanically attached to a metal backing.

12.3 METALLIC COMPOSITES

There are two significant differences between polymeric composites and metallic composites containing molybdenum disulphide. The first is that the processing temperatures of most polymers are below the oxidation temperature of molybdenum disulphide, whereas those of the common bearing metals are above it. This has important consequences for the choice of metal matrix and the techniques available for producing the composites. The second difference is that the structural strengths of the metals are generally much higher than those of the polymers. As a result, incorporation of molybdenum disulphide in a polymer usually strengthens it, whereas incorporation in a metal weakens it.

In spite of these problems, considerable efforts have been made to produce satisfactory composites of molybdenum disulphide in metal matrices, to improve on the rather short lives obtained with solid lubricant films. Many of the products of these studies have showed useful performance, and several have become available commercially.

The problem of avoiding molybdenum disulphide oxidation in the production of composites is illustrated by Table 12.9, which shows the melting-points and other important temperatures for various metals which have been used to produce composites. It can be seen that most of them are higher than the oxidation temperature of molybdenum disulphide (350 - 450°C). Even the softening or sintering temperatures are too high in most cases for processing to be carried out unprotected in air, and manufacture must be carried out in vacuum or an inert atmosphere. This is a complication in research studies, but may be a serious limitation in commercial manufacture, which can make a process uneconomical.

Table 12.9 Melting-Points and Possible Sintering Temperatures of Metals

Metal	Melting-Point	
Aluminium	660°C	
Al Bronze	1040°C	
Brass	930°C	
Cast Iron	1170°C	
Cobalt	1495°C	
Copper	1083°C	
Gold	1063°C	
Gallium	30°C	Eutectic of gallium and indium
Indium	155°C	is liquid at room temperature.
Iron	1535°C	
Lead	327°C	Most metals can be sintered at
Manganese	1244°C	any temperature above half of
Molybdenum	2610°C	their melting point in °K.
Nickel	1455°C	
Silver	961°C	
Steel	ca.1500°C	
Tin	232°C	
Titanium	1660°C	
Tungsten	3410°C	
Vanadium	1890°C	

The weakening effect of incorporating molybdenum disulphide in a metal is often not a serious limitation, since the structural strength required of a bearing is often well below that available, and metallic composites typically contain far higher concentrations of molybdenum disulphide than polymeric composites. It is in any case often possible to support a bearing in a housing which provides the required structural strength. The more important factor in selecting the concentration of molybdenum disulphide in a composite is that the lubricant is only made available by wear. The concentration must be high enough to ensure that the required quantity of lubricant is liberated with an acceptable amount of wear.

The first reports on the incorporation of molybdenum disulphide in a metal matrix were those mentioned previously which were published by Bowden[357] in 1950. He reported a coefficient of friction of 0.13 for a composite in sintered copper. In his other metallic composite the molybdenum disulphide was formed *in situ* by hydrogen sulphide in sintered molybdenum and had a coefficient of friction of 0.06. At about the same time R L Johnson et al[358] at NACA studied the effect of molybdenum disulphide concentration in silver with 5% of copper. They reported coefficients of friction as low as 0.17 and found that the friction decreased with increasing concentration of molybdenum disulphide. Their wear rates were high, around 10^{-5} mm^3/Nm, but this work was the fore-runner of many studies using the same components.

Incorporation of molybdenum disulphide to a concentration of 4% or 10% in sintered bronze, iron or nickel has been achieved by sintering the mixed powders at high pressure[395]. To minimise oxidation the sintering was done in an inert atmosphere, but the internal friction reduction during sintering helped to give more effective sintering, as well as improving geometrical accuracy and die life. Alternative procedures for introducing molybdenum disulphide into sintered components include barrelling and impregnation with a dispersion of the molybdenum disulphide in a volatile liquid.

Powder metallurgy techniques have been used to produce a very wide range of compacts containing molybdenum disulphide in such metals as mixed iron-palladium, iron-platinum[396], tantalum[397], iron-tantalum[397], molybdenum-tantalum[398], and molybdenum-niobium[399]. The concentration of molybdenum disulphide in these compacts has risen to 90% compared with less than 35% in earlier materials. Composites containing nickel were found to be unsatisfactory because of high friction and wear.

Campbell and Van Wyk[396] hot-pressed their iron-palladium and iron-platinum composites at 1090°C (2000°F) and 50MPa. They prepared a variety of composites containing nickel in a similar manner, but none of these was successful. They used graphite dies, and considered that this helped to create a reducing environment during the hot-pressing, which prevented oxidation of the molybdenum disulphide. Hubbell et al[397] used similar pressing conditions to produce their composites with tantalum and iron/tantalum. All these materials gave coefficients of sliding friction between 0.1 and 0.3 when tested at low contact pressure, but at high contact pressures (27.6 MPa) the friction was as low as 0.02 to 0.08. Mecklenburg and Benzing[400-402] studied some of the same materials as Hubbell et al[397], and obtained similar friction values.

Martin and Murphy[403] compared twenty-five different solid lubricant composites for use in small arms. The best performance was obtained with a composite of molybdenum disulphide in molybdenum with niobium and copper. This had a lower wear rate (0.224 x 10^{-6} mm^3/Nm) and lower coefficient of friction (0.05 - 0.15) than any of the twenty-one polymer-based composites.

Suzuki et al[404] made hot-pressed composites of molybdenum disulphide 80%, molybdenum dioxide 10% and niobium 10%. They were pressed at 25 MPa and 1500°C in carbon dies, and the flexural strength was 59 - 63 MPa and the elastic modulus 27.9 MPa. To improve the strength they added 5% of 304 stainless steel, and this gave a flexural strength of 69 - 80 MPa and elastic modulus 43.8 MPa. The coefficient of friction of the latter compact was 0.07 to 0.18 and the specific wear rate 2.2 x 10^{-7} mm^3Nm at 450°C in vacuum.

The compositions and performances of these materials are summarised in Table 12.10. In general the high friction values were at low contact pressure and the low values at high contact pressure, but unfortunately the descriptions given did not always permit the contact pressure to be calculated. The same is true of the wear rates. It is therefore difficult to make useful comparisons between different publications, except to say that in general the compacts containing higher concentrations of molybdenum disulphide have higher wear rates.

Tsuya et al[405] studied the optimum concentration of several different solid lubricant powders in copper composites. In all cases except calcium fluoride they found that the friction of cold-pressed copper composites reached a minimum value at a lubricant concentration of about 7 - 10%, and remained fairly constant up to a concentration of almost 90%.

With composites sintered at 800°C in vacuum the minimum friction was still reached at about 7 - 10% concentration for graphite and tungsten diselenide, but not until about 20% concentration with molybdenum disulphide. When they used a sub-

Table 12.10 Composition and Properties of Some Molybdenum Disulphide/Metal Composites

Composition		Lowest Coefficient of Friction	Ref
MoS_2 content	Other Components		
46.9	Silver 15.6, iron 25, palladium 12.5	0.24	397
63.1	Silver 4.5, iron 21.6, palladium 10.8	0.13	397
5-20	Silver 75-90, copper 5	0.20	358
35	Silver 60, copper 5	0.17	358
80	Iron 16, palladium 4	0.17	396
90	Iron 8, platinum 2	0.29	396
80	Molybdenum, tantalum	0.24(air)	398
		0.02(vac.)	398
80	Tantalum 20	0.17	401
50	Iron 12.5, tantalum 37.5	0.26	400
15	Silver 82.5, copper 2.5	0.25	76
10	Silver 90	0.10	20
20	Silver 80	0.08	20
75	Niobium 10, molybdenum dioxide 10,st.steel 5	0.07	404
80	Niobium 10, molybdenum dioxide 10	0.07	404
x	Copper	0.13	357
x	Molybdenum	0.06	357
x	Silver, PTFE	0.16(air)	398
		0.08(vac.)	398
x	Molybdenum, tantalum	0.04	179
x	Molybdenum, niobium	0.05	399

micronic particle sized molybdenum disulphide in the sintered compact, the minimum friction was not approached until 40% concentration, and they attributed the higher concentrations required to a partial breakdown of the molybdenum disulphide, reacting with the copper. Sintering reduced the minimum wear rates of the molybdenum disulphide composites by a factor of eight or more, and the optimum concentration for minimum wear was increased slightly. They found that the wear rate changes by a factor of over 1,000 as the concentration of solid lubricant changes. There is usually an optimum concentration between 40% and 80% but this is strongly affected by the nature of the test and the composition and state of the counterface.

Yunxin Wu et al[406] also investigated the optimum concentration of molybdenum disulphide in hot-pressed composites with nickel. They found that the optimum molybdenum disulphide content was 60%. This gave a full continuous and homogeneous film on the rubbing surfaces, but the structural strength was still high enough to provide satisfactory lubrication at Herzian contact stresses of 100 MPa and test temperatures up to 250°C. The lowest friction measured was 0.5, and the specific wear rate was rather high, at about 4 x 10^5 mm^3/Nm, so that overall the performance of these compacts was mediocre.

Mizutani et al[407] carried out a review of some of the problems of metal-lubricant composites, and in particular the difficulty caused by the fact that sintering temperatures are limited by the thermal stability of the lubricant. The result of this is that either softer metals must be used, or lower sintering temperatures which result in lower structural strength due to brittleness. This led them to investigate metal-lubricant composites in which the lubricant is synthesised *in situ*. The result[251] was the ternary alloy of iron, molybdenum and sulphur which was described in Chapter 9, and which had the characteristics of a slightly sulphur-depleted molybdenum disulphide composite.

Several publications have reported that composites of molybdenum disulphide with a high nickel content were not satisfactory[396], and nickel alloys are recognised to be susceptible to attack by sulphur or sulphur compounds at elevated temperatures. Al'tman et al[408] found that certain alloying elements, especially molybdenum, can stabilise the composite structures. However, the adverse reports of composites in nickel are not universal, and a few successful examples will be described later.

Molybdenum is generally a satisfactory metal for use in contact with molybdenum disulphide, and there have been several reports of satisfactory performance of composites with molybdenum. Koval'chenko and Yulyugin[409] investigated the effects of temperature, pressure and pressing time on the density of hot-pressed compacts of molybdenum disulphide and molybdenum. They found that full compaction was only obtained at temperatures over 1100°C at 40 MPa and pressing times of 2 mins or more. However, a relative density of about 0.75 was achieved at 1000°C, 40 MPa and 2 mins. Metallographic examination[410] showed a lamellar structure of sintered molybdenum with intermediate layers of molybdenum disulphide.

Tkachenko et al[411] reported that compacts in iron, copper, nickel or cobalt matrices had operating temperature limits between 200° and 600°C, but compacts in molybdenum gave satisfactory friction and wear to 900°C in vacuum. A satisfactory composite for high-temperature aircraft brakes was described as containing 25% molybdenum disulphide, alumina and lead tungstate in a nickel matrix[412]. The composites were pressed at 880 to 1080 MPa and sintered at 1010°C for two hours in vacuum.

Other composites which have been described include sintered bronze filled with PTFE containing dispersed molybdenum disulphide, and molybdenum disulphide in lead, gold, silver, tungsten, silver/5% copper, and iron/tin. When this wide variety of components is considered together with the possible ranges of concentrations and the virtual impossibility of comparing most of the performance data, it becomes very difficult for a potential user to make a sensible choice of composite for a specific application. As with the polymeric composites, a few guidelines can be given:

(i) Low lubricant concentrations will usually give high structural strength and low wear rates.
(ii) High lubricant concentrations favour low friction and effective lubricant transfer.
(iii) Nickel-based composites have often been reported to give poor performance.
(iv) The presence of molybdenum is usually beneficial.
(v) Sintered composites have higher structural strength than hot-pressed composites.

The performance of a composite is strongly influenced by the nature of the counterface and the application conditions, and a development programme is essential before any critical application is attempted. Probably the best indication that a particular composite has a useful performance is when it is available commercially, and commercial products are available under the trade names Bemol, Sinite, Sinitex and Molalloys. Bhushan and Gupta[413] refer to commercially-available compacts of molybdenum disulphide in refractory metals, which can be used at specific loads up to 70 GPa at 500°C and 7 GPa at 800°C in vacuum. These may be the Molalloy compacts, produced by the Pure Carbon Company, Incorporated, one of which was reported[414] to have been used in cargo bay door bearings on the United States Space Shuttle.

An industrial application described by Bessière and Martre[395] was the use of sintered bronze/molybdenum disulphide bushes in a roller table of a system for quench-hardening heavy steel plates. The bearing experienced severe thermal cycles, and this seems the type of application for which metal composites are very appropriate.

12.4 CERAMIC AND INORGANIC COMPOSITES

Although it has been commonplace for over forty years for proponents to talk about the potential value of incorporating molybdenum disulphide in ceramics, there have in fact been very few published examples of the technique. As recently as 1987, Peterson[415] was still referring to the potential value and the actual lack of achievement when he said: "An interesting development to watch will be the development of self-lubricating ceramics. With the low temperature manufacturing processes being developed, it will not be long before someone begins to add solid lubricants."

He went on to say that the main problem is that ceramics wear too slowly to release sufficient solid lubricant. This is certainly a major problem, but another important difficulty is what might be called a mis-match between the temperature capabilities of the ceramics and of the more common solid lubricants. The greatest value of ceramics is for temperatures above 350°C, where steel loses its hardness, many non-ferrous metals become very soft, and only a few polymers can be used even for very short periods. Unfortunately this is also the temperature at which even the most stable of the common solid lubricants have limited lives except in vacuum or inert atmospheres.

As a result, those workers who are trying to develop self-lubricating materials for these high temperatures are looking at much more exotic and much less efficient lubricants, such as the calcium fluoride/barium fluoride eutectic and silver used in the NASA PS200 and PM212 high-temperature composites[416,417].

One early report[418] described the friction and wear performance of a high-temperature compact containing molybdenum disulphide, silica, lead oxide, silver and platinum, but since the relative proportions were not detailed, it is not clear whether the matrix could be described as ceramic. The nature of the constituents would suggest that it was a cermet. It was tested for use in electrical contacts in wire-wound potentiometers for a spacecraft application, and was reported to be satisfactory.

Gangopadhyay et al[419] investigated the use of graphite intercalated with nickel chloride $NiCl_2$ to lubricate silicon nitride or alumina sliding against a steel counterface. The lubricant was quite effective, reducing the coefficient of friction from 0.5 to 0.17 for silicon nitride, and from 0.55 to 0.18 for alumina. There is no obvious reason to expect that the behaviour of molybdenum disulphide would be significantly different.

In fact Van Wyk[210] used something similar to a composite structure for the lubrication of silicon nitride and alumina in plain spherical bearings. He incorporated a 90% molybdenum disulphide/8% molybdenum/2% tantalum compact in holes drilled in the surface of the alumina outer ring, and the details have been described in Chapter 8. The system was very successful, giving a forty times increase in wear life.

It would seem to be technically feasible to produce composites of molybdenum disulphide in ceramics, since bonded coatings have been tested successfully with boric oxide, silica or alumina as bonding agents, and several of them were described in Chapter 11. The extension of these techniques to the production of bulk composites might be tedious but would presumably be possible. Since there is little or no evidence that this has been done, it seems probable that the problems of inadequate lubricant release or temperature mis-match mentioned previously have made the prospect insufficiently promising to justify the effort.

It is less clear why there has been very little use of other types of inorganic matrices. Some inorganic solids would seem to have useful properties for this purpose, but the only one which has been described in any detail is sodium silicate. This was the binder for the bonded coating developed by the US Naval Aircraft Materials Laboratory which was described in Chapter 11, but the same material was used[209] as a bulk composite in certain components in a dry-lubricated engine. It was used to provide transfer lubrication, and presumably it was not capable of use in load-bearing situations.

12.5 TRANSFER LUBRICATION OF ROLLING BEARINGS

The general subject of transfer has been described in Chapter 8. Composites used specifically for transfer lubrication do not need the structural strength which is needed for the construction of load-bearing components. Some structural strength is of course needed to maintain the integrity of the composite reservoir. This may be very low where the reservoir is contained in a recess in a metallic or ceramic component, so that the structural strength is provided by the surrounding material.

Some examples of the practical use of transfer lubrication in a piston engine[209] and in gear sets[207,212] have been described in detail in Chapter 8, but the main use of transfer lubrication is for rolling bearings. Since the earliest days of modern solid lubricant development, their application to rolling bearings has been an area of intensive investigation, because of both its importance and its difficulty. Rolling bearings (ball or roller bearings) are important by virtue of their low friction, the high quality of their manufacture, and the precision with which they can locate and control rotating or oscillating components.

The moving contacts between the rolling elements and the races in rolling bearings are non-conformal, and the contact stresses in those contacts are relatively high. In addition, the clearances and tolerances are small, so that the thickness of dry lubricant films which can be applied is limited. As a result, the life of dry lubricant films is also very limited, and only the use of sputtered films in high vacuum applications has been of practical use. Attempts to develop re-supply techniques have also been of only limited value. Feeding of powder in a gas stream was described in Chapter 9. Supply in the form of dispersions in liquids or greases has been far more successful, but the range of applications is then limited by the properties of the liquids present, so that the benefits of dry lubricants for use in vacuum or at high or low temperatures cannot be realised.

The most promising technique has therefore been transfer lubrication, and a huge amount of effort has been put into the development of suitable transfer systems. Fortunately, the design of rolling bearings lends itself to the incorporation of lightly-loaded lubricant reservoirs, in the form of the separator, also known as a cage or retainer, which separates or locates the balls or rollers. This component is in intermittent sliding contact with the rolling elements and usually with one of the races, but it is not part of the load-bearing path of the bearing, and its contacts with the other bearing components are at relatively low contact stresses. The retainer is therefore commonly used as the solid lubricant reservoir for transfer lubrication of rolling bearings.

The retainer may be completely fabricated from a composite material, or may consist mainly of a composite with reinforcement by metal rings. Alternatively it may be conventionally fabricated of steel or other suitable metals, with composite components bonded, rivetted or pressed onto it, or with holes or grooves filled with the lubricant composite.

The structural strength required of the composite is related to the way in which it is incorporated in the retainer. Because the retainer does not carry the main bearing loads, the structural strength required is not high, even for a fully-fabricated composite retainer. The stresses involved are relatively low contact stresses against the rolling elements and race lands, stresses caused by differential thermal expansion, and centrifugal stresses, since the retainer rotates at approximately half the shaft speed. Possibly an ultimate compressive stress of 40 to 50 MPa is sufficient for most applications. Nevertheless, failures of composite retainers have often occurred in development, and some attention has had to be paid to satisfactory retainer design.

As usual with composites, the formulation is a compromise between the need for adequate structural strength and the need to make available sufficient lubricant. The two-stage transfer path, from retainer to rolling elements, and from rolling elements to races, is inherently inefficient, so that a high lubricant content is necessary. Devine et al[420] found that in the case of one test with a 20 mm bore deep groove ball bearing, there was a twenty-fold increase in bearing life when additional lubricant pockets were inserted in the inner-race lands (i.e. the shoulders of the inner ring adjacent to the ball raceway). The lands in this case were in sliding contact with the retainer, and obviously the additional lubricant supply was effective. Very few subsequent workers have used such additional reservoirs to supplement supply from the retainer.

Many hundreds of different composite compositions have been tested for use in ball or roller bearings, mainly for space use. Some examples of results with polymer-based composites are shown in Table 12.11 and with metallic compacts in Table 12.12. Among the polymeric composites the best results have been reported with composites of PTFE or polyimide. Bearings are commercially available with composite retainers based on either polymer.

Table 12.11 Transfer Lubrication of Ball Bearings with Polymeric Composite Retainers

Bearing Type	Retainer Materials	Load	RPM	Environment	Life	Ref
20mm bore	71% MoS_2, 7% graphite 22% sodium silicate	13N radial, 22N axial	10,000	180°C 400°C 400°C	100hrs 50hrs 1148hrs	421 420
20mm bore	Filled PTFE	50N radial		10^{-5}Torr, room temp.	10^8revs	139
4.76mm angular contact	59% PTFE, 39% glass fibre, 2% MoS2	2.5N radial, 1N axial	8,000	75-90°C	4353hrs+	377
6.35mm deep groove	59% PTFE, 39% glass fibre, 2% MoS_2		2,000	3×10^{-7}Torr	6464hrs	379
40mm bore	80% PTFE, 15% glass fibre, 5% MoS_2	890N axial	20,000	Hydrogen gas -240°C	10hrs+	378
3.175mm bore	605 PTFE, 40% glass fibre, 3% MoS_2	2.4N radial 4.5N axial	3,000	$<5\times10^{-7}$Torr	4630hrs	422
9.6mm deep groove	75% PTFE, 20% glass fibre, 5% MoS_2	20N radial, 20N axial	4,000	Air at 200°C	187	366
3.2mm deep grove	75% PTFE, 20% glass fibre, 5% MoS_2	5N radial	6,000	Air at 250°C	5550hrs	366
20mm deep groove	Polyimide, MoS_2	13N radial, 22N axial	10,000	Air at 316°C	122hrs	423
3.175mm bore	60% PTFE, 40% bronze powder	2.4N radial, 4.5N axial	3,600	$<5\times10^{-7}$Torr	4630hrs	422

Table 12.12 Ball Bearings Lubricated by Transfer from Metallic Composite Retainers

Bearing Type	Retainer Materials	Load	RPM	Environment	Life(hrs)	Ref
4.76mm angular contact	55% bronze, 27% PTFE, 18% MoS_2	2.5N radial 1.0N axial	8,000	78-95°C, 10^{-7}Torr	1984+	7
20mm bore	20% tantalum, 80% MoS_2		5,000	Argon 21-215°C	150	397
20mm deep groove	20% tantalum, 80% MoS_2	34N axial	1,790	10^{-6}-10^{-7}Torr	23,000+	402
Ball thrust	30% gallium/indium, 70% tungsten diselenide	9N	150	10^{-9}Torr	100	424
20mm deep groove	11.8% cobalt, 35.3% silver, 52.9% tungsten diselenide	25N axial	1,790	10^{-6}-10^{-7}Torr	38,000+	402
35mm deep groove	20% gallium/indium, 80% tungsten diselenide	225N radial 225N axial	10,600	Air at 480°C	38	425
20mm deep groove	20% gallium/indium, 80% tungsten diselenide	450N radial 450N axial	3,400	Air at 24°C	1,100	425
20mm deep groove	20% gallium/indium, 80% tungsten diselenide	225N radial 225N axial	10,600	Air at 316°C	215	425
3.175mm bore	20% silver, 12% molybdenum 4% tantalum, 64% MoS_2	2.4N radial 4.5N axial	3,600	$<5\times10^{-7}$Torr	4.63	422
8mm angular contact	Copper, MoS_2, tungsten diselenide	20N axial	9,000	300°C, 10^{-6}Torr	210+	426

The PTFE-based retainers typically contain between 15% and 25% of glass fibre reinforcement and 3% to 5% of molybdenum disulphide. The molybdenum disulphide content increases the structural strength, reduces the wear rate, and improves transfer. The transferred lubricant consists mainly of PTFE and molybdenum disulphide. Bearings of this type have been used successfully in spacecraft, but have also been used in many terrestrial industrial applications.

The availability and extensive applications of these PTFE and polyimide retainer systems led to the production by the Risley National Centre of Tribology of a performance guide for their use in small deep groove ball bearings[366]. It was found that above a certain load threshold life was limited by wear of the balls and races, whereas below that threshold life was limited by wear of the composite retainer. The threshold was at about only 2% of the bearing manufacturer's static load capacity for the bearings. This behaviour is interesting, since with oil or grease lubrication the usual ultimate failure mechanism is rolling-contact fatigue.

Research on metallic compact retainers continues, mainly for long-term applications in spacecraft, and especially where the operating temperatures may be high. Under those conditiions polymers, and especially PTFE, tend to evaporate, or "out-gas". No metal compact system has achieved the same degree of utilisation as the PTFE composites. The excellent results reported for the tungsten diselenide/gallium/indium amalgam would seem to justify extensive use of the material, but it may be that the exotic nature (and therefore high cost) of the components is a limiting factor.

Some indication of the performance and life attainable with transfer lubrication from composite retainers can be obtained from Tables 12.11 and 12.12, but these are just a few examples from a very much larger spectrum of published results. Unfortunately, many of the reports fail to give sufficient data for useful comparisons to be made. In addition, the range of variables involved is so great, especially in respect of the formulation and preparation of the composites, that it is only in limited cases, like the National Centre of Tribology work mentioned previously, that useful life and performance predictions can be made.

12.6 ELECTRICAL BRUSHES AND SLIPRINGS

Transfer of electric signals or power from rotating components is commonly achieved by the use of stationary brushes in sliding contact with rotating rings.

Traditional materials are carbon brushes against copper commutators or sliprings, but carbon brushes rely on graphite content for lubrication of the sliding contact, and graphite requires the presence of moisture or other vapours, or sometimes more exotic materials, to ensure low friction. They cannot therefore be used in vacuum or dry atmospheres[427] or at temperatures above about 180°C, because a rapid wear phenomenon called "dusting" occurs.

Early attempts to overcome this problem included[428] the addition of hygroscopic compounds, liquid lubricants, metal halides[429] and solid lubricants, including PTFE[429] and molybdenum disulphide[430,431] to the carbon to provide lubrication. However, one of the basic reasons for using carbon for brushes is its good lubricant properties, so that when these are lost and must be supplemented by the addition of solid lubricants, there is no dominant reason for continuing to use carbon.

For space use, the voltages and current densities are usually low, and there has been extensive study of systems in which the brush and slipring materials are metallic, and lubrication is provided by molybdenum disulphide or other dichalcogenides.

Molybdenum disulphide itself is most reliably considered as a semi-conductor (see Chapter 4), and if used as a continuous film over the contact surfaces it forms a high-resistance layer. Attempts to overcome this problem have included coatings containing metal powder and molybdenum disulphide[432] and very thin molybdenum disulphide films formed *in situ*[433], by burnishing[434] or by sputtering[163]. The *in situ* and burnished films gave reasonable performance, but in general it has been found that if sufficient molybdenum disulphide is present to provide effective lubrication electrical noise is too high. Thus one series of experiments with burnished molybdenum disulphide gave electrical noise levels of 2.1 to 2.7 mV compared with a total output of 11.9 to 13.5 mV[434]. Conversely, if the quantity of molybdenum disulphide is low enough for satisfactory electrical performance, the friction and wear life are unacceptable.

Surface films containing molybdenum disulphide have therefore been very little used for electrical contacts, and the usual technique is to use conducting compacts containing molybdenum disulphide or other dichalcogenides. Table 12.13 lists some of the compacts which have been used.

Composition (%)		Slip Ring	Brush Pressure (Pa)	Speed (m/s)	Time (hrs)	Wear Rate (mm^3/Nm	Current Density ($amps/m^2$)	Electrical Noise (mV)	Atmos. Pressure (Torr)	Friction	Ref
Lubricant	Metal										
15 MoS_2	82.5 silver 2.5 copper	Silver	7×10^4	0.25	1200	3×10^{-6}	5×10^5	ca.0.3	10^{-8}	ca.0.2	432
12.5-15 MoS_2	82.5-85.0 silver, 2.5 copper	Silver or gold	$4\text{-}7\times10^4$	0.18	700	10^{-5}	3×10^5		10^{-9}		435
15MoS_2	82.5 silver 2.5 copper	Silver	7×10^4	8×10^{-4}	600	10^{-6}	4×10^5	5-20	10^{-9}	ca.0.2	76
15$NbSe_2$	82.5 silver, 22.5 copper	Silver	7×10^4	8×10^{-4}	600	10^{-5}	4×10^5	5-80	10^{-9}	ca.0.2	76
15MoS_2	85 silver	Silver	3×10^4	2.6×10^{-5}	1019	2×10^{-3}	10^5	0.15	10^{-8}		436
15$NbSe_2$	85 Silver	Silver	2.5×10^4	2.6×10^{-5}	1035	3×10^{-3}	10^6	0.08	10^{-8}		436

Table 12.13 Performance of Some Lubricating Compact Brush Materials in a Vacuum of $10^{-8}$to 10^{-9}Torr (0.14 to 1.4μPa)

There is a curious anomaly in the performance of these materials which may be related to the complex effects of purity and temperature reviewed in Chapter 4. Niobium diselenide (Nb Se_2) is a better conductor[74] than molybdenum disulphide, but apart from one report[436] in which the test conditions were unconventional[437], compacts containing molybdenum disulphide have generally performed better than those containing niobium diselenide, in terms of wear, electrical resistance and electrical noise. Apart from the compacts listed in Table 12.13, the superiority of molybdenum disulphide was also confirmed in contacts with silver and graphite[438].

The compression of the Ag/Cu/MoS_2 compacts during forming causes flattening of the silver particles, so that the grain structure of the compact is anisotropic. Studies of the friction and wear behaviour have shown that the wear rate is lowest when the flattened silver particles are at right angles to the contact surface and parallel to the direction of sliding[439]. In this configuration the wear rate was lower by a factor of 2 to 5 than when the flattened particles were parallel to the counterface.

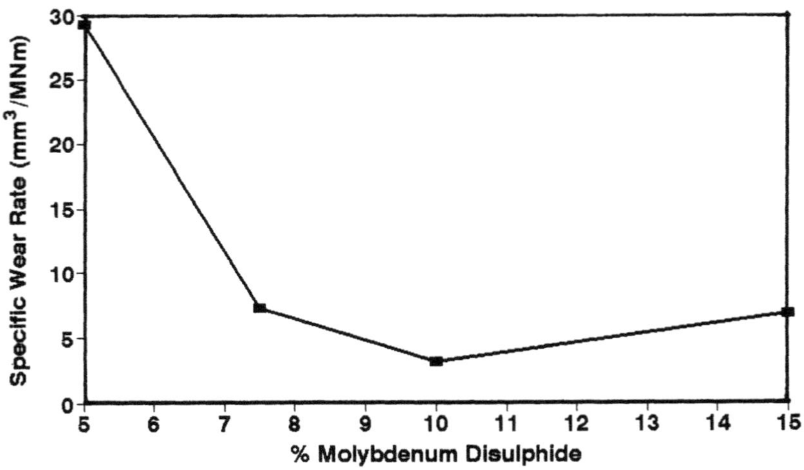

Figure 12.5 Variation of Brush Wear Rate with Molybdenum Disulphide Content (Ref.442)

The friction level changed with continued rubbing. The static coefficient decreased from 0.45 - 1.35 to 0.21 - 0.36 after one day, and the dynamic coefficient increased from 0.03 - 0.25 to 0.14 - 0.30 after two days, and remained at 0.13 -

0.23 after that. The exact composition of this material was found by analysis to be 90.6% Ag, 2.5% Cu and 6.9% MoS_2.

Other successful combinations have been reported to include molybdenum disulphide in dispersion-hardened silver, 15% molybdenum disulphide with 80% silver and 5% nickel, and 50% molybdenum disulphide, 33.3% silver and 16.7% nickel[440]. Tanaka and Kimura[441] studied the effect of electric current on the wear of the last of those compacts. They found that the amount of wear increased progressively with increasing amperage, but that the increase represented a change in the size of wear particles rather than an increase in the number of particles.

A survey of European Space Tribology Laboratory vacuum testing on satellite slip-rings reported that slip-rings designed for satellite solar-array mechanisms had operated satisfactorily over long periods with a variety of Ag/Cu/MoS_2 compacts[442]. The relationship between brush wear rate and molybdenum disulphide content is shown in Figure 12.5 for some hot-pressed composites in silver with 1% of copper.

Table 12.14 Some Composites Tested by Christy for Small Actuator Motors (data from Ref.443)

1. Carbon graphite, copper, processed to produce copper sulphide.
2. As 1, but with higher sintering temperature.
3. Carbon graphite with silver.
4. Molybdenum disulphide 80%, tantalum/tungsten 20%.
5. Natural graphite, copper/barium fluoride 45%.
6. Graphite, plastic with molybdenum disulphide

Apart from the various Ag/Cu/MoS_2 compacts, Christy[443] reported successful tests in air and vacuum for six different composites used as brush materials in small electric motors. The materials are listed in Table 12.14, and only two of them contained molybdenum disulphide. It is interesting that four of the materials contained graphite with a variety of additives. This suggests that considerable progress has been made in recent years in resolving the problem of graphite's poor lubricating behaviour in vacuum. At the present time the preferred material for brushes for space use would probably still be a compact of silver, copper and molybdenum disulphide with approximately 5-10% of molybdenum disulphide.

On theoretical grounds it might be better to use metallic brushes against lubricating compact slip rings, because the faster wearing surface would have the greater surface area, but attempts to do so have given results inferior to those obtained with compact brushes against metal slip rings[432].

~ ~ ~ ~ ~ ~ ~ ~ ~ ~ ~ ~ ~ ~ ~ ~ ~

CHAPTER 13.

USE IN OILS AND GREASES

13.1 INTERACTION BETWEEN MOLYBDENUM DISULPHIDE AND LIQUIDS

In general the presence of liquids reduces the effectiveness of molybdenum disulphide lubrication. The effect on bonded films varies from a drastic reduction in wear-life[444,445] to almost immediate failure [446]. The load-carrying capacity is also reduced, as shown in Table 13.1 [445] and the coefficient of friction increased[444,447]. Similar effects have been found with burnished films [448], as shown in Table 13.2, and a marked increase in friction was reported[444] for an *in situ* film. In the case of the burnished film, scanning electron microscopy showed that after the application of a mineral oil, renewed sliding caused immediate cracking and bulging of the film. The appearance of the film suggested that intermittent adhesion or stick-slip might have occurred. These publications all related to various types of lubricating oil, either mineral, ester or silicone, but Gänsheimer[449] reported similar results with water, ethylene glycol, glycerol, dodecanol, dodecylamine, dodecane, hexadecane, octadecane, and paraffinic white oil.

The reason or reasons for the adverse effects have not been definitely established. DiSapio and Moloney[444] suggested that the deterioration is caused by weakening of the bonds between molybdenum disulphide and substrate. This may be partly true, but cannot be the only cause, since such an effect would seem likely to lead to rapid film stripping rather than increased but more or less steady friction and wear. Hopkins and Campbell[445] listed three ways in which liquids could affect bonded films: by creating hydraulic forces under load which would tend to break up the film, by softening a binder and reducing structural strength, and by increasing the friction between mating surfaces. It is possible that in fact all four mechanisms contribute to the situation to varying extents and with different liquids or films.

Table 13.1 Load-Carrying Capacity and Wear Life of Molybdenum Disulphide in the Falex Tester With and Without Mineral Oil (Ref.445)

Lubricant	Load-carrying Capacity (Jaw-Load, lb)		Wear Life at 1000lb Jaw Load	
	Clean	Wet With SAE 30 Oil	Clean	Wet With SAE 30 Oil
Air-cured film	1,500	1,250	-	-
Heat-cured phenolic-bonded film	2,600	1,600	110	3.0
Heat-cured phenolic-bonded film	3,000	1,625	248	3.0
Sodium silicate-bonded film	1,500	1,125	10	2.5
Inorganic air-dried film	-	-	406	11.0
High-temperature polyimide binder	4,500	2,950	502	4.5
Sputtered MoS_2 film (no binder)	-	-	90	3.0

Gänsheimer observed a difference between the effects of polar and non-polar liquids. Polar liquids caused a much greater and more rapid increase in friction than non-polar liquids, and he suggested that the adverse effects were due to adsorption. Adsorption would be expected to be greater with polar molecules, both on the edge-sites of the molybdenum disulphide and on the metal substrate.

There may also be more specific effects of certain liquids, such as the softening of the binders in bonded films suggested by Hopkins and Campbell[445]. Any adsorption of a liquid on the surface basal planes, or any intercalation between lamellae would interfere with the ideal charge distributions and inter-lamellar spacing found with uncontaminated molybdenum disulphide. These effects would in fact be similar to the adverse effects of water vapour and other vapours on the friction of burnished films.

Table 13.2 Effect of Mineral Oils on the Friction of a Burnished Film (Ref.448)

Oil	Coefficient of Friction	
	Dry	Oil added
Base Oils		
Spindle extract (3.17cSt)	0.05	0.09
Light extract (9.72cSt)	0.035	0.09
Interneutral extract (21.09cSt)	0.035	0.085
Diesel Oil		
MIL-L-2104A	0.045	0.085
In all cases the film was disrupted after the introduction of oil		

One apparently beneficial interaction was described by Braithwaite and Greene[95]. They found that the wettability of steel and cast iron surfaces to a lubricating oil was markedly increased by the presence of a burnished molybdenum disulphide film, in terms of both rate of spread of oil droplets and the area wetted. Both effects are illustrated in Figure 13.1. The improvement in wettability was influenced by the extent of oxidation of the powder used to produce the burnished film. The least oxidised powder was the most effective, but even the most highly oxidised powders had some beneficial effect.

The implication of this work is that the presence of a burnished film on a metal surface should help to improve lubrication by a mineral oil, especially where there might otherwise be some tendency for partial oil starvation. It should be remembered, however, that this film was produced by burnishing molybdenum disulphide powder. Commercial dispersions, bonded films, composites or greases, as well as fully formulated lubricating oils all contain other substances which may significantly affect wetting behaviour.

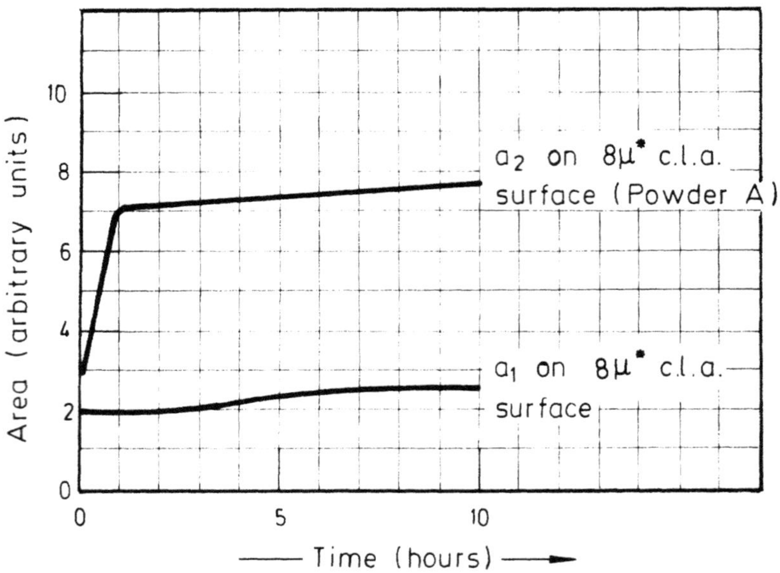

Figure 13.1 Effect of Burnished Film of Molybdenum Disulphide Powder on Wetted Area (Ref.95)

While molybdenum disulphide lubrication is adversely affected by liquids, liquid lubrication can be improved by molybdenum disulphide. There is in fact a considerable industry based on the use of molybdenum disulphide in liquids, in dispersions, lubricating oils, greases, anti-seizes and pastes. There is an important contradiction in the fact that in many situations the lubricating action of molybdenum disulphide is either partly or completed destroyed by the presence of liquids, whereas in other situations it can provide useful lubrication benefits in a liquid medium. A great deal of effort has been applied in attempts to establish the mechanism and the conditions by which lubrication by molybdenum disulphide can occur in the presence of a liquid.

There is no doubt that the presence of a liquid affects the ability of molybdenum disulphide to form a good adherent film on a bearing surface. It is often not possible to observe a deposited film on surfaces which have been lubricated by oils or greases containing molybdenum disulphide. Black et al[447] found that steel balls which had been "rumbled" for 8 days at low load in a 10% dispersion of molybdenum disulphide in oil acquired a maximum film thickness of 9.4 $\mu g/cm^2$. A similar treatment in dry powder gave a film thickness of over 200 $\mu g/cm^2$. At much higher load the formation of an adherent film has been demonstrated[450] but the dispersion in that case was in a paraffinic white oil, and such oils are not representative of lubricating oils, since they are virtually free of the polar compounds which most easily react or adsorb on surfaces.

Rolek et al[451] tried to distinguish between viscosity, temperature and composition in their influence on the effects of molybdenum disulphide dispersions in oils. They used a series of white oils, mineral oils, and the same mineral oils with some polar additives removed. The results were not entirely clear, but they supported Tsuya's findings (see below) on the effect of viscosity. However, they also seemed to indicate a more specific effect of temperature and the presence of polar additives. It seemed that there was a specific inhibiting effect of polar additives in suppressing any friction reduction by the molybdenum disulphide. In addition they identified a temperature effect distinct from its effect on viscosity, and suggested that this might be related to a transition temperature, possibly associated with desorption of polar compounds.

Under boundary or mixed lubrication conditions there is usually a reduction in friction when molybdenum disulphide is dispersed in an oil[153,452] but it can have no useful effect under fully hydrodynamic lubrication conditions, because there is no contact between the bearing surfaces.

Rylander[453] showed by a combined theoretical and experimental study that in a full fluid film hydrodynamic bearing the friction could only be increased by the addition of dispersed molybdenum disulphide to the oil. In the convergent zone of the bearing prior to the point at which the oil film thickness is equal to the effective particle size, there is a small increase in viscous friction due to the influence of the dispersed particles in increasing the effective viscosity. In the convergent zone between the point where the oil film thickness is equal to the particle size and the point of minimum separation there is a greater increase in friction associated with shear of the particles. In the divergent zone there is again a small increase in viscous friction. In the most unfavourable case he studied there was an increase of 70% in the total friction with a 1% dispersion of molybdenum disulphide in the oil.

Similar results were obtained by Aoki and Nakahara[454], except that they concluded that the main increase in friction arose from rubbing friction between the particles and the bearing surfaces, and not from shear of the particles. It is possible that both mechanisms could contribute to the friction, but the extent of the individual contributions would depend on whether the stresses imposed on the particles were such as to cause them to shear or slip. It seems probable that the friction between film and substrate will normally be greater than the inter-lamellar friction, so that shear of particles is inherently more likely.

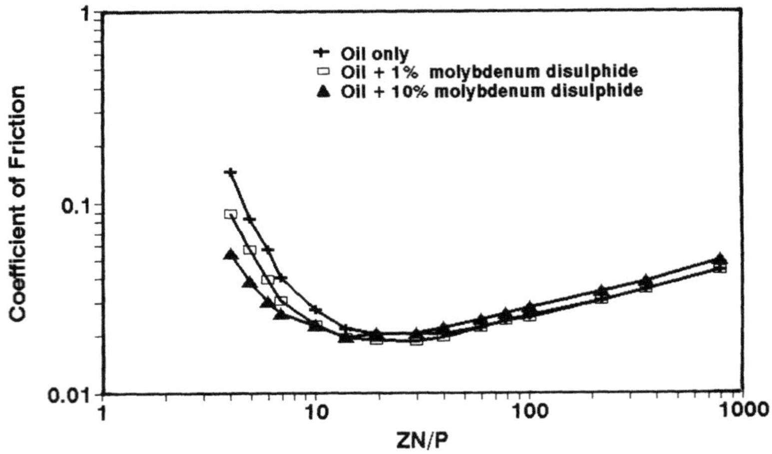

Figure 13.2 Variation of Friction with Sommerfeld Number for a Series of Dispersions of Molybdenum Disulphide in Mineral Oil (Ref.455)

Herzig[455] also studied the effect of dispersed molybdenum disulphide in a hydrodynamic bearing, and showed that, where full fluid film lubrication is not present, dispersed molybdenum disulphide can decrease friction. The coefficient of friction in an oil-lubricated journal bearing is related to the Sommerfeld Number ZN/P where Z is the viscosity, N the rate of rotation, and P the bearing pressure or specific load. Figure 13.2 shows some empirical relationships between coefficient of friction and the Sommerfeld Number for an oil with different concentrations of dispersed molybdenum disulphide in a foil bearing[455].

It can be seen that at low Sommerfeld Number, that is in the mixed or boundary region with high load, low viscosity or low speed, the molybdenum disulphide dispersions give a decrease in friction varying from 40% with a 1% dispersion to 60% with a 10% dispersion. At higher Sommerfeld Number with full fluid film separation, there is an increase in the coefficient of friction of 14% with the 10% dispersion, while the 1% dispersion has the same friction as the base oil. However, extrapolation of the curves to Sommerfeld Numbers of 10^{-3} or 10^{-2} suggests that the 10% dispersion could give an increase in friction of over 25%. This represents the condition in a lightly-loaded high speed bearing.

Thus any advantage of molybdenum disulphide in an oil in reducing friction will be limited to the boundary and mixed lubrication regions, where the reduction in friction may be considerable. One theoretical analysis suggested that the influence of the dispersed powder depends only on particle shape, size and concentration[455], or in other words that the dispersed powder is simply forming a physical barrier between the interacting surfaces. It would follow that the same effect could be produced by other dispersed solids, and this was confirmed by studies with zinc sulphide, zinc pyrophosphate and calcium hydroxide.

Apart from the frictional effects, two other performance characteristics need to be considered, namely load-carrying capacity and wear. Most of the available information on these subjects has been obtained from the practical use of commercial materials and will be discussed later, but Groszek and Witheridge[225] found that molybdenum disulphide ground conventionally in air gave a small increase in Mean Hertz Load and Weld Load in a Four-Ball Test as 5% dispersions in a mineral oil. However, oleophilic molybdenum disulphide, ground in n-heptane, gave a 170% increase in Mean Hertz Load and a 260% increase in Weld Load.

Tsuya et al[456] studied the effect of oil viscosity on the contribution of added molybdenum disulphide to the load-carrying capacity of the oil. They found that a given concentration of molybdenum disulphide was more effective in increasing the load-carrying capacity of a low-viscosity oil than a high-viscosity oil. These results are consistent with Groszek's evidence [457] that higher molecular weight (i.e. more viscous) paraffinic hydrocarbons are more strongly adsorbed than lower molecular weight paraffins on the non-polar adsorption sites (i.e. the basal planes) of molybdenum disulphide. As a result, the more viscous oil would interfere more strongly with the ability of the molybdenum disulphide to adhere to the load-bearing surfaces. Tsuya et al also found that the wear rate with 1.5% of added molybdenum disulphide was lower than that with most of the base oils, by a factor of roughly ten or more.

One curious effect of molybdenum disulphide in lubricating oils or greases is an apparent increase in the rolling-contact fatigue (L_{10}) life. Popinceanu et al[458] used up to 1.5% concentration in oils or 5% in greases in a total of over seven hundred deep groove ball bearings, and found increases in L_{10} life of up to 100%. Similarly, Scott and Blackwell[459] tested 5% or 10% of molybdenum disulphide in a variety of synthetic oils and greases, and found increases in mean life for all except a di-ester oil. Kuhnell and Stecki[460] found a 60% increase in life for a 1% dispersion in an unspecified lubricating oil.

These results appear to be contradicted by those of Ishibashi et al[461], who reported reduced lives, but their tests used a combined rolling/sliding motion with separate rolling elements, and were therefore not directly comparable. Later work by Soda and Yamashita[462] found that the effect on fatigue life depended on the degree of slip in the contacts. Under slip conditions there was an increase in fatigue life, but in the absence of slip the particles of molybdenum disulphide appeared to act as stress raisers, and fatigue life was reduced. These various results show that in practice the effect depends on operating conditions, and also on the type of rolling bearing. For example, the degree of slip varies considerably between deep groove ball or cylindrical roller bearings and angular contact ball or tapered roller bearings. However, the results of Soda and Yamashita do not explain the deleterious effects reported by Ishibashi et al, who had incorporated slip in their tests.

A further complicating fact is that the presence of solid particles in general is now recognised as having a detrimental effect on the life of rolling-element bearings[463]. Thus the improvements reported by Popinceanu et al, Scott and Blackwell, and Kuhnell and Stecki differ from the usual effect of solid particles, and must represent some specific property of molybdenum disulphide.

Rolling-contact fatigue typically takes the form of pitting or spalling of the surface of a rolling element or race. It originates in sub-surface cracking at the point of maximum stress and propagates into a pattern of cracks with the eventual loss of a particle from the surface. It tends to be accentuated by any factor which promotes cracking, such as oxidation, aqueous corrosion or hydrogen embrittlement. There have been reports that rolling-contact fatigue is inhibited by factors which tend to heal micro-cracks, such as vacuum or a reducing environment, but it is not easy to imagine how molybdenum disulphide fits into this picture. One possibility is presumably that by reducing friction in any slipping contacts, molybdenum disulphide reduces the total stress in the contact.

Hisakado et al[464] suggested that benefits from the use of molybdenum disulphide can only arise when the bearing geometry is such that a large number of particles is carried into the zone of minimum oil film thickness. To some extent this view was supported by the findings of Casano and Sliney[465], who carried out tests with molybdenum disulphide and graphite in two different super-refined mineral oils. They used a ball-on-flat test geometry in which different combinations of rolling and sliding could be applied. They found that in pure rolling or combined rolling and sliding a film of packed solid lubricant was formed. At higher speeds it was more difficult to form a film, while in pure sliding no visible film at all was formed, and there was a tendency for the powdered lubricant to pack into the entry zone and prevent ingress of oil.

Gänsheimer and Holinski[466] showed that a certain specific load is required on the contact zone in order for a molybdenum disulphide film to be deposited on a metal surface from a dispersion in a mineral oil. Holinski[467] subsequently showed that molybdenum disulphide films were formed on tappets, camshaft, valves and probably cams in a Chevrolet engine operating with a 0.5% dispersion in the oil.

Although there is no generally-accepted explanation of the mechanisms occurring between liquids and molybdenum disulphide, there is a possible theoretical explanation which is consistent with most of the published evidence. The subject of surface energy and wettability is basic to such an explanation. Braithwaite and Greene[95] showed that a burnished molybdenum disulphide film is more readily wetted than a steel or cast iron substrate by a typical lubricating oil. In addition they showed that the extent of oxidation of the coating had a significant effect on its wettability, the least highly oxidised surfaces being more readily wetted.

This is of course consistent with the work of Groszek and others[224,225] on oleophilic molybdenum disulphide. They showed that molybdenum disulphide which had been ground in a non-polar liquid (n-heptane) was more readily wetted by non-polar liquids than similar powders which had been ground in air and were therefore, by implication, more highly oxidised.

Where a molybdenum disulphide film has been formed on a surface in air, the nature of the interface between coating and substrate is not fully established, but it is probable that the initial adhesion, either chemical or mechanical, is preferentially at edge sites. This will inevitably result in discontinuities, flaws and gaps in the film adjacent to the substrate surface.

The high degree of wettability of molybdenum disulphide by oils would be likely to generate capillary pressure within any such gaps or flaws, so that some weakening of the film attachment would result. At the same time, previous evidence shows that in the presence of contaminating liquids an increase in sliding friction would certainly arise. The resulting combination of increased stresses and weakening of the film attachment provides a sufficient explanation of the deterioration in coating performance, even if no further effects such as binder softening or intercalation arise.

The situation is quite different when molybdenum disulphide powder is used in a liquid. As has been shown, friction reduction and film formation only arise when the geometry permits particles of the powder to be trapped between bearing surfaces, and probably sheared. Such break-up of particles within a non-polar liquid is directly comparable with the procedure used by Goszek for the production of oleophilic molybdenum disulphide, so that the resulting fractured particles will presumably also be oleophilic.

Groszek and Witheridge[225] showed that such oleophilic powders are highly effective in increasing load-carrying capacity. They provided some evidence that the effectiveness is due to the fact that oleophilic particles preferentially attach to bearing surfaces in a basal plane orientation. In that orientation the molybdenum disulphide would provide less opportunity for capillary penetration, and would be less susceptible to chemical reactions.

In summary, therefore, it seems at least possible that molybdenum disulphide coatings preformed in air are adversely affected even by non-polar liquids because the non-basal-plane orientation adjacent to the substrate surface renders them liable to capillary penetration and chemical effects. Conversely, the breakdown of molybdenum disulphide particles dispersed in an oil when sheared between bearing surfaces will generate oleophilic particles which can attach to the bearing surfaces in a basal plane orientation. This provides greater stability against physical or chemical degradation by the oil, as well as giving the optimum condition for effective wear resistance and load-carrying capacity.

The relative inefficiency of film formation in a liquid may be due to either one or both of two different factors. The first is that since the molybdenum disulphide originally present in a dispersion will normally have been processed or ground in air the proportion of oleophilic particles present will be small until the dispersion has been subjected to considerable further comminution in the oil. This possiblity is supported to some extent by Holinski's work, since the conditions at cams, tappets, camshaft and valves in an engine would encourage a high level of particle fracture.

The alternative possible contributory factor is that there is competition between the liquid and the oleophilic particles for attachment to the bearing surfaces. This is supported to some extent by Tsuya's work, since the more viscous oil would presumably compete more successfully than a less viscous oil for attachment to the metal surface. It would also explain the effect of polar compounds and a possible desorption transition temperature described by Rolek et al, as well as the fact that the reliable reports of film formation all refer to non-polar liquids. Polar compounds and liquids would normally adsorb more strongly to metal surfaces, and thus compete effectively with the oleophilic particles.

These suggestions represent only a possible explanation for the apparent contradiction between the degradation of pre-formed coatings in a liquid and the strong evidence that effective films can be formed in liquids. In the absence of any fully-proven explanation of the interactions, at least the following empirical conclusions can be reached.

(i) Film formation is more difficult in liquids. It only takes place where the geometry encourages entrainment of lubricant particles into the contact zone, the operating speed is low, and the contact region enables particles to be loaded with sufficient pressure to be sheared or to attach to the bearing surfaces.

(ii) Pre-formed films are less durable in liquids and are readily disrupted.

(iii) Friction may be increased by added molybdenum disulphide in full fluid film situations or decreased under boundary lubrication conditions.

(iv) There is usually an increase in load-carrying capacity as a result of using molybdenum disulphide dispersed in an oil.

13.2 USE IN LUBRICATING OILS

The concentration of molybdenum disulphide dispersed in an oil may be as high as 60% or as low as 0.1%. At the higher concentrations the dispersions become pastes, and these will be considered in Section 13.4.

A high proportion of the molybdenum disulphide used in lubrication is used in the form of dispersions in oil. In many cases the concentrations are less than 3%, and Johnson[7] questioned whether any improvement was obtained at such low concentrations, since many of the published data showed improvements less than the experimental error. More recent work suggests that in many cases such small improvements are real.

Quantitatively the greatest volume use is probably in engine oils, where the incentive for its use is improved fuel economy. Early tests produced conflicting reports about the effect of molybdenum disulphide dispersions on fuel consumption, and in 1975 Risdon and Gresty reviewed the published results of dynamometer, track and fleet tests over the twelve-year period from 1963 to 1974. The tests compared the performance of vehicles using ordinary engine oils with the same oils containing different concentrations of dispersed molybdenum disulphide from 0.2% to 7%. They found[468] that the optimum concentration was 1%, and that this gave an improvement of between 2.3% and 6.4% in fuel consumption at the 95% confidence level, the average improvement being 4.4%.

They also reported a general reduction in sludge, varnish, wear and oil thickening due to oxidation, with no adverse effect in any of the test engines. The improvement in deposits was also confirmed by Müller and Bartz[469], who reported improved piston cleanliness. Further investigation suggested that this was due to an improvement in oxidation stability, but the way in which molybdenum disulphide can improve oxidation stability is not clear. Other beneficial mechanisms which have been suggested are improved dispersion of solid contaminants by the dispersant used to stabilise the molybdenum disulphide dispersion, and prevention of surface deposits by the formation of a molybdenum disulphide film.

A later study by the Ford Motor Company[470] investigated the use of a 1% dispersion of molybdenum disulphide in the engine oil in standard Ford cars. The results showed fuel economy improvements between 0.8% and 4.7%, and subsequent more extensive studies confirmed improvements of over 3% with no adverse effects.

Following such reports, molybdenum disulphide-containing engine oils became commercially available in the late nineteen-seventies. However, there is as yet no indication that any major vehicle manufacturer recommends the use of engine oils containing molybdenum disulphide. Almost all manufacturers recommend the American Petroleum Institute categories such as CF or SJ or their European ACEA equivalents, and none of these yet includes molybdenum disulphide. It seems probable that the greatest use is by trucking or bus fleets, where a 3% fuel saving could be very significant.

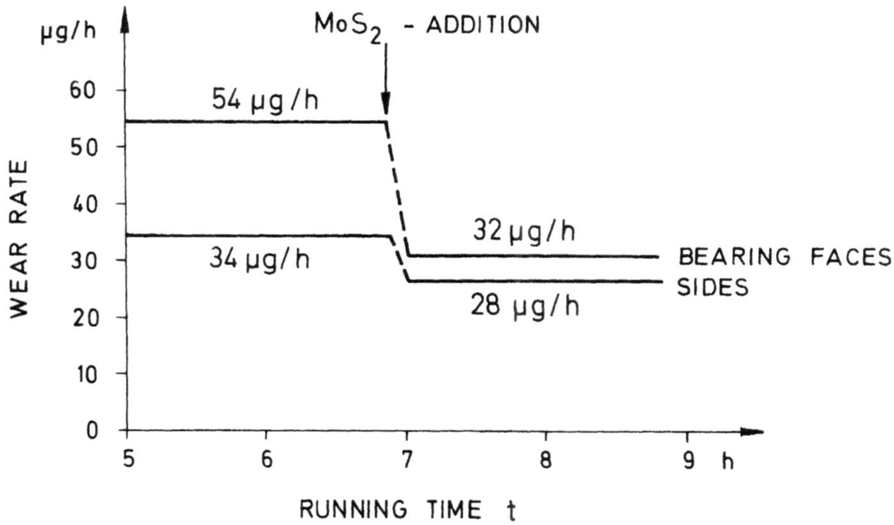

Figure 13.3 Effect of Molybdenum Disulphide Addition on Wear Rate in a Single-Cylinder Diesel Engine (Ref.471)

Overall, similar improvements in wear in engines have also been found[468]. Shadow et al[471] used radioactive tracers to study the wear in piston engines. They found a consistent decrease in piston wear in engines when a dispersion of molybdenum disulphide in the engine oil was used. Figure 13.3 shows the effect on wear rate when the molybdenum disulphide was added, and the reduction in wear rate of the bearings was over 40%.

There have also been several laboratory studies which showed[464,472,473] significant reductions in wear in a variety of tests at molybdenum disulphide concentrations between 0.5% and 10%, but unfortunately most of these studies used a paraffinic white oil, so that no useful practical inferences can be obtained from them. White oils are not typical of mineral lubricating oils in their friction and wear behaviour, as they have inferior anti-wear properties and are known to be untypical in their response to additives.

Practical vehicle engine oils invariably have quite high concentrations of complex additive packages, and this raises two important factors in assessing the effect of molybdenum disulphide. The first is the possibility of beneficial or antagonistic interaction between the molybdenum disulphide and one or more constituents of the additive package. The second is whether any beneficial effect of molybdenum disulphide might be equalled by some more conventional additive, while avoiding the increased friction at high Sommerfeld numbers or the risks of sedimentation or blockage associated with dispersions of insoluble additives.

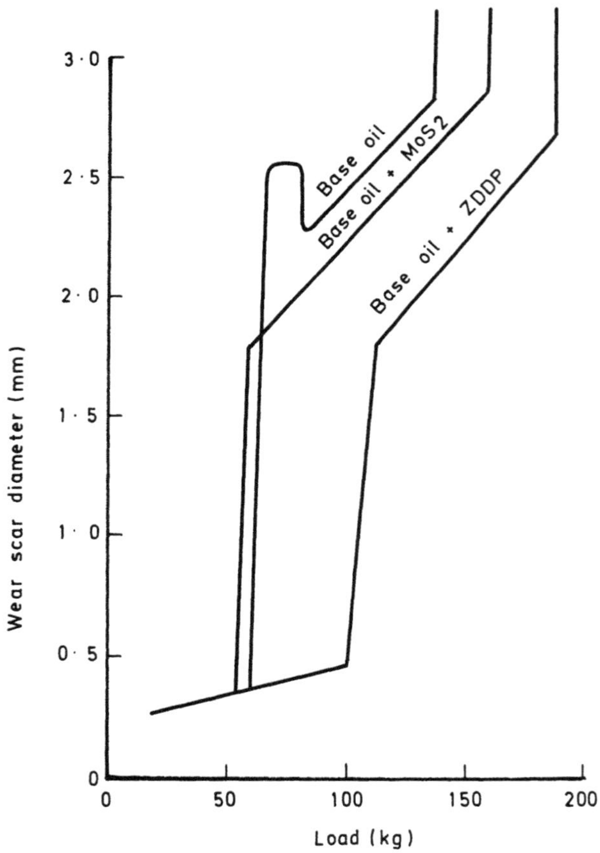

Figure 13.4 Four-Ball Machine Load/Wear Scar Relationships for Oil with Molybdenum Disulphide or Zinc Dialkyldithiophosphate (Data from Ref.474)

Several studies have been performed to investigate both these aspects. Thorp[474] carried out four-ball tests in accordance with the standard IP239 Test for Extreme Pressure Properties on a mineral base oil and samples of the same oil containing either a commercial zinc dialkyldithiophosphate (ZDDP) additive, or a suspension of molybdenum disulphide, or both ZDDP and molybdenum disulphide. Typical results are shown in Figure 13.4 in the form of the standard graph of mean wear scar diameter against load. They show that 1% of molybdenum disulphide gave an improvement of approximately 24 kg in the weld load, with a slight deterioration in the Initial Seizure Load. However, the ZDDP (whose concentration was not stated) gave an increase of approximately 50 kg in the Initial Seizure Load. The ZDDP was therefore far more effective than the molybdenum disulphide in increasing the load-carrying capacity of the oil.

These results must be treated with caution. ZDDP's are anti-wear additives, as well as anti-oxidants and corrosion inhibitors, but are not particularly effective in increasing load-carrying capacity, so that the comparison between the two additives in this case is not a very demanding one. On the other hand, the highest concentration of molybdenum disulphide used was only 1%, well below the concentrations normally used for load-carrying performance, which would typically be greater than 5%.

The more interesting results obtained were for the use of both additives together, all of which showed a further increase in load-carrying capacity, so that any interaction in these tests was beneficial. The greatest improvement was approximately 39 kg increase in Initial Seizure Load compared with the solution of ZDDP alone. Curiously, 1% of molybdenum disulphide gave only about 18 kg improvement over the ZDDP solution. There was virtually no increase in weld load compared with the ZDDP, and this again suggests that the concentrations of molybdenum disulphide were too low to be very effective. Thorp explained these results on the basis that molybdenum disulphide cannot compete with the base oil for adsorption on the steel surfaces, but can adsorb on top of an adsorbed ZDDP film, but there is no real proof of this explanation.

Bartz[475] carried out similar tests, and his results are shown in Figure 13.5. They show that 1% of molybdenum disulphide gave a significant improvement in the initial seizure load compared with the base oil at the expense of higher wear scar diameters at low load. When 1% of molybdenum disulphide was used in conjunction with a ZDDP there was little if any improvement. Bartz[476] also used the standard

Four-Ball Machine later to study the interaction between molybdenum disulphide and several anti-wear and extreme-pressure additives and detergent/dispersant additives in a mineral oil. Unfortunately these results are difficult to compare directly with those of Thorp because he only reported wear scar diameters at two load levels. He found that at high load (1000N) with 1% of molybdenum disulphide, the combination with a ZDDP gave a wear scar diameter higher than either additive separately, and comparable to that of the base oil, and he described this as an antagonistic effect between the two additives.

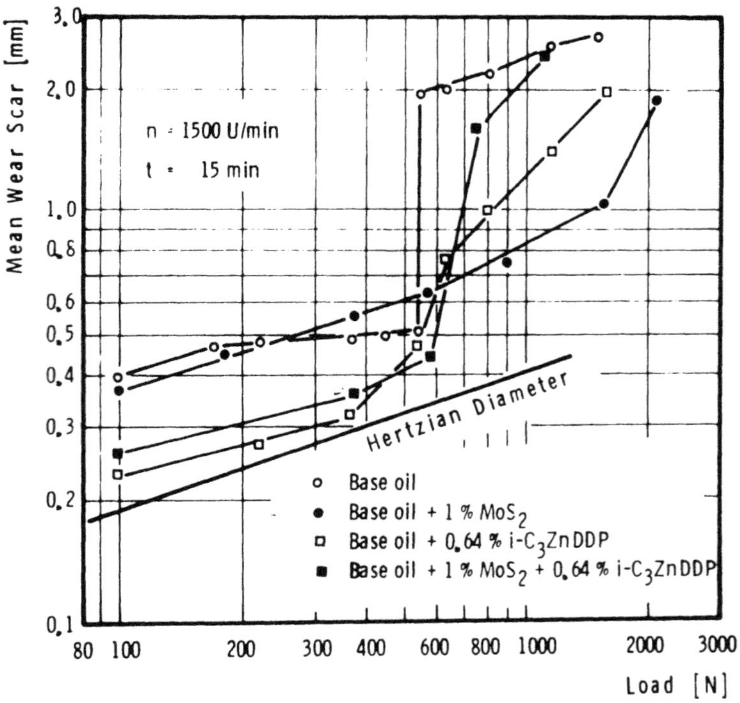

Figure 13.5 Four-Ball Machine Test Results for Base Oil Containing Molybdenum Disulphide and Zinc Di-Isopropyldithiophosphate (Ref.475)

This is certainly contrary to Thorp's result (Figure 13.4) but unfortunately the 1000N (ca 100 kg) load is close to the transition load, and rapid changes in wear scar diameter with load occur at that point, which could be significantly affected by the use of a different mineral oil or a different ZDDP. It may be reasonable to sum up these sets of results by saying that the combination of molybdenum disulphide and a ZDDP appears to give increased load-carrying capacity but may increase wear.

A similar result was found by Bartz for the combined use of molybdenum disulphide and a lead/sulphur/chlorine gear oil additive in an FZG Gear Test. He found that the combination gave higher wear rate but higher failure load than the lead/sulphur/chlorine additive alone. Tests with other additives gave a less clear picture, the results varying with the type of additive and the relative concentrations. Bartz and Oppelt[477] also investigated the interaction with oil-soluble additives and the resulting effect on wear rate in several different test machines, and concluded that overall the effect was beneficial.

On the whole, no clear picture emerges of the interaction between molybdenum disulphide and the conventional engine or gear oil additives. Both beneficial and deleterious interactions have been observed, but when the level of experimental error is considered, no dramatic or synergistic effects have been proved. The potential for fuel saving in motor vehicles is obviously significant, and it seems a pity that no major lubricant manufacturer has apparently made a determined effort to establish the optimum conditions for using molybdenum disulphide in engine oils.

One well-established benefit of a molybdenum disulphide dispersion is that if there is total oil loss from the system, the seizure of bearings, pistons or gears can be delayed by the residual molybdenum disulphide film on bearing surfaces. This effect was confirmed dramatically by Crump[478], who described tests in which cars completed up to 130 miles without damage after oil containing molybdenum disulphide had been drained from the sump. Less dramatically, Holinski[479] found indications that the beneficial effects of molybdenum disulphide in a gear oil in reducing friction continued after the oil was replaced by a normal oil. These tests were carried out in a Bartel Lubrimeter, and showed consistent reductions in friction from 0.1 for the standard gear oil to 0.06 when 0.5% of molybdenum disulphide was added.

It might be expected that molybdenum disulphide would be more effective in gears than in many other types of mechanism. Gears typically involve combined rolling and sliding action in elastohydrodynamic conditions, with either very small oil film thickness or some degree of solid/solid contact. Obviously the extent of any benefit would depend on the type of gear and the loading conditions.

Gresty et al[480] reported that direct measurement of the effect of 3% of a molybdenum disulphide concentrate in the oil on the efficiency of a rear axle test gear gave results varying from about 1.5% decrease to about 2% increase in efficiency. However, parallel tests with a towing dynamometer had shown that 1% of molybdenum disulphide reduced the running temperature of the axle by 8°C. Since oil temperature has a significant effect on gear efficiency, because of the influence on viscous drag, they corrected the test rig results for temperature, and obtained revised results indicating efficiency increases between 2% and 5%.

Because of the high degree of sliding between worm and wheel, friction is particularly important in worm gears, and it is common practice to incorporate friction-reducing additives in worm gear oils. The addition of molybdenum disulphide to the oil in one type of worm gearbox[481] was reported to reduce the starting current to the drive motor by 50%. However, the factors contributing to starting current are complex, depending on the type of drive motor, inertia of various system components, the effectiveness of the retained oil film on the gear surfaces, and so on. A more meaningful indication is the efficiency in steady-state operation, and Smith and Marshek[482] found that in one worm gear speed reducer a dispersion of 1% of molybdenum disulphide in the lubricating oil increased the average efficiency from 68.8% to 72.2%, a relative improvement of 5%, with a corresponding reduction in temperature.

This application illustrates the fact that friction will often be reduced by the addition of molybdenum disulphide to an oil where the geometry is such that the particles can enter the oil film and be loaded against the bearing surfaces. This is the case when boundary lubrication is occurring, and frequently when elastohydrodynamic lubrication is taking place. It can also arise where hydrodynamic lubrication is marginal, and it is often in such cases that frictional heating problems and premature failures occur.

An example of this type of trouble-shooting application was the addition of a molybdenum disulphide dispersion to the lubricating oil in a 400 ton metalforming press, which was reported[483] to have eliminated overheating, increased unit production rate and reduced downtime.

Excessive noise in gears and transmissions is often related to frictional problems, especially friction-induced oscillations or stick-slip. Addition of molybdenum disulphide can sometimes reduce the friction problems. An example was the addition of 10% of a molybdenum disulphide dispersion to the lubricating oil in a gearbox transmitting 75 kW in a tube mill. This was reported to have eliminated excessive noise, improved the condition of pitted gear teeth, and improved the quality of the products. Presumably all three benefits derived from the elimination of friction-induced oscillations, and the resulting overall reduction in machine vibrations.

When reduced wear is associated with the addition of molybdenum disulphide to a lubricating oil, the type of wear is always likely to be adhesive. The effects of molybdenum disulphide in such a case would be a direct reduction in the degree of adhesion by the interpolation of a molybdenum disulphide film, as well as a reduction in the overall stress levels due to lowering of friction.

This discussion has so far been related to oils such as mineral oils and esters, which are themselves inherently good lubricants. Where oils such as silicones, halocarbons, polyphenyl ethers or perfluoropolyethers are used, the potential for using molybdenum disulphide would appear to be much greater, but in fact with the exception of silicones there has been no evidence of much use in such oils. There have been several commercial products consisting of molybdenum disulphide in silicone oils or greases, and a United States military specification, M1l-L-25681 (now DOD-L-25681D) was issued to cover a molybdenum disulphide-containing silicone oil for use at temperatures up to 390°C on slow-speed sliding surfaces.

Dispersions of molybdenum disulphide in polyglycols are a special case. On heating to decomposition temperature, polyglycols decompose without leaving a deposit. Dispersions in polyglycols can therefore be used in high-temperature applications such as oven conveyors as a convenient means of feeding lubricant into the bearings. When the polyglycol carrier decomposes, it leaves a clean layer of molybdenum disulphide to continue the lubrication. A graphite dispersion in polyglycol is often preferred in such applications since on greater heating the graphite will also be removed without leaving a deposit, but in fact the deposit of molybdenum trioxide which would be left by molybdenum disulphide under the same conditions is relatively harmless and has some lubricating properties.

Oleophilic molybdenum disulphide (see Chapters 7 and 9) has been reported as giving particularly effective wear reduction[225] when used as a 5% dispersion in a mineral oil.

The use of molybdenum disulphide dispersions in metalforming was mentioned in Chapter 9. Similar dispersions in water, light solvents or oils have been tested for use in many different metalcutting processes, including drilling[484], milling, turning, broaching, boring, reaming[485] and grinding[486]. The results were generally favourable in terms of cutting speed, accuracy, surface finish and tool life. The improvements in tool life in machining of steels with one fluid containing oil, sulphonates, corrosion inhibitors and dispersed molybdenum disulphide in water are shown in Table 13.3.

Table 13.3 Improvement in Tool Life with a Molybdenum Disulphide-Containing Cutting Fluid (Ref.485)

Operation	Surface speed (ms^{-1})	Feed rate	Tool Life (No. of parts machined)	
			Water emulsion	MoS_2 fluid
Slot milling	0.5	0.6	40	80
Drilling	0.23	0.03	40	80
Groove turning		0.25	40	375
Spline broaching	0.05		60	120
Boring	76	2×10^{-4}	50	120
Broaching	0.05		50	100
Reaming	3	4×10^{-4}	60	110

One problem in metalcutting is molybdenum disulphide contamination of the workpiece, which requires additional processing to remove it. The general trend has therefore been to use dilute oil-in-water emulsions ("soluble oils") and aqueous solutions ("synthetics") for preference in metalcutting operations. Dispersions are more widely used in metalforming operations, especially in cold or warm pressing and extrusion, where their advantages are greater, and this type of application is described in Section 13.4.

Some commercial dispersions of molybdenum disulphide in oils are listed in Table 13.4.

Table 13.4 Some Commercial Dispersions of Molybdenum Disulphide in Oils

Concentrated Dispersions for Blending into Mineral Oil Systems at Low Concentration
Acheson DAG Dispersion 707
Acheson DAG Moly 725
Molykote M55 Plus
MolyPaul Molyphide Dispersion
Rocol ASO
Less Concentrated Dispersions in Mineral Oil, Intended for Use Without Further Dilution
Acheson DAG Moly 724
Molykote 1231LN
Rocol MO Grades
Dispersions in Synthetic Oils for Special Applications, Usually at High Temperatures
Molykote M30
MolyPaul Chainlife SM
Rocol CL 280M
Rocol MU-F

13.3 MOLYBDENUM DISULPHIDE IN GREASES

Lubricating greases are oils which have been thickened to a semi-fluid consistency by dispersing in them a colloidal solid. They therefore differ in several respects from other semi-fluid materials such as heavy petroleum oil, bitumen, petrolatum (petroleum wax) or low-molecular-weight polymers. One of the most important of these differences is that under certain circumstances they may separate into their two main constituents, thickener and base fluid. Another important difference is that although they behave as soft solids when stationary or moving slowly, they can be made to flow readily at high shear rates. As a result their lubricating behaviour is similar to that of an oil when they are used in mechanisms such as rolling contact bearings.

When molybdenum disulphide is added to a grease, the effect on the flow properties depends critically on the concentration of molybdenum disulphide. Where the quantity added is high enough to cause a major change in the flow properties, the result is a product which no longer behaves as a grease, but behaves like a paste. Such materials are generally used as anti-seize compounds rather than lubricants, and they will be described in Section 13.4.

Two disadvantages of using molybdenum disulphide in oils are the risk of the solid particles agglomerating or sedimenting out, and the probability of increased torque, noise, friction and wear at high speed and low load. These disadvantages are far less likely to arise in greases. Except in the softest greases the powder is extremely unlikely to agglomerate or sediment out, while greases tend not to be used in high-speed bearings or other high-speed components. Molybdenum disulphide greases therefore have considerable advantages with only minor technical limitations in most situations. As a result they are used in large quantities, and as early as 1971 the world consumption was quoted[487] as over fifty thousand tonnes.

There have been several detailed comparisons of the performance of greases with and without molybdenum disulphide, including an intensive survey carried out[488-495] by Climax Molybdenum Company between 1967 and 1977. The results of those comparisons confirmed that the molybdenum disulphide greases usually performed better, and rarely worse, than similar greases without molybdenum disulphide.

Since 1977 several new thickener types have become widely used, including complex (sodium, lithium, calcium or aluminium), polyurea, modified clays, and other non-soap thickeners. To bring the picture up-to-date, Risdon[496] carried out an evaluation in 1986 of three of the newer grease types not included in the previous programme. They were lithium complex, aluminium complex and polyurea. They included one EP (extreme pressure) and one non-EP grease for each type of thickener, and all were of the same consistency grade, NLGI No 2.

The tests used were all ASTM standard grease tests, Four-Ball Wear (ASTM D2266), Four-Ball EP (ASTM D2596), Falex EP (ASTM D3233), Ball Joint Torque Stability (ASTM D3428), Timken EP (ASTM D2509), Corrosion Preventive Properties (ASTM D1743) and Oxidation Stability (ASTM D942). The molybdenum disulphide addition at 1%, 3% or 10% by weight caused small changes to the Four-Ball Wear Scar, the Timken OK Load and the Oxidation Stability, but the results were all within acceptable limits and there were no significant incompatibilities. There were improvements in load-carrying properties in the Falex and Four-Ball EP tests.

The GMR (General Motors Research) Ball-Joint Tester is a device designed to assess the suitability of lubricants for use in vehicle suspensions and steering systems. It evaluates a lubricant in terms of torque, torque stability, extended duration and thin film operation. In the Climax tests Risdon[493] used it to test greases containing molybdenum disulphide. The performance in all cases was satisfactory, but improved as the concentration increased from 1% to 3%. Later tests[494] with the same tester were used to assess the potential for reducing energy consumption. Addition of Technical Fine molybdenum disulphide to the test greases was found to give a reduction in torque (i.e. energy dissipated) after only one minute of operation. Effective torque reduction was found with addition of 5% of molybdenum disulphide in several greases, but the greatest and most consistent reduction in torque and operating temperature was obtained with 20%.

One surprising finding[495] was that 5% of molybdenum disulphide reduced the amount of abrasive wear when greases containing abrasive contaminants were tested in the Falex Tester and the GMR Ball-Joint Tester. The mechanism involved is not clear, and it is generally unusual for any lubricant to reduce abrasive wear. One possibility is that, as reported by Gänsheimer and Holinski[466], under high contact stresses and high temperature, molybdenum disulphide reacts with iron to release molybdenum into a steel surface and increases the surface hardness. On the other hand, such a process seems unlikely in the conditions present in this study.

Antony et al[497] similarly investigated the effects of molybdenum disulphide and graphite on the performance of a lithium-based grease and an organo-clay-thickened grease. They found that either of the additives at a concentration of between 1% and 5% w/w improved the EP and anti-wear properties of both greases. When both solid lubricants were used together, there was a synergistic improvement in weld load and wear scar diameter with the lithium-based grease, and in the EP properties of the organo-clay-based grease, but there was an increase in wear scar diameter with the latter grease. They also found no significant change in the other physical or chemical properties of the greases, other than a slight increase in oil separation. The results for molybdenum disulphide are summarised in Table 13.5 and show that apart from a marked increase in weld load there were only slight changes in properties.

Table 13.5 Effect of Molybdenum Disulphide on Properties of Lithium-Based and Organo-Clay Based Greases (Ref.497)

Characteristics	Concentration of MoS_2, wt.%					
	0	1.0	2.0	3.0	4.0	5.0
Lithium-Based Greases						
Cone penetration (unworked)	260	262	262	263	262	262
Cone penetration (worked)	262	263	264	263	263	262
Drop point °C	195	194	193	193	194	193
Cu corrosion at 100°C, 24hrs.	Pass	Pass	Pass	Pass	Pass	Pass
Heat stability, oil separation, wt.%	0.2	0.6	0.6	0.5	0.4	0.4
Weld load, kg.	160	225	225	225	225	280
Wear scar diameter, mm.	1.0	0.85	0.85	0.90	0.90	0.90
Organo-Clay-Based Greases						
Cone penetration (unworked)	262	260	258	260	255	252
Cone penetration (worked)	267	273	272	268	271	268
Drop point °C, greater than	330	330	330	330	330	330
Cu corrosion at 100°C, 24hrs.	Pass	Pass	Pass	Pass	Pass	Pass
Heat stability, oil separation, wt.%	Nil	0.1	0.02	Nil	Nil	0.2
Weld load, kg.	160	180	225	225	225	225
Wear scar diameter, mm.	0.6	0.55	0.55	0.55	0.55	0.55

The indications are therefore that the incorporation of molybdenum disulphide in greases will usually give an improvement in the load-carrying performance with little or no adverse effect on other grease properties at concentrations up to 10%.

Table 13.6 shows typical load-carrying capacity figures for lithium greases with and without molybdenum disulphide, as given by the Four-Ball EP, Falex, and Timken procedures. The Mean Hertz Load figures are based on the ASTM D2596 test method. The similar IP239 method will give values generally between 10% and 20% higher. The table shows that the load-carrying capacity of molybdenum disulphide greases is consistently higher than that of equivalent greases which do not contain molybdenum disulphide. Similar improvements are often obtained with conventional EP additives, but very high load-carrying capacities are often obtained by using both molybdenum disulphide and another EP additive.

Table 13.6 Typical Load-Carrying Capacity Figures for Lithium Soap Greases With and Without Molybdenum Disulphide
(Mean figures from several publications.)

Grease	4-ball EP Tester		Timken Tester		Falex Machine
	Mean Hertz Load	Weld Load (kg)	Load capacity (lb)	Film Endurance (mins.)	Load Capacity
Lithium grease	24	156	4-13	1-5	650
+ 3% MoS_2	37	200	5-12	118-327	733
+ 10% MoS_2	50	316	7-17	112-155	933
Lithium EP grease	38	316	6-22		800
+ 3% MoS_2	52	316	9-23		1033
+ 10% MoS_2	70	400	7-23		1200

The best performance found in the Climax Molybdenum Company test programme was for a lithium hydroxystearate soap-thickened grease with a conventional EP additive and 10% by weight of molybdenum disulphide with a nominal particle size of 7.0 μm. This grease had a Timken OK Load of 23 lbs (10.5 kg), Falex load capacity of 1450 lbs (658 kg), Mean Hertz Load of 90.5 and Weld Load of 630 kg.

There is a general tendency for wear to be reduced in molybdenum disulphide greases, but the improvement is not as great nor as consistent as for load-carrying capacity. Friction is also generally lower. The effects on physical and chemical properties of greases are more complicated, and addition of molybdenum disulphide has been shown to affect the stability of the gel structure, the oxidation resistance and the corrosion resistance.

It is a fairly common phenomenon for additives or contaminants to reduce the stability of the colloidal structure of a grease, and even the mixing of two similar stable greases can cause de-stabilization. For this reason, approval of a particular brand of grease against an official specification will sometimes require proof of compatibility on mixing with other brands previously approved. In an extreme case the reduction in stability can be so great that the grease loses its semi-solid nature and liquefies. The effect of added molybdenum disulphide, and many other relatively inert solids is far less extreme. Typical effects might be a reduction in the Drop Point, which is the temperature at which a grease becomes liquid on heating, or an increased tendency to soften when heavily worked.

The effect on oxidation resistance or corrosion resistance may be a slight improvement, or more often a small deterioration. All these effects can be offset by proper re-formulation. Fully formulated molybdenum disulphide greases can thus have performances as good as those of similar greases without molybdenum disulphide, while having higher load-carrying capacity, better wear prevention and reduced friction. However, it is generally recommended that non-specialist users of greases should not make their own additions of molybdenum disulphide to greases unless they have facilities for adequate testing of the products.

In the early days of incorporation of molybdenum disulphide in greases concern was often expressed that the dispersed powder would lead to jamming of ball or roller bearings because of accumulation of the powder in small clearances. Caution was perhaps understandable, although graphited greases had already been used successfully for many years. There is of course always a possibility that a badly-formulated or badly-made grease might suffer agglomeration of the solids incorporated, but many tests and applications have been performed without any jamming problems. Risdon and Binkelman[490] reported the results of ASTM D1741 tests on a series of lithium greases containing up to 50% of added molybdenum disulphide. This test uses a ball-bearing running at 3500 rpm, and one of the failure criteria is stalling of the drive motor due to high bearing torque.

In Risdon and Binkelman's tests the functional life of the molybdenum disulphide greases was in all cases longer than that of the base grease. Kitchen[498] also successfully operated rolling bearings for 100,000 hours with a grease containing 50% of molybdenum disulphide. On the whole, however few people would recommend the use of a grease containing more than 20% of molybdenum disulphide in a high-speed rolling bearing, and the subject of optimum concentration is considered later.

Table 13.7 Increase in Load-Carrying Capacity of a Di-Ester Grease With Molybdenum Disulphide Content (Data from Ref.173)
(Mean Hertz Load figures obtained by ASTM D-2596; the IP239 method will give results between 10% and 20% higher)

MoS_2 Content	Mean Hertz Load
0%	18
3%	40
5%	50
10%	52
20%	59

The comparisons described so far all related to mineral oil greases, but Devine et al[173] made a similar comparative study of diester oil greases. The results were generally comparable to those for mineral oil greases. There was an increase in load-carrying capacity, and Table 13.7 shows the progressive increase in Mean Herz Load with molybdenum disulphide content. There was also an improvement in wear prevention with no adverse effect on storage stability, oil separation, high-temperature performance or rust-preventive properties. There was some decrease in oxidation resistance, especially with the finer particle sizes, but with the larger 7.0μm particle size the decrease was only slight. Overall the optimum concentration of molybdenum disulphide was 5%.

Molybdenum disulphide has been used in greases based on other oils, as well as mineral oils and esters, but probably the only other common type is silicone. Silicone greases are usable to over 250°C, but they are poor lubricants under boundary lubrication conditions. Addition of molybdenum disulphide gives a useful improvement in friction, wear prevention and load-carrying capacity, while maintaining the high temperature capabilities. Many such greases have been produced commercially, and are used for applications such as bearings, slides and valves which operate under high loads and temperatures.

Some commercial greases have also been produced which are based on a perfluorinated polyether oil with added molybdenum disulphide. They can be used for even higher temperatures than the silicone-based greases, but are much more expensive and would normally only be used on small components. They have the further advantage of excellent resistance to chemical attack.

The subject of the optimum concentration of molybdenum disulphide in greases was a matter of controversy for several years. Early British practice was to use at least 10%, whereas early United States practice was to use less than 5%. Comparative tests of several greases showed that in a range of applications there was little difference in performance between 5% and 10%, as is also shown in Table 13.7, and for military and aviation standardization purposes a concentration of 5% was adopted in the early nineteen-sixties. The resulting greases were found to be suitable for many highly-loaded rolling bearing applications.

Greases are available which contain anything from 1% to 60% of molybdenum disulphide. The choice of concentration is a compromise which depends on the bearing loads and the importance of good flow characteristics. The greater the concentration the better are the load-bearing properties, but above 20% there would normally be a progressive deterioration in the flow properties. At 50% or more the flow properties are likely to be poor, and such products behave more like pastes than greases. The greases with less than 3% of molybdenum disulphide have relatively low load-carrying capacity, but help to maintain lower friction in thin-film conditions, and give some residual lubrication after running dry.

Some of the highest volume applications are in the automotive industry. According to McCabe[89], the earliest automotive users were Rolls-Royce and Chrysler. He reported that Chrysler used molybdenum disulphide greases in all passenger car applications, although it seems unlikely that he meant this to inlude wheel bearings. Rolls-Royce were said to be using them in leaf springs, steering linkages, brake expander units, door locks and window winders. He further reported that General Motors, Ford and British Motor Corporation started to use them widely during the early nineteen-sixties, although in the regular surveys of United States car lubricants in the NLGI Spokesman, only Ford has been shown as specifying molybdenum disulphide greases consistently over the past thirty years.

The Caterpillar Company has also been quoted[499] as specifying a multipurpose (usually NLGI No 2) molybdenum disulphide grease for its heavy duty products from about 1973. Such greases have also been reported to be more widely used in off-road vehicles, and equipment in the construction industry[500]. In both cases the molybdenum disulphide concentration is between 3% and 5%.

Another major user is the steel industry, and Jost and Hicks[501] and Forsythe[502] described a wide variety of applications, including crane slewing plates, slideways, run-out table bearings, gears, work-roll bearings and motor bearings. In other industries the variety of applications is indicated by the list in Table 13.8.

Table 13.8 Some Applications of Molybdenum Disulphide Greases

1. Gears of a jack-up drilling rig.
2. Main "X" bearing in a single-point mooring.
3. Swivel bearing in DC-10 cargo handling.
4. Cams, followers and bearings in paper plate stamping machines at 150°C.
5. Railway rolling-stock wheel mounting.
6. Wire rope lubricant.
7. Extrusion of steel billets.

The use of oils containing molybdenum disulphide to solve lubrication problems was mentioned previously. Similar use of greases containing molybdenum disulphide can be even more effective, because the benefits of molybdenum disulphide in greases are not offset by any serious disadvantages.

Table 13.9 Some Commercial Molybdenum Disulphide Greases
(Some of these may now be obsolete)

Product	Description
Achesons Multipurpose No 2	Mineral oil base
Molykote BR2	Mineral oil, lithium soap
Molykote Longterm	Mineral oil, lithium soap
Molykote TTF52	Mineral oil, inorganic thickener, low temperature
Molykote 165LT	
Molykote FB180	Mineral oil, bentonite thickener, high temperature
Molykote 1121	
Molypaul Grade 1110	
Molypaul Easymesh	
Molypaul Thermo-Paul 1 and 2	
Rocol MG	
Rocol Sapphire Hi-Load 2	
Rocol MTS 1000	
Molykote 1132	Synthetic oil, inorganic thickener, high temperature
Molypaul Molyrace HT	
Aeroshell Grease 17	
Molykote ET 300	Silicone oil base
Aeroshell Grease 15	
Rocol MX 33,44,66,550	

T.I.Fowle[503] has pointed out the potential for molybdenum disulphide to reduce 'wire-wool' failures of plain bearings. This type of failure can occur when a hardened adhesive wear particle becomes embedded in a relatively soft bush, causing severe abrasive wear of the counterface. The formation of the initial wear particle can arise from a transient failure of the hydrodynamic lubricant film, caused, for example by temporary interruption of a drip feed or coolant flow. Addition of a small amount of molybdenum disulphide, say 3% or more, to the lubricant can prevent the short-term occurrence of adhesive wear, and thus the formation of the initiating particle.

He gave an interesting description of an industrial lubrication problem in crusher roll bearings in the Carribbean sugar industry. The unhardened steel rollers were supported by phosphor bronze half-bearings and were water-cooled, but severe adhesive wear took place which was caused mainly by over-heating combined with high bearing loads. The problem was cured by the use of 3% of molybdenum disulphide in a grease produced from a bentonite clay-thickened high-viscosity mineral base oil. The reduction in friction coupled with high load-carrying capacity was sufficient to eliminate the adhesive wear which had previously led to failures.

One type of molybdenum disulphide grease which is technically interesting is the type in which oleophilic molybdenum disulphide is used as the thickener[224]. This eliminates the need for a non-lubricating thickener, and thus has the potential for reducing the total solids content, especially for high-temperature greases, in which the thickener content is often particularly high. There have been some reports of satisfactory testing for aerospace applications, but it is not clear whether any major production or use has occurred.

Table 13.9 lists some of the many commercial molybdenum disulphide greases.

13.4 PASTES AND DISPERSIONS

The oils and greases containing molybdenum disulphide which have been described so far still behave primarily as oils and greases. In other words they are liquid or semi-liquid lubricants with good rheological properties. When the solids content increases to a stage where the materials no longer flow smoothly, especially in small channels or orifices, it becomes more convenient to differentiate them by calling them dispersions and pastes.

Dispersions have previously been described in Chapter 9 in the context of volatile liquids designed to supply powder to a system. But dispersions in non-volatile liquids are also used as a means of feeding a high concentration of molybdenum disulphide into a system while still maintaining some degree of flow.

There are three main types of application for which molybdenum disulphide pastes and concentrated dispersions are used. They are used as anti-seize compounds for threaded fittings, pipes, flanges, and so on. They are also used to assist assembly, and to a lesser extent running-in, of many different types of mechanical euqipment. Finally, they are used in high-load metalforming processes.

Almost any liquid can be used as the carrier for the dispersions, and the choice is largely determined by cost, viscosity, any adverse effects on the equipment or products, and adequate wetting of the molybdenum disulphide powder. The most common carrier liquids are probably mineral oils because of their low cost, wide range of viscosity grades, and relative chemical inertness. When it is important to avoid carburisation of metals at high temperatures, polyglycols, vegetable oils such as rapeseed oil, and low-molecular-weight polymers such as polyethylene or polybutene can be used. For low flammability chlorinated diphenyls have been used, but it is more usual to use an aqueous dispersion and allow the water to evaporate, relying on the residual dry film for friction reduction.

Pastes can be made with the same carrier liquids by simply increasing the concentration of powder to achieve a paste-like semi-solid consistency. More stable pastes are obtained by using a carrier which is inherently semi-solid, such as a grease, petrolatum (soft petroleum wax) or a semi-fluid polymer. The concentration of powder in a paste or a dispersion may be anything between 35 and 75%, depending on the application. The British military specification Def Stan 80-81/1 requires not less than 50% of molybdenum disulphide in a mineral oil grease for an anti-seize and anti-scuffing compound for use up to 250°C. This is probably a fairly typical level for anti-seize use.

The Government of the German Democratic Republic published a standard TGL 10596/03 in 1977 for a product containing not less than 40% in a stabilised paste with mineral oil and adhesive components[504]. This material was recommended for a wide range of applications, including:-

Assembly of equipment.
Installation and removal of bearings and seals.
Loosening of threaded joints.
Reducing stick-slip motion of slideways.
Lubrication in hostile environments.
Lubrication of sliding under extreme pressure.
Lubrication of metal stamping, pressing, drawing, etc.

In other words, this one material was intended to meet the requirements of anti-seize and anti-scuffing compounds, pre-assembly, and metalforming. A primary objective of the standard was obviously to simplify supply in a relatively tightly-controlled economy. In more competitive economies a very large number of alternative commercial products are available, and a few of these are listed in Table 13.10.

Table 13.10 Some Commercial Dispersions and Pastes for Anti-Seize, Assembly and Metalforming
(Some of these may now be obsolete)

Product	
Pastes	
Molykote G-n plus	Mineral oil
Molykote G-Rapid plus	Mineral oil
Molykote U-n	Polyglycol
Molykote X	Mineral oil
Acheson Gredag YP	Petrolatum Anti-Seize
Molykote M77	Silicone oil
Rocol MT-LM	Assembly and Running-In
Rocol Anti-Seize J166	Anti-seize
Molypaul Easyrun	Assembly, Anti-Seize
Molypaul Polypaste 300	Grease-based Anti-Seize
Viscous Dispersions	
Molykote M30	Synthetic oil
Rocol CL280M	
Rocol ASP	Mineral oil, anti-scuffing
Molypaul Chainlife SM	Polyglycol

The lower concentrations would often be used for pre-lubrication in machine assembly. For these and for anti-seize applications, the dispersions or pastes are commonly applied by brushing. For heavy metalforming operations the highest concentrations would be suitable, in order to maintain sufficient coverage as the surface area of the workpiece increases during deformation. For small items brush application would again be satisfactory, but dipping and spraying are also used and become essential for very large workpieces.

Several publications have described the use of molybdenum disulphide pastes for the lubrication of screw threads. This is an application in which the paste assists assembly and subsequently acts as an anti-seize. The critical factors in screw or bolt assembly are the friction between male and female thread and between bolt-head and workpiece, and their effect on the relationship between assembly torque and bolt tension (clamping force). The stiffness and integrity of a bolted assembly are determined by the bolt tensions, but since it is impracticable to measure bolt tension directly, the required bolt tension is controlled by the torque applied to the bolt or nut. However, the relationship between assembly torque and bolt tension is determined by head friction and thread friction. If the friction is reduced, the required bolt tension will be obtained at a lower assembly torque.

Table 13.11 Effect of Lubricants on Thread Friction (Ref.505)

Material and Conditions	Lubricant	Starting Torque (Nm)	Spread (Nm)	Friction μ
Blackened 8.8 steel (constant clamping force 34kN)	Mineral oil	70	5	0.13
	MoS_2 paste	45	5	0.07
Zinc-plated 8.8 steel (constant clamping force 34kN)	None	93	53	0.17
	MoS_2 paste	45	5	0.07
	Copper paste	52	0	0.10
A2 steel (constant clamping force 27kN)	None	101	17	0.25
	MoS_2 paste	40	5	0.09
	Copper paste	53	6	0.12
	MoS_2 bonded film	36	0	0.07
	Graphite bonded film	25	5	0.05

Gänsheimer and Wessely[505] examined the value of several different lubricants for lubrication of screw or bolt threads. Some of the torque and friction values they measured are shown in Table 13.11. They concluded that mineral oils were of little value for screw or bolt assembly, and recommended the use of pastes or bonded coatings based on graphite, copper powder, aluminium powder or molybdenum disulphide as solid lubricants.

Newnham et al[506] also found that oil lubrication was relatively ineffective for assembly of titanium bolts, but a bonded molybdenum disulphide film was effective for the whole period of the test. Birger and Iasilevitch[507] also found that molybdenum disulphide coatings gave a considerable reduction in threaded assembly torques. However, both they and Gänsheimer and Wessely pointed out that poor surface finish on threads was a major cause of difficulty in assembling screw fittings, and this results in failure of pre-assembly coatings in repeated assemblies.

An important concern when using molybdenum disulphide pastes for pre-assembly lubrication of screw threads is that the resulting thread and head friction can be very low. If the assembly torque which is applied is based on an unlubricated friction value of perhaps 0.2, or even an oil-lubricated value of 0.15, then the bolt tension with molybdenum disulphide will be two or three times as high as it should be. This can result in thread stripping, or bolt stretching. An even more dangerous result if the equipment is subject to vibration or cyclic loading may be fatigue failure of the bolts if the resulting tensile stress in the bolts exceeds the critical fatigue stress. This has been known to lead to major premature failures in heavy equipment such as full-face tunnelling machines. It is important to ensure that if molybdenum disulphide is used to assist assembly or as an anti-seize in threaded connections, the torque applied is properly calculated to take into account the considerable reduction in thread and bolt-head friction which can result.

One other area in which the use of molybdenum disulphide may be undesirable for pre-assembly or anti-seize use is when nickel alloys are used in critical applications, especially at high temperatures. Such alloys are extensively used in aviation gas turbine engines, and there have been cases in which serious failures of gas turbines have been attributed to the effect of molybdenum disulphide-based lubricants on high-temperature components[508]. Provided these two hazards are kept in mind, molybdenum disulphide dispersions and pastes are of great value as anti-seize compounds and as assembly aids.

Pastes and dispersions have also been recommended for running-in ("breaking in") equipment after assembly, but there are limitations on the extent to which this can be applied. Because by definition pastes have very poor flow properties, they cannot be used as substitutes for the conventional lubricants when running-in new equipment. To a lesser extent the same is true of viscous dispersions. If care is taken during assembly to ensure that all the interacting surfaces of new equipment are coated to some degree with a paste or dispersion, then operation of the equipment for a brief period at low speed and lightly loaded may be useful in eliminating roughness and interferences. Operation at anything approaching normal operating speed and load is much more likely to be unsuccessful, if not disastrous.

An alternative is to assemble equipment with a paste or dispersion, and then to introduce the normal operating lubricants before operating the equipment. Kawamura et al[509] investigated this procedure using a paste of 50% molybdenum disulphide and 10% of a lithium grease in an ISO 32 grade turbine oil. They found that the procedure was of little benefit because the paste rapidly dispersed into the normal lubricating oil. In most cases better results are likely to be obtained by the use of a small concentration (5% to 10%) of molybdenum disulphide dispersed in the normal operation lubricants during the running-in period. However, great caution should be used in attempting this in any complex or critical systems because of possible adverse effects.

The metalforming industries use large volumes of molybdenum disulphide dispersions and pastes, and Nittel[226] described molybdenum disulphide dispersions as indispensable for cold extrusion. Many of these dispersions are aqueous, but he reported the increasing use of "black soaps" in which the aqueous dispersions are added to reactive soaps, giving a thicker, almost paste-like consistency. Similar black soaps have also been used in wire-drawing[510], and Yunusov et al[511] reported that the resulting steel cables had improved tensile strength and ductility, although they found that the concentration of molybdenum disulphide was critical and had to be between 4% and 10%.

Bhattacharya et al[512] studied the effect of molybdenum disulphide concentration on friction in metal deformation. Their results (Figure 13.6) show an optimum at 3 - 4 wt.% and a steady decrease at higher concentrations for unannealed aluminium alloy, and two separate optimum concentrations for annealed alloy. Nandi[513] also reported a drop in extrusion load for lead billets at certain specific concentrations.

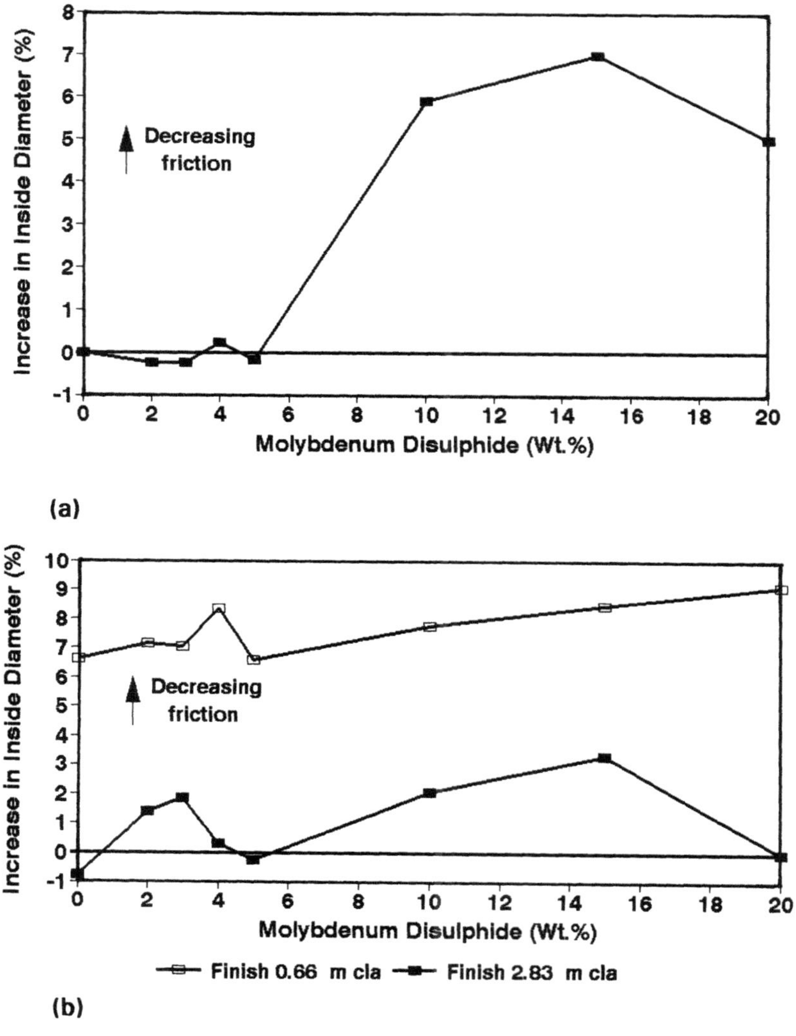

Figure 13.6 Effect of Molybdenum Disulphide Content in a Mineral Oil on the Friction in Deformation of Aluminium Rings (Ref.512)
(a) Annealed aluminium (b) Unannealed aluminium

High-solids dispersions and pastes are used in many different forming operations, especially in cold or warm forming of steel by drawing, pressing or extrusion. The critical problem in such operations is that high proportions of freshly-exposed surface are created, especially at high reduction ratios. This freshly-exposed surface has a strong tendency to adhere, or pick up, on the surfaces of the forming tools or dies. The reduction ratios attainable, and therefore the whole economics of the forming process, are limited by the ability of the lubricant to reduce friction between tool and stock, and to prevent pick-up, at the very high contact pressures required. Molybdenum disulphide gives high load-carrying capacity with coefficients of friction as low as 0.03 to 0.04 under such high load conditions, and spreads very effectively to maintain a film on freshly-exposed surface.

This remarkable ability to lubricate deformation of steel can be readily demonstrated in the laboratory with the two common test machines, the Shell Four-Ball Machine and the Falex Pin and V-Block Tester. As described previously in Chapter 5, lubricants with more than about 35% of molybdenum disulphide will result in extrusion of steel test pieces at temperatures little higher than room temperature. In both cases the surface area of the specimens is increased by about 200% at low power smoothly and without pick-up on the counterfaces.

~~~~~~~~~~~~~~~~
~~~~~~~~~~~~~~~~

CHAPTER 14.

OTHER LAMELLAR SOLID LUBRICANTS

14.1 OCCURRENCE AND PROPERTIES

The lamellar crystal structure is found in many other inorganic chemical substances, and several of these have lubricating properties. Naturally-occurring lamellar materials include various micas, talc and graphites, and both talc and graphite have been used as lubricants for thousands of years. Many similar synthetic compounds also have useful lubricating properties, including poly (carbon monofluoride), usually called graphite fluoride, molybdenum diselenide, and the disulphides and diselenides of niobium and tungsten.

During the nineteen-fifties and nineteen-sixties a much wider range of similar materials was investigated, including the disulphides, diselenides and ditellurides of titanium, zirconium, hafnium, thorium, vanadium, uranium, chromium, rhenium and tantalum. All of them, as well as the ditellurides of molybdenum, niobium and tungsten, were found to be unsatisfactory either in friction or film-forming properties, or both. Other synthetic lamellar substances include cadmium chloride, cadmium iodide and boron nitride. The low stability of the cadmium compounds has generally ruled them out as lubricants, and boron nitride, like talc, will not form satisfactory lubricating films.

As a result, the number of lamellar solids which are of interest as lubricants is reduced to eight including molybdenum disulphide, and the density and resistivity of the other seven are listed in Table 14.1. These materials, together with PTFE, are the main alternatives to molybdenum disulphide for solid lubrication. Some of the more important aspects of PTFE have been discussed in Chapter 12, and the properties and performance of the seven other lamellar solid lubricants will be presented briefly in this chapter.

Table 14.1 Physical Properties of the Lamellar Solid Lubricants

Substance	Density ($g.cm^{-3}$)	Volume Resistivity (Ω.cm)	Ref.
Graphite	1.4-1.7	0.005	518
Graphite fluoride	2.4-2.8	>3,000	525
Tungsten disulphide	7.5	14.4	73,74
Niobium disulphide	4.4	0.0031	73,74
Molybdenum diselenide	6.9	0.0186	73,74
Tungsten diselenide	9.0	144	73,74
Niobium diselenide	6.25	0.00054	73,74

The lubricating performance of the lamellar crystal structure is strongly dependent on the inter-lamellar separation and bonding, and these can be modified by the intercalation of other substances between the lamellae. The general subject of intercalation will therefore be discussed before proceeding to descriptions of the individual lamellar solid lubricants.

14.2 INTERCALATION

Although a lamellar crystal structure is favourable for solid lubrication, the inter-lamellar spacing and the nature of the inter-lamellar bonding are of major importance in determining the resistance to inter-lamellar shear, and therefore the sliding friction, of lamellar compounds.

Strong electronic (covalent or electrovalent) bonding is desirable within a crystal lamella, to provide structural strength and to ensure that when shear forces are applied to the crystal, shear takes place between lamellae, and not within them. Conversely, strong bonding between lamellae is undesirable, as it leads to high inter-lamellar shear resistance and high friction. Ideally, the inter-lamellar forces are limited to weak van der Waals forces, and the inter-lamellar space is often called the "van der Waals gap". In general wide inter-lamellar spacing favours low inter-lamellar shear resistance.

It has been known since the middle of the nineteenth century that atoms or molecules can be introduced between the carbon layers in graphite, and more recently it has been recognised that many other lamellar crystalline materials will behave in the same way. This phenomenon of "intercalation" results in modification of many of the properties of the basic crystalline material. In particular it leads to an expansion of the inter-lamellar gap, but the effect on inter-lamellar shear resistance will also depend on any effects on bonding which may arise.

Table 14.2 Part of the Periodic Table Showing the Transition Elements Whose Dichalcogenides Have Lamellar Crystal Structures and Good Lubricating Properties

GROUP NUMBER			
4	5	6	7
Ti	V	Cr	Mn
Zr	Nb	Mo	Tc
Hf	Ta	W	Re

In 1970 Jamison and Cosgrove[514] reported an interesting study of the relationship between the crystal structure and the lubrication performance of the sulphides and selenides of several of the transition metals in Groups 4, 5, 6 and 7 of the Periodic Table (see Table 14.2). They showed experimentally that satisfactory film formation and low friction were only obtained within certain closely-defined limits of crystal structure. It is not proposed here to attempt to present full details of the rather complex crystallographic considerations, but only to give a simplified version of the essential aspects of their findings.

The required crystal structure consists of the lamellar stacking of hexagonal crystals in which the repeating unit (unit cell) has hexagonal symmetry and includes two adjacent lamellae. This is conveniently described as a 2H arrangement of the unit cell. However, the frictional properties are also strongly dependant on whether the chalcogen (sulphur or selenium) atoms and metal atoms in one lamella are located vertically above those in the adjacent lamella, or displaced laterally.

Three theoretical arrangements are shown in Figure 14.1. Arrangement (a) shows the chalcogen atoms in one lamella riding completely on top of those in the adjacent lamella. Arrangement (c) shows them packed as closely together as an assembly of spheres permits, while arrangement (b) shows an intermediate situation. (It should be emphasized that these are only illustrative, and do not necessarily represent actual crystal structures). It can be seen that, in sliding, arrangement (a) will give the least interference between adjacent lamellae, while arrangement (c) will give the greatest interference.

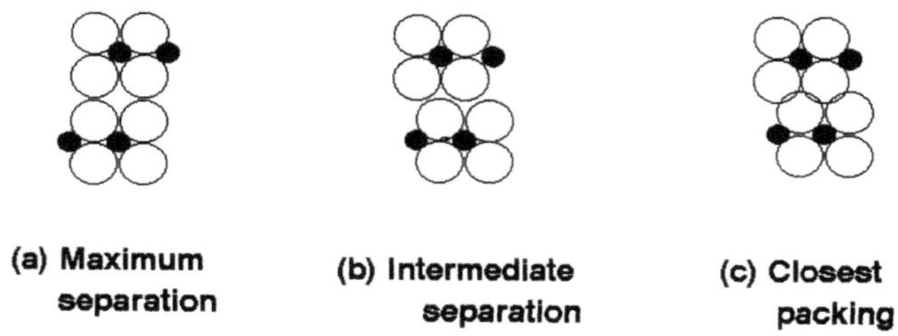

Figure 14.1 Three Theoretical Geometries for the Interaction of Dichalcogenide Molecules

The various arrangements can be defined in terms of the dimensions of the unit cell. In all three cases the cell edge length parallel to the basal plane (A) is equal to the diameter of a chalcogen atom, but the cell height normal to the basal plane (C) varies from 2.0 atomic diameters in (a) to approximately 1.82 atomic diameters in (c).

The importance of Jamison and Cosgrove's work was in showing that the experimentally-determined lubricating properties can in fact be explained in terms of these cell parameters. They found that good film formation and low friction were only obtained when the ratio C:A was greater than 1.93. This represents very good agreement with the theoretical analysis based on rigid spheres, in view of the fact that the effective shape of the atoms is distorted by chemical bonding and other aspects of electron distribution.

Jamison subsequently suggested[67] that in the case of molybdenum and tungsten disulphides and diselenides, with unbonded electron orbitals which intersect the (002) planes of the crystal lattice, the electrons are reflected and exchange momentum with the lattice, causing an effective pressure which expands the lattice. In addition, because the chalcogen electrons are largely involved in bond formation within the lamellae, each lamellar surface has a net positive charge, and the resulting repulsive force also tends to increase the inter-lamellar spacing.

In the case of the equivalent niobium compounds, the same electronic effects are not present. He postulated that in pure stoichiometric niobium disulphide this results in poor lubrication. When good lubrication behaviour is observed, it is probably caused by additional niobium atoms intercalated between the lamellae, which contribute non-bonding electrons. On the basis of this theory, non-bonded atoms intercalated between the lamellae can increase the inter-lamellar spacing, whereas bonded intercalated atoms increase the resistance to inter-lamellar shear, and therefore the friction. However, an alternative interpretation is that certain intercalated atoms alter the interaction between the niobium atoms, allowing rearrangement to the 2H structure of molybdenum disulphide, and it is the favourable structure which provides good lubrication performance.

Jamison[515] found that low concentrations of intercalated copper or silver in niobium disulphides and diselenides promoted good lubricating performance. Higher concentrations increased the resistance to inter-lamellar shear, and therefore the friction, but improved high temperature performance due to the reduced intracrystalline shear and some sacrificial oxidation of the intercalated metals.

In general the effects of intercalated elements or compounds are variable and complex, and may be advantageous or disadvantageous.

14.3 GRAPHITE

Graphite has been known as a lubricant for over 2000 years, and until about 1960 was the most extensively-used solid lubricant. In terms of annual consumption, it may still be the leading solid lubricant, because of its major application in carbon brushes in electrical equipment. Its technology is highly developed, and there are several comprehensive books about it[516,517]. A comparison between graphite, molybdenum disulphide and PTFE was made by Stock[518].

Graphite is a dark grey-black crystalline material very similar in appearance to molybdenum disulphide, and its confusion with molybdenum disulphide over many years has been described in Chapter 1. Its major characteristics and lubricating properties are listed in Table 14.3. It occurs naturally in large veins or flakes, and is recovered by surface or subterranean mining. It is also manufactured commercially, mainly from petroleum coke, by a process known as graphitizing, by heating in a non-oxidising atmosphere to 2600° to 3000°C.

Table 14.3 Main Characteristics of Graphite as a Lubricant

Advantages
Low friction, in the range 0.05-0.15
Maximum PV about 0.7MN/ms when dry
Good adhesion
Usable at low temperatures and to 540°C in air
Good performance in presence of liquids
Good thermal and electrical conductivity
Disadvantages
Poor performance in vacuum or when very dry
Black, and therefore unacceptable for certain processes
Very complex and variable materials

Since 1950 graphite has been increasingly supplanted for lubricant use by molybdenum disulphide, and in some types of application by PTFE, and the amount of research and development effort applied to graphite has been relatively small. There are three main reasons for this. Probably the most important is the great variability in composition and behaviour of the natural products. Their purity varies from 80% to 90%, although this can be increased in refining to 98%. Even at the highest purity there are variations in crystallinity and behaviour, depending on the original source and the resulting variations in the nature of impurities. Synthetic graphite is more consistent in crystallinity and purity, but still shows some variability depending on the nature of the carbon source.

Graphite is also more limited by temperature and environment. It only gives low friction in the presence of moisture or certain other contaminants, and this is discussed in more detail later. As a result, without special treatment, its use in air is limited to temperatures below 160° or above 350°C and to pressures above about 350 mbars, (35 kPa). Finally, the load-carrying capacity of graphite is significantly lower than that of molybdenum disulphide.

Graphite has several corresponding advantages over molybdenum disulphide which account for its continuing large-scale use. The most important of these is its high electrical conductivity. It also has useful thermal conductivity. A third advantage in some applications is that when heated to over 540°C in air it oxidises to carbon dioxide (or sometimes carbon monoxide) leaving no solid residue.

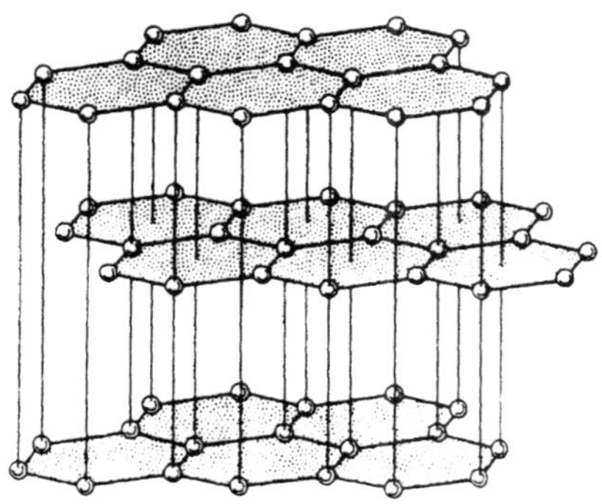

Figure 14.2 Crystal Structure of Graphite

The crystal form is a hexagonal lamellar structure, as shown in Figure 14.2. This has a general resemblance to that of molybdenum disulphide and the other transition metal dichalcogenides, but the main difference lies in the fact that each layer consists of a repeated hexagonal arrangement of carbon atoms, whereas in molybdenum disulphide and its analogues the layers consist of molecules, or assemblies of two different atoms.

The bonding between carbon atoms within a layer consists of strong covalent chemical (electronic) bonds, and this provides the high structural strength. The bonding between layers was originally assumed to consist only of weak van der

Waals forces. The inter-layer separation is much greater than the intra-layer separation of the carbon atoms. This led Bragg in 1928 to suggest[519] that the cohesive energy between layers would be low and that this would permit easy inter-layer sliding and the observed low friction. This theory became suspect after it was recognised in the nineteen-forties[520] that graphite exhibited low friction only in the presence of moisture or other adsorbed vapours. In 1952 Brennan[521] calculated the interlayer energy from the interaction of the molecular orbitals, regarding each layer as a single large molecule. He found in considering the overlapping influences of the Π-electrons that in fact the first order interaction was strongly attractive.

Bryant et al[522] subsequently showed that in vacuum the interlayer bond strength was from six to ten times as high as values reported in air. When substances such as water vapour or oxygen are present, they prevent Π-electron bonding, and the interlayer forces are reduced effectively to van der Waals forces. When those substances are removed, as in vacuum, the result is strong interlayer bonding. Thus graphite is not inherently a good lubricant, but becomes one when certain contaminants are intercalated between the carbon layers. Water vapour is very effective in this situation, but water and other volatile compounds cease to be effective at temperatures above about 160°C. There has therefore been a great deal of effort to identify alternative intercalates to improve lubrication over a wide temperature range.

Several metal oxides and salts have been reported[523] to be effective over the temperature range from 65° to 550°C, but there is little evidence that such intercalated compounds have any advantages over molybdenum disulphide, and they do not seem to have been used to any significant extent in practical applications. Above 350°C graphite again gives low friction in contact with certain metal surfaces, and this is believed to be caused by interaction with surface oxides. Friction starts to increase again above 540°C when graphite begins to oxidise, but it can be used for short periods at higher temperatures because the oxidation products are gaseous and it leaves no undesirable residues. It has therefore been used to lubricate many hot metalworking processes, but with steels at high temperatures it can cause carburization and embrittlement.

Graphite has been used in most of the forms which have been described for molybdenum disulphide. It is most widely used in carbon brushes, where it provides effective lubrication of the brush-commutator contacts, and in a variety of dispersions, such as those listed in Table 14.4.

Table 14.4 Some Graphite-Containing Dispersions

Liquid	Graphite Concentration %	Applications
Water	20-30	Mould lubricant (release agent), tool lubricant, rubber lubricant, electrically conducting coating
Mineral oil	10 35-40	Mould and tool lubricant Antiseize, metalforming
Castor oil	10	Mould lubricant for natural rubber
Isopropyl alcohol	10-20	Dry film mould lubricant, anti-seize, electrically-conducting coating
Polyglycol	10-20	High temperature lubricant

14.4 GRAPHITE FLUORIDE

Although it is formally described as poly(carbon monofluoride), graphite fluoride is a much more variable material than that name implies, and the name graphite fluoride is probably a more accurate description of it. Its general formula is $(CF_x)_n$, and x can vary from about 0.25 to 1.15. It is prepared by direct fluorination of graphite at elevated temperature, usually between about 420° and 550°C, although fluorination takes place at any temperature between 315° and 900°C[524]. The composition of the product varies with the reaction temperature and the fluorine pressure. The general crystal structure is shown in Figure 14.3.

The lubricating and other properties vary with the composition. The friction decreases with increasing fluorine content up to a content of roughly x = 0.6 and then remains constant, and friction values have been quoted as low as 0.02[525]. The film-forming properties also improve with increasing fluorine content up to a level of x = 1.0. Graphite fluoride appears to be an intrinsic lubricant, and unlike graphite it is effective in vacuum[526] and dry argon[525]. Reports of its maximum operating temperature vary enormously, from 200°C to over 500°C. The reasons for this seem

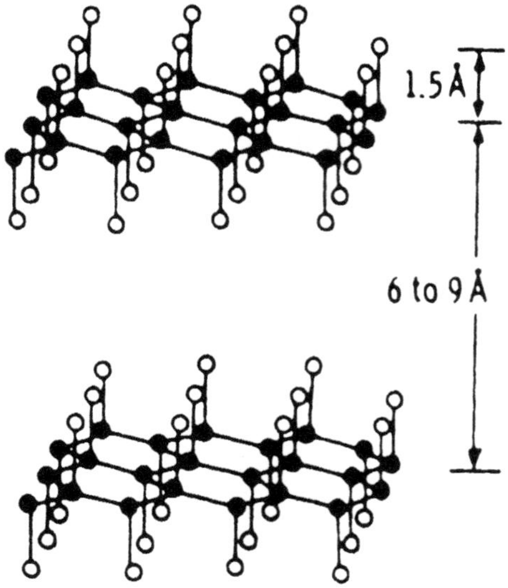

Figure 14.3 Crystal Structure of Graphite Fluoride (Ref.525)

not to have been firmly established, although the fluorine content appears to have some influence on the thermal stability.

Fusaro[527] studied the mechanisms of lubrication and failure of rubbed films of graphite fluoride in comparison with molybdenum disulphide. He found that in general the lubrication mechanisms were similar, but graphite fluoride showed a greater tendency to flow under stress. This was an advantage in the early stages of operation, as it facilitated the formation of a satisfactory lubricating film. Subsequently it also led to flow away from the sliding contact, and ultimate failure took place due to the resulting depletion of the film. On the other hand, Atkinson and Waghorne[528] reported that graphite fluoride films failed by chemical decomposition, so both mechanisms can presumably occur under different conditions.

Fusaro[527] found that graphite fluoride was less adversely affected by moisture than molybdenum disulphide. This was confirmed by Kinner et al[132], whose results are shown in Table 14.5. However, these results also show a considerable increase in wear life for graphite fluoride in dry nitrogen compared with dry air, and this tends to confirm the findings of Atkinson and Waghorne[528] that chemical decomposition is a factor in film failure.

Table 14.5 Effect of Gaseous Environment on the Wear Lives of Molybdenum Disulphide and Graphite Fluoride Films (Ref.132)

Conditions	Wear Life (minutes)			
	Molybdenum disulphide		Graphite fluoride	
	Room temperature	200°C	Room temperature	200°C
S96 annulus on S96 disc				
Laboratory air	300	51	305	86
Wet air	135	83		
Dry nitrogen	1350	185		
Wet nitrogen	>1500	510		
Three-line contact, EN31 riders on S96 disc				
Laboratory air	680	72	705	86
Dry air	300		840	
Wet air	150		800	
Dry nitrogen	1750		7000	
Sliding speed 0.2m/s, load 340N				

Kinner's results for tests at atmospheric pressure tend to indicate that there is no consistent advantage in wear life for graphite fluoride compared with molybdenum disulphide, but he reported that R D Arnell had found in unpublished investigations that the wear life of burnished graphite fluoride films was considerably reduced in high vacuum.

Graphite fluoride has been tested satisfactorily as a rubbed film or bonded film[529], in polymeric[526] and metal[530] composites, and as a dispersion in oils. In general, it is superior to graphite in many respects, but its advantages over molybdenum disulphide are slight. As a synthetic material, it is more expensive, and in spite of many promising early reports, its long-term impact so far has been relatively small.

14.5 TRANSITION METAL DICHALCOGENIDES

Apart from molybdenum disulphide, the only transition metal dichalcogenides which have shown real promise for lubricant use are the disulphides of tungsten and niobium and the diselenides of molybdenum, tungsten and niobium. Their properties are similar in many respects to those of molybdenum disulphide, but there are some differences which affect their suitability as lubricants. Most publications state that none of these five compounds occurs naturally, but in fact small quantities of hexagonal tungsten disulphide do occur naturally, and Graesar[61] reports hardness measurements on natural hexagonal tungsten disulphide from the Italian Alps. However, this material was found to contain 6.0% of molybdenum, equivalent to about 10% of molybdenum disulphide, and separation of this from the tungsten disulphide would almost certainly be uneconomical, if not technically impossible.

It is therefore probably true to say that the only practical source of any of the five compounds is by synthesis, with the result that they are all more expensive than molybdenum disulphide. The synthesis in each case is by heating the powdered metal with sulphur or selenium. The product in most cases is non-stoichiometric and the crystals which form initially from the walls of the reaction vessel do not have the desired hexagonal lamellar structure. The hexagonal crystal structure can usually be obtained by "annealing" or holding the product at 200°C or more in an inert atmosphere or vacuum.

Their crystal structures have been mentioned briefly in connection with intercalation in Section 14.2. All five compounds can be obtained in the layered hexagonal crystal form, and most are also found in rhombohedral or trigonal form. The compounds of the Group 6 metals, molybdenum and tungsten, as well as niobium diselenide, have a hexagonal form similar to that of molybdenum disulphide, in which the metal atoms in one layer are displaced sideways from those in the layers immediately above and below. This structure results in the widest interlamellar spacing, the easiest interlamellar shear, and the lowest friction.

However, pure niobium disulphide has a hexagonal structure in which the metal atoms in each layer are located directly above or below those in adjacent layers. Jamison[67] showed that with this configuration niobium disulphide is not a good lubricant. He postulated that when niobium disulphide behaves as a good lubricant, additional niobium atoms intercalated into the structure will have resulted in a change in electron bonding to favour the molybdenum disulphide structure.

Compound	In air					Vacuum		
	Ref. 531	Ref. 514	Ref. 73	Ref. 74		Ref. 84	Ref. 532	Film formation
Molybdenum disulphide	0.05	0.04-0.045	0.18	0.23	0.17	0.45	0.07	Good
Molybdenum diselenide	0,07	0.057	0.17	0.22	0.17	0.45		
Tungsten disulphide	0.10	0.051-0.053	0.17	0.17	0.15	0.45		Good
Tungsten diselenide	0.05	0.037-0.047	0.09	0.17	0.10	0.04	0.15	Good
Niobium disulphide			0.08	0.16	0.07			Poor
Niobium diselenide	0.29	0.058-0.075	0.12	0.17	0.17			Weak
Test Conditions	Slow speed press fit test	Ball on flat, 2m/s, 3-500g load	Flat/flat 0.25 m/s, 6.9×10^5Pa	Compact on flat: .04m/s, 5.5×10^5 Pa	Compact on flat: 0.36m/s 11.7×10^5Pa	Sphere on flat, 10^{-6} to 10^{-8} Torr	Flat on copper, <0.75 m/s	

Table 14.6 Some Reported Coefficients of Friction for Transition Metal Dichalcogenides

Subject to this complication, the friction properties of the group are generally similar to those of molybdenum disulphide. The actual coefficients of friction vary with load, speed, temperature and humidity, but some reported figures are shown in Table 14.6. The chemical properties are also similar to those of molybdenum disulphide. They are resistant to attack by water, alkalis and most acids, but are attacked by aqua regia and hot concentrated hydrochloric, nitric or sulphuric acids. The most significant differences are in their electrical conductivity and their oxidation resistance.

Table 14.7 Limiting Temperatures for Dichalcogenides in Air and Vacuum

Compound	Decomposition Temperatures (°C)					
	In air			In vacuum		
	Ref. 73	Ref. 534	Ref. 139	Ref. 73	Ref. 84	Ref. 535
Molybdenum disulphide	350	435	350	1350	1090	930
Molybdenum diselenide	400	479		1350	980	760
Tungsten disulphide	440	485	400	1350	1040	870
Tungsten diselenide	350	540		1350	930	705
Niobium disulphide	420	449		1050		
Niobium diselenide	350	410		1350		

Table 14.7 shows some reported limiting temperatures in air and in vacuum. The oxidation temperatures quoted are, as always, rather arbitrary and variable because the oxidation is a multiphase process and is strongly affected by the particle size of the solid and the pressure and circulation of air or oxygen. However, the enormous variation in the decomposition temperatures in vacuum for the first four compounds is impossible to explain, and the best that can be said of them is that the order of ranking is similar.

Tungsten disulphide has been generally reported as having better oxidation resistance than molybdenum disulphide, and this is confirmed by the limiting temperatures listed in Table 14.7. The higher oxidation resistance is thought to be due to the tungsten dioxide which is formed initially providing better protection against further oxidation than that given by molybdic oxide. This argument would presumably apply equally to the two diselenides.

Sliney[533] indicated that the comparison between tungsten disulphide and molybdenum disulphide is more complicated. He showed that both compounds oxidise at temperatures as low as 300°C, and that below 340°C tungsten disulphide oxidises more rapidly than molybdenum disulphide. The oxidation rates in that temperature range are very low, and with loosely-compacted powders of 1μm average particle size the time required to oxidise 50% of the materials at 300°C was approximately one to two weeks. Above 340°C molybdenum disulphide oxidises more rapidly than tungsten disulphide, and it is at these higher temperatures that the relative oxidation resistance is more important. The relationships are shown in Figure 14.4.

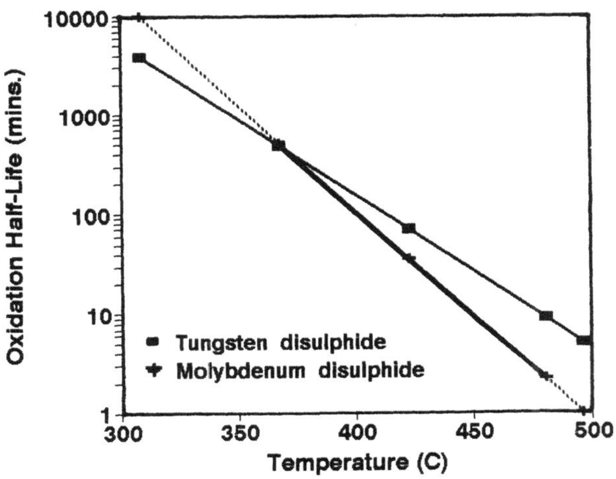

Figure 14.4 Oxidation Rates of Molybdenum Disulphide and Tungsten Disulphide (Ref.533)

The changes in friction and wear behaviour with change in temperature in air are presumably determined by oxidation. Figure 14.5 shows that at low temperatures both molybdenum disulphide and tungsten disulphide have similar coefficients of friction of about 0.07 to 0.08, but a marked increase in friction occurs at 400°C with molybdenum disulphide and 600°C with tungsten disulphide. The corresponding temperatures for increasing friction reported by Tsuya[136] were 300°C and 450°C.

The effective temperatures in the sliding contact will of course, be significantly affected by frictional heating and therefore by sliding speed and load, but the important conclusion is that the increase in friction with rising temperature occurs at a temperature at least 100°C higher with tungsten disulphide than with molybdenum disulphide.

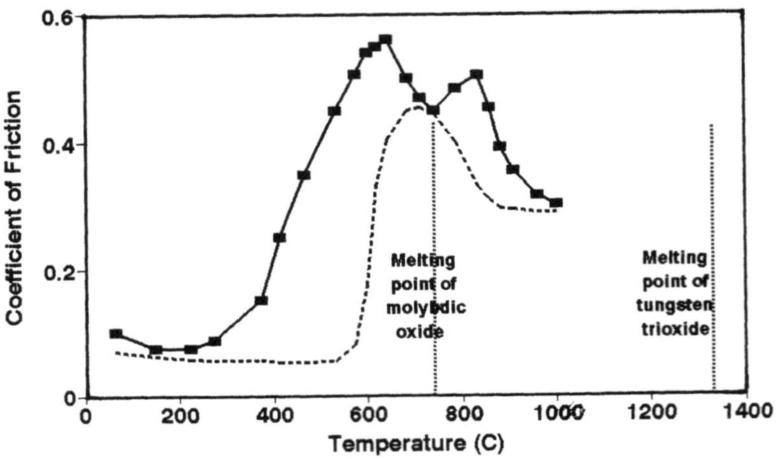

Figure 14.5 Variation of Friction with Temperature for Molybdenum Disulphide and Tungsten Disulphide in Air (Ref.533)

Both Tsuya[136] and Ducas[21] found that in moist air the friction of tungsten diselenide rose more than that of molybdenum disulphide. More surprisingly Ducas found that the friction of other dichalcogenides fell in moist air. His results are listed in Table 14.8.

The higher possible operating temperature with tungsten disulphide, due to its greater oxidation resistance, and the resulting maintenance of low friction to higher temperature, have been the main reasons for its use in practical engineering applications. For example, Radcliffe and Parry[536] reported in 1979 that tungsten disulphide, having better oxidation characteristics than molybdenum disulphide, "is an extremely important lubricant in current British gas-cooled nuclear reactors". Similar studies of the other dichalcogenides have given less consistent results, and none of them has been used to any significant extent because of any reported advantage in high temperature performance.

Table 14.8 Change in Friction of Dichalcogenides Tested in a Steam Atmosphere (Data from Ref.21)

Material	Change in friction when tested in steam
Molybdenum disulphide	+76%
Molybdenum diselenide	+122%
Tungsten disulphide	-48%
Tungsten diselenide	-40%
Niobium diselenide	-46%

The film-forming properties of the different dichalcogenides also vary considerably, and were mentioned briefly in the description of intercalation in Section 14.2. Salomon et al[534] rated molybdenum disulphide, tungsten disulphide and tungsten diselenide as being superior to both molybdenum diselenide and niobium diselenide in forming burnished films. They reported that the two latter compounds failed to burnish. On the other hand, Giltrow[248] studied the endurance of burnished films of several dichalcogenides on mild steel and tool steel, and some of his results are shown in Table 14.9. He obviously succeeded in producing burnished films of molybdenum diselenide and niobium diselenide. However, their endurance was inferior to the other materials, and to that extent his ratings confirmed those of Salomon.

The other major difference between the various synthetic dichalcogenides lies in their electrical conductivities, as shown in Table 14.1. These figures should be considered relative rather than absolute, since values quoted by different investigators have differed by factors of over two hundred[537]. Overall the lowest resistivity is that of niobium diselenide, and this has led to many investigations of its potential for use in situations, such as high vacuum, where graphite cannot be used. In brush compositions, however, molybdenum disulphide has generally been more successful, and this subject has been considered in more detail in Chapter 12.

The poor film-forming properties of some of the synthetic dichalcogenides can of course be overcome by incorporation in bonded films, but apart from tungsten disulphide there seems to be little incentive for doing so. Stupp[562] compared the performances [24] of tungsten disulphide and molybdenum disulphide in phenolic-bonded films at room temperature, and found very little difference between them, but he apparently did not study any of the other dichalcogenides. Bonded films containing tungsten disulphide in an inorganic binder are also available commercially.

Table 14.9 Endurance of Some Burnished Dichalcogenide Films (Based on data from Ref.267)

Dichalcogenide	Endurance (revs.)	
	On mild steel	On tool steel
Molybdenum disulphide	3.7×10^3	3.8×10^3
Molybdenum diselenide	7.2×10^2	4
Tungsten disulphide	2.6×10^3	2.6×10^3
Tungsten diselenide	2.8×10^3	3.5×10^3
Niobium disulphide	30	9
Niobium diselenide	74	20

Spalvins[274,538] has reported that molybdenum diselenide, tungsten disulphide and niobium diselenide have all been sputtered satisfactorily. Sputtered tungsten disulphide performed well in vacuum, with a coefficient of friction of 0.05 and a life of over a million test cycles. In air the coefficient of friction rose to 0.22 and film failure occurred rapidly. The behaviour was therefore very similar to that of molybdenum disulphide. Bergman et al[539] later reported the results of sputtering a series of transition metal dichalcogenides onto steel surfaces. They found that only molybdenum diselenide had a fine crystalline structure, the others all having a turbostatic or rag structure. These included tungsten disulphide and niobium diselenide, so presumably their results differed from those of Spalvins because of differences in the sputtering conditions. Their results for niobium diselenide and tungsten disulphide tested on a pin-on-disc machine show early friction peaks suggesting that considerable energy was required to re-orient the films to a low-friction configuration.

Kirner[540] reported the successful use of plasma spraying for niobium diselenide, niobium ditelluride, and a mixture of tungsten disulphide and silver, but the performances in high vacuum and high temperature were inferior to those obtained with molybdenum disulphide. There has been a great deal of Russian work on the transition metal dichalcogenides, mainly produced by the Self-Propagating High Temperature Synthesis procedure described in Chapter 2[37,541]. There have also been descriptions of *in situ* coatings[253,542], as an alternative method of obtaining improved film formation, but most of this work showed that the performance was inferior to that of natural molybdenum disulphide.

Davis and Presland[543] reported the formation of a surface film of tungsten disulphide on the worn surfaces of the steel balls when four-ball testing was carried out with an oil containing a long-chain tertiary alkyl primary amine tungstate and a sulphur additive such as zinc dialkyl dithiophosphate. Thus every technique which has been used to produce molybdenum disulphide films has also been used for the other lubricating dichalcogenides, but with the possible exception of tungsten disulphide, the extent of practical application in film form seems to have been very limited.

The poor film-forming properties of some of the synthetics, and the occasional reports of inferior film performance even for the best of them, are presumably less critical when they are incorporated in composites. The ability to form a satisfactory transfer film on a counterface would still be important, but there seem to have been no fundamental studies of this aspect of their performance. Whatever the reason, published papers indicate that far more effort, both research and commercial, has been devoted to the development and use of the synthetic dichalcogenides in composites than in film form.

As early as 1971 the Bemol company in the United States supplied data on sixteen different available composites of the synthetic dichalcogenides, and these are listed in Table 14.10, together with four molybdenum disulphide composites for comparison. It is interesting to analyse the performance figures quoted, which are all apparently determined experimentally. The friction values show no major differences between the various dichalcogenides, although it is significant that the poorest compound, niobium disulphide, is not used in any of the Bemol products. Otherwise the lowest friction occurs with the highest concentration of the lubricant component.

The temperature limits are more curious. There seems to be no evidence from other published literature for the much higher temperature performance in vacuum of the niobium diselenide and tungsten diselenide than tungsten disulphide, and the very inferior performance of molybdenum disulphide. Similarly, there is no obvious reason for the very low limit in air of 180°C for tungsten diselenide in silver, or for the low limit of 260°C for molybdenum disulphide in either silver or nickel.

Table 14.10 Properties of Some Dichalcogenide Composites (Data supplied by Bemol Inc.,)

Composition wt. %	Density (g/cm^2)	Coefficient of friction	Max. service temperature, °C	
			Air	Vacuum
Ag 90;$MoSe_2$ 10	8.82-9.31	0.10-0.18	350	400
Ag 85;$NbSe_2$ 15	8.57-9.04	0.08-0.17	350	400
Ag 80;$Nbse_2$ 20	8.32-8.78	0.07-0.15	350	400
Ag 70;$NbSe_2$ 30	7.85-8.28	0.06-0.14	350	400
Ag 80;$NbSe_2$ 15;C 5	7.35-7.76	0.08-0.17	350	
Ag 90;MoS_2 10	8.45-8.92	0.10-0.19	260	400
Ag 80;MoS_2 20	7.63-8.06	0.08-0.18	260	400
Ag 85;WSe_2 15	9.22-9.73	0.10-0.18	180	400
Ag 75;WSe_2 25	9.07-9.57	0.07-0.15	180	400
Ag 70;$NbSe_2$ 20;PTFE 10	6.14-6.48	0.06-0.15	260	260
Ag 70;WSe_2 20;PTFE 10	5.29-5.59	0.06-0.15	260	260
Ni 90;WS_2 10	7.87-8.31	0.12-0.20	350	540
Ni 80;WS_2 20	7.72-8.14	0.08-0.15	350	540
Ni 65;WS_2 35	7.52-7.94	0.05-0.12	350	540
Ni 90;$NbSe_2$ 10	7.69-8.12	0.12-0.20	350	650
Ni 80;$NbSe_2$ 20	7.38-7.79	0.08-0.18	350	650
Ni 85;WSe_2 15	8.02-8.46	0.11-0.20	350	650
Ni 75;WSe_2 25	8.04-8.48	0.09-0.17	350	650
Ni 90;MoS_2 10	7.24-7.64	0.08-0.15	260	400
Ni 85;MoS_2 15	7.10-7.50	0.07-0.15	260	400

Twelve other representative composites described in research publications are summarised in Table 14.11. There are also many variations in the relative concentrations of the various components within these twelve types, and most represent the final selection from a wide range of compositions which were investigated.

One of the earliest studies was by Boes and Bowen[544]. They incorporated PTFE in their composites to improve transfer and film-forming behaviour, and a metal binder, or matrix, to provide adequate structural strength. They found that the optimum ratio of PTFE to dichalcogenide was 3:1 by volume. A higher proportion of PTFE resulted in significantly increased wear rate. They also found that silver was generally better than copper, allowing higher metal content and giving lower friction.

Table 14.11 Some Composites of Synthetic Dichalcogenides

Composition		Coefficient of friction	Refs.
Dichalcogenide	Matrix materials		
Molybdenum diselenide	Silver, PTFE		425,544,548
Tungsten disulphide 22%	Copper		83
Tungsten disulphide 22%	Silver		83
Tungsten disulphide 20%	Copper, tin		136
Tungsten disulphide 20%	Copper, silver		136
Tungsten disulphide 52.9%	Cobalt, silver	0.2	397,402,549
Tungsten diselenide 7.5%	Silver 70%, PTFE	0.1	544
Tungsten diselenide	Gallium/indium eutectic	0.15 to 200°C	546,547
Tungsten diselenide	Gallium/indium eutectic, carbon fibre/polyimide		193
Tungsten diselenide	Silver, PTFE		212,398,544
Tungsten diselenide	Silver, polyimide		545
Niobium diselenide 10%	Copper 60%, PTFE 30%	0.16	544

Bowen[212] tested nine different composites of 10% tungsten diselenide, PTFE and either silver, copper, or a copper bronze in gears and bearings for possible use in a space vacuum application. The composite was used in the form of an idler gear for the gears and as retainers for the 25mm. bore ball bearings. The tests were conducted in a vacuum of 10^{-5} to 10^{-6} Pa (10^{-7} to 10^{-8} Torr), including one test of 100 hours in which the temperature was -100°C for 40 hours and then rose slowly to 149°C for 40 hours. Most of the tests survived for 100 hours, but one premature failure occurred in which one idler gear consisted only of 7% tungsten diselenide and 93% PTFE with no metal matrix.

The composite of 52.9% tungsten disulphide in a cobalt and silver matrix, described as AF-SL 14, was one of a series of composites developed by the USAF, and was tested satisfactorily by several investigators[548,549].

Boes et al[425] chose a composite of molybdenum diselenide with PTFE and silver for testing for the retainers of ball bearings which operated successfully at 316°C (600°F). Most other investigators concentrated on tungsten disulphide and tungsten diselenide because of their better high temperature performance, although Rempe[545] also found a composite of tungsten diselenide and silver in polyimide satisfactory for ball bearing retainers in liquid hydrogen.

Boes[74,546] also developed a new type of compact using tungsten diselenide as the lubricating component. The tungsten diselenide in powder form was impregnated with liquid gallium or liquid gallium/indium eutectic, and Boes found that the powder would absorb up to 30% of the liquid metal. The so-called "amalgam" was then compacted in a die at room temperature and up to 350 MPa applied pressure. The result was a hard non-porous compact which could be machined, drilled or threaded and showed very little deterioration at temperatures up to 800°C in air. The compact was subsequently used successfully for retainers for 20 mm and 35 mm bore ball bearings operating at over 10,000 rpm in air at 316°C, and for periods of 35 - 38 hours at 480°C[425,193]. It was also selected[178] after a long-term study as the lubricant component in carbon fibre-reinforced polyimide for use in gas turbine ball bearing retainers. Kiparisov et al[547] studied its performance, and reported a coefficient of friction less than 0.15 up to 200°C.

The important potential applications which have been highlighted for these composites are in electrical brushes for space use and in retainers for rolling bearings, but for both purposes the most widely-used materials are still composites of molybdenum disulphide. Overall it is difficult to sum up the status of the synthetic transition metal dichalcogenides. There have been enthusiastic reports of their performance and use, and certainly individual compounds have specific advantages, namely the oxidation resistance of tungsten disulphide and the electrical conductivity of niobium diselenide. Nevertheless, apart from tungsten disulphide, none has yet formed the basis for any important commercial or technological application, and the inference must be that use of the others is limited to a few special situations.

~ ~ ~ ~ ~ ~ ~ ~ ~ ~ ~ ~ ~ ~ ~ ~

CHAPTER 15

CORROSION AND FRETTING

15.1 BACKGROUND

Probably the most controversial of all subjects associated with molybdenum disulphide has been that of corrosion. After the rapid increase in its application between about 1955 and 1962, there were widespread reports of corrosion arising from its use. In Europe thia was serious enough to result in a loss of confidence and a bad reputation from which it has probably still not completely recovered almost forty years later.

In the United States it was already much more widely used than in Europe, including important applications in military aircraft and in the space programme. As a result there was a greater emphasis on analysing and overcoming the problem, but the association between molybdenum disulphide and corrosion still exists in many people's minds, in spite of the fact that the problem has been understood and controlled for well over twenty years.

It is fair to say that some of the early reports were ill-founded, and magnified by rumour. There were also categorical reports from some users that molybdenum disulphide, even as free powder or burnished films, positively prevented or cured corrosion problems. These must in most cases have been equally ill-founded, although they provide an interesting parallel with the comments of Cramer in 1764, which were quoted in Chapter 1.

One important factor in those early days may well have been the fact that designers were used to dealing with lubricating oils and greases which gave very effective corrosion protection to bearings and adjacent surfaces. They often failed to realise that when the oils and greases were replaced by solid lubricants the corrosion protection was lost. Subsequent operation in corrosive environments, such as marine equipment or vehicle and aircraft wheel assemblies, then found the system inadequately protected, and serious corrosion could occur. Other factors which confused the situation were the variable purity of different molybdenum disulphide supplies, the use of excessive ball milling, which caused oxidation and metal contamination, and the use of sub-micronic powders which were more susceptible to atmospheric oxidation.

Nevertheless, there were reliable reports of corrosion occurring with molybdenum disulphide in films and in greases. Several such reports arose from the US Army[189], originating with a salt fog test of a missile launcher in which all parts coated with solid film lubricant rusted badly. Subsequent reports described galvanic corrosion of various metals with molybdenum disulphide in moist atmospheres.

Laboratory tests confirmed that in the absence of effective corrosion inhibitors molybdenum disulphide could cause corrosion in humid environments. Kay[550] showed in tests with different steels that corrosion was accelerated in the presence of loose molybdenum disulphide powder, especially ball-milled or micronated powder. Her test conditions were realistic, namely 20°C and 90% relative humidity for six days, but the use of loose powder was not representative of practical use, and it subsequently became clear that burnished films were less active in promoting corrosion. Calhoun et al[189] also showed that molybdenum disulphide in a bonded film actively promoted corrosion, although their test conditions were severe, consisting of salt fog and salt spray tests. They found that corrosion was more severe when graphite was present, but that molybdenum disulphide also clearly caused corrosion.

Groszek and Witheridge[225] compared the corrosivity of three molybdenum disulphide powders in conditions similar to those used by Kay. They placed equal quantities of each on part of the surface of an AISI 52100 steel disc and exposed the discs to air saturated with water at 20°C for 120 hours. The three powders were a sample meeting the British military specification DEF-2304, a sample of the same material further ground in a soft vacuum, and referred to as "polar MoS_2", and a sample of oleophilic powder ground in n-heptane. They found that the "polar" powder produced very heavy corrosion, but the other two powders gave far less corrosion. The difference was considered to be due to the far higher proportion of edge sites in the "polar" material. This would be logical in view of the greater reactivity of edge sites compared with the basal planes of the crystals.

Apart from the corrosivity of the molybdenum disulphide itself, molybdenum disulphide films give relatively little physical protection to metal surfaces. Burnished films are fairly permeable, and the crystal structure is itself permeable to vapours and gases. Gresham[316] reported that in general bonded coatings with high proportions of binder gave excellent corrosion resistance, but other reports, such as that of Calhoun et al[189] conflict with this view. Even bonded films, in spite of their superficial resemblance to paint films, are far less effective in preventing access of corrosive substances, for two reasons. In the first place they are formulated primarily for lubrication, whereas paints are formulated for protection as well as decoration. In the second place, bonded molybdenum disulphide films are ideally only about 4 to 10 μm thick, compared with 50 to 200 μm for paint. In fact Gresham suggested that for optimum corrosion resistance, a thickness of 12.5 to 50 μm might be needed, but obviously this may conflict with the optimum thickness for lubrication in certain circumstances, such as under high loads.

It follows that if a corrosive environment is present, it seems unlikely to be physically prevented from reaching the surface of the substrate. DeLaat[551] reported cases of serious corrosion with bonded films, although in some at least of these the films had not been properly used. On the other hand, Peterson and Finkin[343] carried out a very thorough survey of US Naval Air Rework Facilities in 1970-71. At that time the system would certainly have still included coatings of older formulations, without improved corrosion inhibition, but they found very few definite cases of corrosion, and in several of those the use of bonded films was probably not responsible.

Gabel and Peterson[552] later concluded that most bonded films provide some degree of corrosion protection as compared with unprotected metal surfaces. Once films have been burnished or run in, the degree of protection is very much reduced, although the fact that burnished films present a high proportion of inert basal surface to the atmosphere presumably explains the fact that such films experience less corrosion than loose powder.

15.2 THE CHEMICAL ENVIRONMENT

The complex possible reactions of water and oxygen with molybdenum disulphide have been discussed in Chapter 4, but as Lancaster[102] pointed out, the oxidation mechanism which is thermodynamically favoured is the one which leads to the production of sulphuric acid. Other proposed mechanisms can yield sulphur dioxide or hydrogen sulphide. There is therefore clearly a potential for corrosion by the normal products of oxidation, which is itself promoted by the presence of moisture. For this reason molybdenum disulphide powder is sometimes stored in an inert gas atmosphere or treated with oxidation inhibitors, and some specifications include pH limits on aqueous extracts.

Even in the virtual absence of moisture, molybdenum disulphide can react with iron or other metals to produce the corresponding sulphide, and Heinicke[553] has reviewed the tribochemical activation of such reactions. Other workers have described similar reactions in the presence of moisture[47] or heat, and the further reaction in the presence of moisture or especially dilute acids to produce hydrogen sulphide. Gänsheimer has reviewed such reactions[142], and especially their activation in sliding contacts.

The formation of iron sulphide, or any other metal sulphide, and subsequent hydrolysis to release hydrogen sulphide, represents a corrosion process. The various oxidation processes discussed all involve the production of hydrogen sulphide, sulphur dioxide or sulphuric acid. In the absence of effective protection, any one of these is a potential corrodent, especially in association with any wear which takes place.

Apart from simple corrosive attack on a metal surface, Gabel[337] reported a more specific situation leading to corrosion. This occurred when a dry-film-lubricated metal surface was in contact with an unlubricated anodized aluminium surface. Laboratory investigation at 95% relative humidity and 49°C (120°F) confirmed that in this situation the corrosion resistance of the anodized aluminium was reduced from 1100 hours to fewer than 200 hours. It was not clear whether dissimilar metals were involved, but the occurrence suggested that corrosion might be due to the development of an external potential in the presence of an electrolyte.

15.3 CORROSION PROTECTION

Several techniques were investigated in order to reduce the extent of corrosion problems. They fell into two categories, either to protect metal surfaces against attack or to incorporate inhibitors in the lubricant formulation.

Several different pre-treatments were reported to give improved corrosion resistance to metal substrates. Calhoun et al[189] reported that zinc phosphating alone or followed by zinc plating or cadmium plating gave a marked improvement in corrosion resistance of steel. The best results were obtained with zinc phosphating and cadmium plating, but unfortunately this gave the shortest wear life. The longest wear life was obtained with zinc phosphating alone, and this gave a useful improvement in corrosion protection, but much less than plating. Black[554] reported that a bonded molybdenum disulphide coating with a ferro-manganese phosphate pre-treatment gave better corrosion resistance to steel fasteners than was given by galvanising, although presumably much of the protection was given by the phosphating rather than the bonded coating.

The biggest single improvement in formulation was almost certainly the virtual elimination of graphite from bonded coatings and composites. Both Calhoun et al[189] and Gabel[316] had found that removal of graphite brought about a marked reduction in corrosion. Unfortunately there was also a significant reduction in wear life, but this was restored when the graphite was replaced with antimony trioxide and there was no corresponding increase in corrosion.

Gabel[337] also investigated the use of corrosion inhibitors. She found that antimony trioxide and dibasic lead phosphite appeared to have a synergistic effect in reducing corrosion with a phenolic-bonded coating. She also found that sodium nitrate, which acts as an oxidising passivator on steel, was effective, but less so than dibasic lead phosphite.

Niederhäuser et al[290] used the alternative approach of coating the molybdenum disulphide in a sputtered coating with PTFE to protect it from attack by moisture. They found that there was some reduction in corrosion of a steel substrate, although a greater reduction was obtained when the coating was sputtered onto a rhodium interlayer. Zhai et al[555] also reported some improvement in the stability of a composite plating when cerium was incorporated to coat the molybdenum disulphide and reduce moisture attack, but they did not directly assess corrosion resistance.

The benefits of incorporating a corrosion inhibitor were shown by a report in 1980[556], which showed that coatings to the US specifications MIL-L-8937C and MIL-L-23398, which contained corrosion inhibitors, survived the four test cycles in salt spray cabinet and sulphurous acid corrosion testing, whereas a MIL-L-81329 coating without a corrosion inhibitor survived only one cycle.

The degree of improvement in a few years is clearly indicated by the marked reduction in the number of publications about corrosion after **1976**. With the elimination of graphite and of excessive ball-milling, and the incorporation of corrosion inhibitors, it should now be possible to use molybdenum disulphide in bonded films, composites, greases or anti-seizes with confidence that no corrosion problem will arise, although commonsence precautions should always be taken:

(i) Where possible, manufacturers' data sheets should be examined to ensure that corrosion inhibitors are used, or that the results of corrosion testing are quoted.
(ii) If the presence of significant moisture, or more corrosive materials such as acids or brines, is unavoidable, use corrosion-resistant substrate materials as much as possible.
(iii) In very critical situations, test for corrosion under simulated service conditions.

All these are sensible precautions, whether molybdenum disulphide is used or not.

There is very little information available about corrosion risks with the other lubricating dichalcogenides. Spalvins[274] mentioned that the friction and endurance of a sputtered tungsten disulphide film deteriorated considerably when tested in the atmosphere instead of in vacuum, and the deterioration was associated with formation of sulphuric acid and corrosion of the substrate. These results are not very different from those obtained with molybdenum disulphide in the same period, and there seems to be no reason to expect the problem to be any greater or any less with the other dichalcogenides.

15.4 FRETTING

Fretting is of course not a form of corrosion, but it is often confused with corrosion, and can sometimes be associated with serious corrosion. It is in fact a form of oxidative wear which occurs when solid surfaces, usually metallic, rub together in a low-amplitude oscillation. The damage can be very severe. With metals it is characterized by pitting which resembles corrosion, and with ferrous metals there is a characteristic production of red ferric oxide.

When fretting occurs in an environment which is potentially corrosive, such as acidic or moist atmospheres, quite severe corrosion can often result. This is encouraged by the removal of protective oxide surface films, and by the production of surface pitting by the fretting process. The resulting corrosion is often referred to as fretting corrosion, although the term is often misunderstood and misused.

The occurrence of fretting is strongly dependent on the amplitude of the relative motion, and its severity depends on the frequency of the oscillation and the contact pressure. Since in most fretting situations one surface is driving and the other driven, the amplitude of the relative movement depends to some extent on the friction in the contact. In general, lower friction often leads to higher amplitude, and in different cases this can either take the amplitude into or out of the regime in which fretting is most severe. The lubricant can also prevent metal-to-metal contact, and can change the contact temperature and affect oxygen access, both of which will change the rate of oxidation.

Overall, the effect of any lubricant in a specific fretting situation tends to be rather unpredictable, and this is true of molybdenum disulphide. In a liquid or grease it has been reported[556] not to be particularly effective in reducing fretting damage, but this is also likely to depend on other parameters of the system. On the other hand, bonded films are likely to prevent direct metal-to-metal contact and reduce oxygen access as well as reducing friction, and in one set of tests bonded films were reported[557] to have produced a million-fold improvement in fretting life compared with a 2000-fold improvement with molybdenum disulphide powder. However, as an illustration of the unpredictability of lubricant performance in fretting situations, in another set of tests free powder was shown[558] to give comparable improvement to the best bonded films, while Müller[223] found free powder to be ineffective except when used together with tricresyl phosphate.

~ ~ ~ ~ ~ ~ ~ ~ ~ ~ ~ ~ ~ ~ ~ ~

CHAPTER 16

SELECTION AND USE

In certain respects this final chapter will serve as a summary of practical aspects described in previous chapters. In addition it seems desirable to put the whole subject of molybdenum disulphide lubrication into its proper context by trying to indicate when it's desirable to use solid lubricants, when molybdenum disulphide is the best choice, and how to use it.

16.1 SELECTING THE CLASS OF LUBRICANT

Oil has been the first choice for lubrication for hundreds of years, and seems likely to continue as first choice for the foreseeable future. The second choice is grease, and that too is likely to continue virtually unchallenged. It is only when these two classes run into limitations that other possible forms of lubricant are normally considered.

Perhaps the most obvious problem which counts against oils or greases is the need to operate in hostile environments. These include:

16.1.1 **Very High Temperatures:** The commoner oils and greases are limited to maximum temperatures in air of perhaps 160° to 180°C for any extended operation, and to 240°C for very brief periods. More exotic (and expensive) liquids such as silicones, polyphenylethers and perfluoropolyethers can be used in turn up to about 350° - 400°C, but they are poor lubricants in boundary conditions. In inert atmospheres they can be used to higher temperatures, but in vacuum or eventually in inert gases they are limited by high vapour pressure. At higher temperatures still it is theoretically possible to use liquid metals or liquid oxides or salts, but these are also poor lubricants and present severe handling difficulties. Solid lubricants become serious contenders for use at any temperature above about 180°C.

16.1.2 **Very Low Temperatures:** Conventional oils and greases can be used down to perhaps -75°C. At lower temperatures liquid lubrication can be provided by cryogenic liquids down almost to absolute zero (-273°C), but their viscosities are extremely low, and they are technically difficult to handle. Solid lubricants probably become serious contenders at any temperature below -50°C.

16.1.3 **High Vacuum:** With proper shielding oils and greases can be used in high vacuum applications, but solid lubricants are always potential alternatives.

16.1.4 **Nuclear Radiation:** All the more common oils and greases are organic compounds and have limited resistance to nuclear radiation. The same applies to most polymers. Liquid metals have been successfully used in intensive radiation environments. The inorganic solid lubricants have very high resistance to radiation.

16.1.5 **Aggressive Chemical Environments:** Mineral oils and their inorganic-thickened greases can have useful resistance to chemical attack. The possible use of solid lubricants depends on the specific nature of the chemical environment.

Apart from their potential for use in hostile environments, there are also many situations in normal environments in which solid lubricants are useful. Their advantages and disadvantages stem mainly from their lack of flow properties, and are listed in Table 16.1.

Specific load and speed have a major influence on the choice of the class of lubricant, as shown diagrammatically in Figure 16.1. More specific limits of load and speed are shown in Figure 16.2, but even these should be considered only as approximate guidelines. The actual limits will be influenced by many other factors,

Table 16.1 Advantages and Disadvantages of Solid Lubricants

Advantages
1. Practically no tendency to flow, creep or migrate, so that they can be relied on to remain in place for long periods.
2. Little tendency to contaminate adjacent systems or products.
3. Very low volatility makes some of them usable in high vacuum.
4. Usable at very high or very low temperature.
5. Chemical inertness to other chemicals.
6. Generally more stable to radioactivity.
Disadvantages
1. Difficult or impossible to feed or replenish.
2. Finite wear rate limits their life.
3. Poor thermal conductivity limits sliding speed.
4. Thermal expansion often very different from metals, so that clearances can be lost with temperature change.
5. Generally higher friction.
6. More complicated to apply.
7. No coolant properties.
8. No corrosion protection.

such as the type of lubricant in each class, the life required, re-lubrication, and environmental factors such as temperature, pressure and vibration.

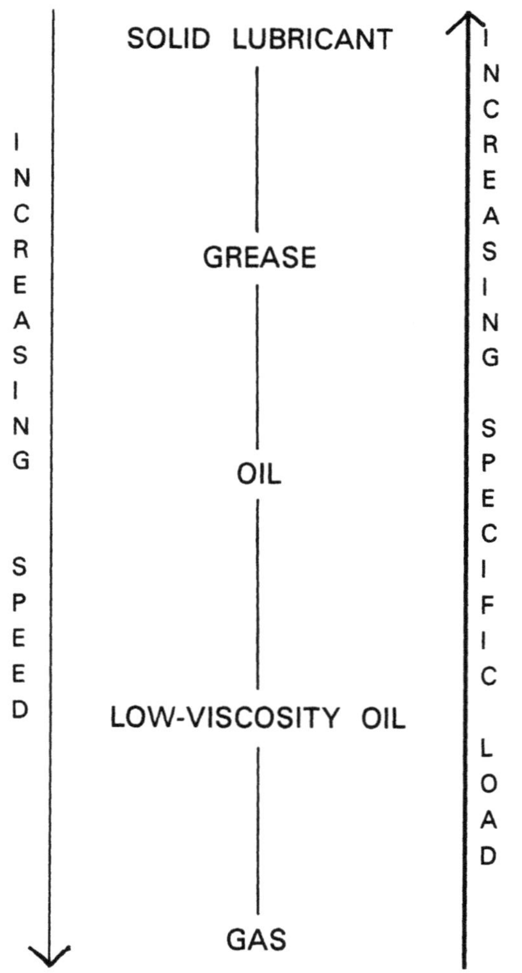

Figure 16.1 Effect of Speed and Load on Choice of Lubricant Type

The maximum speed for using solid lubricants is determined mainly by heat build-up, and this in turn is affected by friction, thermal conductivity and other forms of heat loss. The maximum specific load is determined by three principal factors. One is the ability of the particular lubricant to withstand the applied stresses. The second is the effect of load on friction and heat generation. The third is the effect on wear rate.

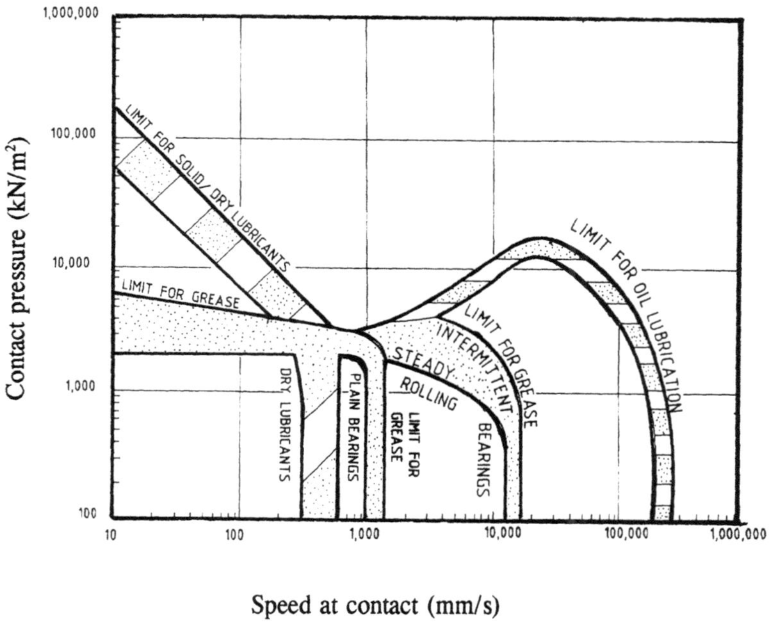

Figure 16.2 Approximate Speed and Load Limits for Different Classes of Lubricant

The load-carrying capacity of solid lubricants is in fact higher than that of any other class of lubricant, and at low speed they can withstand loads up to about 500 - 700 MPa, or the maximum which can be supported by the substrate. There is also a limit on the combination of specific load and sliding speed represented by a PV factor of about 1 to 3.5 MPa.m/s.

A major advantage of solid lubricants is their very low tendency to cause contamination of industrial products such as foods, pharmaceuticals and textiles. Many of them, including PTFE, graphite and molybdenum disulphide, are also virtually non-toxic, so that they do not constitute a hazard in small concentrations in food. On the other hand, the dark colour of graphite and molybdenum disulphide makes them unattractive for use in some industries, because even very small amounts of contamination are visually unacceptable. (It should in any case be noted that lubricants for use in the food and drugs industries are generally strictly controlled by law.)

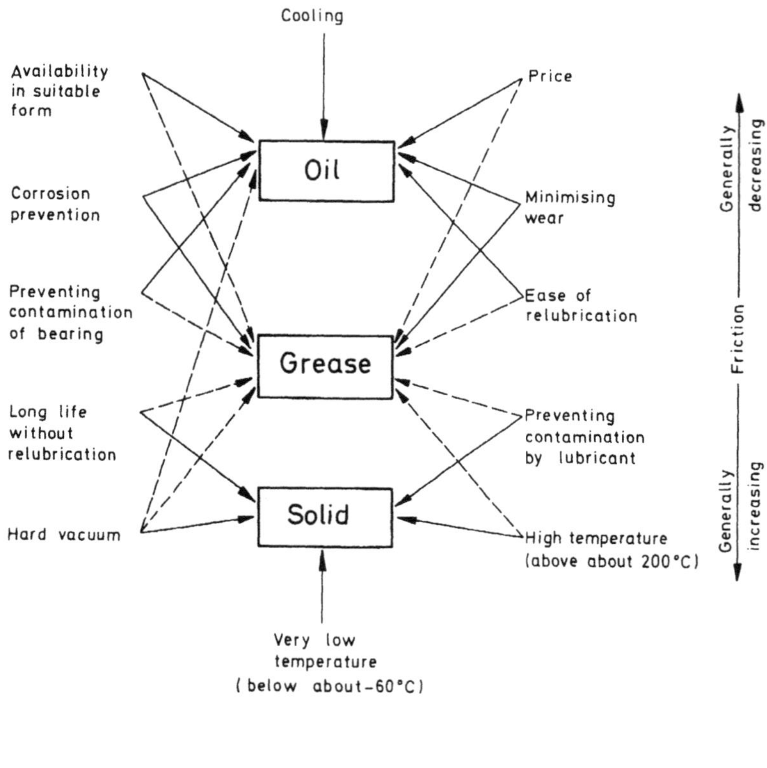

Figure 16.3 Factors Affecting the Choice of Lubricant Class
(1st choice —— ; 2nd choice - - - -)

Finally, it may be useful to recall that even when the conditions favour the use of an oil or a grease, there can be significant advantages in dispersing a solid lubricant in them where increased load-carrying capacity is beneficial.

The factors listed in this section give some indication of the situations in which the use of solid lubricants should be considered and some of the most important are brought together diagrammatically in Figure 16.3. The next section will try to show how to select the best type of solid lubricant for a particular application.

16.2 SELECTING THE TYPE OF SOLID LUBRICANT

The factors involved in choosing the best solid lubricant for a particular application are complex. Where the application is in an important mechanical system, it will be necessary to study the properties and performance of the various candidate materials in detail, and then to carry out realistic evaluations to ensure that the chosen material is suitable.

Table 16.2 Comparative Properties of Molybdenum Disulphide, Graphite and PTFE

Property	Solid Lubricant		
	Molybdenum disulphide	Graphite	PTFE
Friction coefficient	0.002-0.3	0.05-0.15	0.03-0.1
Maximum PV	3.5MPa.m/s	0.7MPa.m\s	~60kPa.m/s*
Thermal conductivity	Poor	Good	Very poor
Electrical conductivity	Poor to fair	Good	Very poor
Max.temp. in air,°C	350-400	540	300
Max.temp. in vacuum	650+	Not usable	Low
Adhesion	Very good	Good	Very good
Colour	Dark grey/black	Dark grey/black	White
Chemical resistance	Resistant	Resistant	Very resistant

* Increased by reinforcement to about 3MPa.m/s, but friction is then higher

For less critical applications the choice of solid lubricant will also be less critical, and will often be made on a cost basis. The three most commonly-used lubricants in low-technology applications and consumer products are PTFE, molybdenum disulphide and graphite. Their characteristics are compared in Table

16.2, but for many simple applications such as hinges, locks, latches, swivels, etc., all three may be suitable. The choice will then often be made on the basis of handling convenience and availability, and all three materials are readily available in dispersions in aerosol form.

For high-technology applications, a much wider range of solid lubricants needs to be considered. Table 16.3 compares the limiting temperatures for use of fourteen of these materials, and Table 16.4 gives some additional information about seven of them, together with applications for which they have been reported to have advantages. However, even for high technology applications the extent of use of most of them is very limited.

Table 16.3 Approximate Temperature Limits for Some Solid Lubricants

Solid Lubricant	Temperature Limit °C	
	In air	In vacuum
Molybdenum disulphide	350-400	650-1000
Molybdenum diselenide	400-480	760-1000
Tungsten disulphide	400-480	970-1050
Tungsten diselenide	350-480	700-1050
Niobium disulphide	420-450	1000
	350-410	1050
Graphite	540	Not usable
PTFE (unfilled)	290	200(outgasses)
PTFE (reinforced)	300	250(outgasses)
Graphite fluoride	470	470
Calcium fluoride	900	900
Boron nitride	850	1700
Lead	Unusable,oxidises	300
Silver	900	900

These temperatures represent best estimates since for most materials reported temperature limits vary widely.

Table 16.4 Some Characteristics and Applications of Less Common Solid Lubricants

Lubricant	Characteristics	Applications
Tungsten diselenide	Similar to molybdenum disulphide but better oxidation resistance at high temperatures	Nuclear plant
Niobium diselenide	Poorer lubricant than molybdenum disulphide, but better electrical conductivity	Electrical contacts
Graphite fluoride	Performance generally similar to molybdenum disulphide. Less adversely affected by moisture, but reported to have shorter life in vacuum	
Calcium fluoride	Not usable below 400°C	
Boron nitride	Poor film former	
Lead	Not usable in air because oxidation leads to film failure	Lubrication of rolling bearings in vacuum
Silver	Friction quite high. Used with calcium fluoride to provide lubrication from 0°C to 900°C	

16.3 Use of Molybdenum Disulphide

Where molybdenum disulphide is the preferred lubricant for a particular application, it is necessary to select the appropriate form to use and determine exactly how to use it. Information about each form is summarised in the following paragraphs.

16.3.1 **Loose Powder** can be used to lubricate many simple mechanisms provided that the geometry of the contacting surfaces enables the powder to enter the contacts, and that there is no danger of powder jamming the mechanism. An effective film can be formed on the rubbing surfaces, but the technique is generally inefficient and can be very dirty. Loose powder can also be used for forming burnished films.

16.3.2 **Dispersions** in water or volatile solvents are a more convenient way in which to introduce molybdenum disulphide into simple mechanisms, but they are more important as a way of forming a uniform coating on a bearing surface, by brushing, dipping or spraying.

16.3.3 **Burnished Coatings**, when fully consolidated and highly burnished, have high load-carrying capacity, very low friction, and low wear rate. For the longest wear life, the optimum thickness in the fully burnished form is probably about 3 to 5 μm thick. Would be one of the best techniques for lubricating bearings with very small clearances in air. Burnished films can be produced from loose powder, dispersions in volatile solvents or soft bonded films.

16.3.4 **In Situ Films** are complicated to produce and there is little information about the results of practical applications. They would appear to have advantages for producing films on surfaces with complex geometry such as gears, and like thin burnished films they could be used for systems with small bearing clearances.

16.3.5 **Sputtered Films** require specialised equipment and experience for their production. They have excellent performance in vacuum, fair performance in dry or inert environments, and poor performance in air. However this is a technology which is rapidly developing, and improved performance in air is likely to be achieved reliably in future.

16.3.6 **Bonded Films** are the most convenient way of producing reliable, durable films for use in normal atmospheres. Many different compositions are commercially available, with different binders and additives, and different concentrations of molybdenum disulphide. With the softer films, or under high sliding loads, the films become consolidated and behave in a similar manner to other burnished films, with little or no wear,and eventual failure is normally by oxidation and blistering. With harder films or under low sliding loads the friction remains higher and there is steady wear, the eventual failure normally being by depletion of the coating.

16.3.7 **Composites** are used in two different ways. They may be used for the manufacture of low-friction components, in which case the structural strength needs to be high and the molybdenum disulphide concentration may be low. Alternatively they may be used to provide transfer lubrication, in which case the lubricant content needs to be high, and high structural strength is less important. PTFE is often incorporated to improve lubrication and transfer. The matrix material needs to be selected to suit the operating conditions, and composites with metallic matrices may be more suitable for higher temperatures or nuclear radiation.

16.3.8 **Dispersions in Lubricating Oils or Greases** are generally used to provide increased load-carrying capacity, but they also reduce friction under boundary conditions and may delay seizure in the event of oil loss. In general, extreme-pressure additives are recommended when the lambda ratio, the minimum film thickness in a bearing divided by the combined surface roughness, is less than 1.2. They are also recommended when the calculated dynamic load P is greater than 0.25 times the manufacturers' dynamic load rating C for a ball bearing or greater than 0.15C for a roller bearing. Similar recommendations might be appropriate for the use of molybdenum disulphide in an oil or grease, especially when the running speeds are low. In automotive engines such conditions do not generally apply, except in cams and tappets, and the main reason for incorporating molybdenum disulphide in an engine oil is to reduce friction and consequently fuel consumption.

16.3.9 **Concentrated Dispersions and Pastes** can give exceptionally high load-carrying capacity, as well as very low friction under high loads. They can be used in metalforming to obtain high reductions and high deformation rates. They are also effective anti-seize compounds, but when they are used in bolted assemblies, care must be taken not to overtension the bolts.

Table 16.5 Some Techniques for Using Molybdenum Disulphide

Application	Appropriate Lubrication Technique
Plain bearings in air to 350°C	Bonded, burnished or sputtered films or filled polymer bushes
Plain bearings in vacuum	Polyimide or inorganic bonded, or burnished or sputtered films, or filled polymer bushes
Heavily-loaded rolling bearings in air to 200°C	Molybdenum disulphide grease
Lightly-loaded rolling bearings in air to 350°C or in high vacuum	(a) Burnished or sputtered films on races, cages and rolling elements (b) Composite retainers or land inserts, with lead on rolling elements or races
Screw threads, rivets, or fasteners in air to 150°C	Molybdenum disulphide paste anti-seize
Screw threads, rivets, or fasteners in air to 350°C or in vacuum	Burnished or sputtered films or powder
Chains, sprockets, conveyor chains or rollers	Dispersions in oils or greases (polyglycol for high temperatures)
Universal couplings	Molybdenum disulphide greases or pastes
Open gears	Molybdenum disulphide grease

Table 16.5 lists a number of engineering components with the most appropriate form of molybdenum disulphide to use for lubricating them. Table 16.6 summarises some of the important factors to consider in designing for the use of molybdenum disulphide films.

Table 16.6 Important Factors in Designing for the Use of Molybdenum Disulphide Films

1. Hardness of counterface is not critical, but soft counterfaces will wear and add to the problem of loss of stiffness. There is no inherent advantage in using materials over about 800 VPN.

2. Surface finish of counterface should be about 0.1 to 0.3μm c.l.a.

3. Stiffness can be improved by assembling with zero initial clearance or even slight interference, provided that thermal or other dimensional changes can be satisfactorily accomodated.

4. Where required machine life is greater than anticipated lubricant coating life, there should be facility for easy replacement of coated components.

5. Solid films are not generally as effective in corrosion prevention as either lubricating oils or paints, so that careful attention should be paid to corrosion-resistant materials and treatments and design features.

6. If temperature range is wide, proper attention must be paid to thermal expansion or contraction and to temperature stabilisation of materials of construction.

7. Since the performance of molybdenum disulphide films is adversely affected by most liquids, care should be taken to prevent liquid contamination.

8. Since films tend to be weakest at their edges, the whole of the rubbing contact area should be coated.

9. Oils, greases or temporary corrosion preventives should not be used to prevent corrosion of coated components in storage, because such materials will decrease the wear life.

10. The presence of substances such as PTFE or oils may prevent adequate film formation by molybdenum disulphide because of preferential wetting.

Table 16.7 Guide to Dynamic Friction Coefficients for Different Forms of Molybdenum Disulphide

Form of Lubricant	Load	Environment		
		Vacuum	Dry air	Moist air
Unconsolidated film	High	0.08-0.12	0.08-0,12	0.15-0.2
	Low	0.10-0.15	0.10-0.15	0.20-0.40
Highly-burnished film	High	0.02-0.04	0.03-0.06	0.08-0.15
	Low	0.05-0.10	0.06-0.12	0.15-0.20
Unburnished bonded film	High	0.06-0.12	0.08-0.15	0.15-0.25
	Low	0.10-0.25	0.10-0.30	0.20-0.30
Sputtered film	High	0.01-0.04	0.02-0.06	0.12-0.18
	Low	0.04-0.10	0.04-0.12	0.12-0.20
Composite	High	0.05-0.20	0.06-0.20	0.10-0.25
	Low	0.07-0.25	0.08-0.30	0.12-0.35
These figures are only rough estimates, intended for general guidance, since the friction values vary so much with purity or composition, as well as all the various environmental factors				

In many situations the actual value of the coefficient of friction is not a critical design factor. Torque-limited designs can occur in spacecraft and to a lesser extent in aircraft, but more often the requirement is that friction should be reasonably predictable and consistent, without seizure or stick-slip. The friction of a solid lubricant is generally smooth, but its value varies with load, speed, temperature and environment.

It is difficult to predict the actual coefficient of friction in a particular situation because of the many factors which affect it, but Table 16.7 lists some typical coefficients of friction in different situations.

The final requirement, having selected the solid lubricant and the form in which it should be used, is to establish how to apply it and use it to achieve the required performance. With the great expansion in the availability of different commercial products in recent years has come a greater maturity in the technical documentation and the technical advice provided by suppliers. In most applications the suppliers' recommendations should be followed in using their products. Divergences from suppliers' recommendations should only be made on the basis of sound testing and experience.

Hopefully this book, and especially the last eight chapters, will have provided information to supplement and clarify the suppliers recommendations, and to provide a sounder basis for the testing and experience.

~ ~ ~ ~ ~ ~ ~ ~ ~ ~ ~ ~ ~ ~ ~ ~

REFERENCES

1. Dowson, D., History of Tribology, Longman Group Limited, London, 1979.
2. Pliny, Natural History, Books 33 and 34. Rackham, H.,Loeb Classical Library, Heinemann, London, 1952.)
3. Gunther, R.T., The Greek Herbal of Dioscorides, Hafner, New York, 1959.
4. Agricola, De Re Metallica,(Hoover, H.C. and Hoover, L.H.,trans.), The Mining Magazine, London, 1912.
5. Cramer, J.A., Elements of the Art of Assaying Metals, L. Davis and C. Reymers (publ.), London, 1764.
6. Northcott, L., Molybdenum, Butterworths, London, 1956.
7. Johnson, R.L., A Review of the Early Use of Molybdenum Disulfide as a Lubricant, NLGI Spokesman, **32**, 298, (1968).
8. Dickinson, R.G. and Pauling, L., The Crystal Structure of Molybdenite, J.Amer.Chem.Soc., **45**, 1466, (1923).
9. Koehler, W., Antifriction and Antiabrasive Metal, US Patent 1,714,564 (1927).
10. Cooper, H.S. and Damerell, V.R., Lubricant, US Patent 2,156,803 (1939).
11. Wilkey, J.E., "Frictional Material and Method of Making Same", US Patent 2,284,785 (1939).
12. Bell, M.E. and Findlay, J.H., Molybdenite as a New Lubricant, Phys.Rev. **59**, 922, (1941).
13. Boyd, J. and Robertson, B.P., The Friction Properties of Various Lubricants at High Pressures, Trans. ASME, **67**, 51, (1945).
14. Norman, T., Molybdenite as a Die Lubricant, Metal Progress, **50**, 314, (1946).
15. Johnson, R.L., Godfrey, D. and Bisson, E.E., Friction of Solid Films on Steel at High Sliding Velocities, NACA Tech Note No. 1578, (April, 1948).
16. Hood, J.H. and Campbell, M.E., Boeing Experience with Bonded Solid Film Dry Lubricants, WADC Tech Report 59-244, p.162, (October, 1959).
17. Van Wyk, J.W., "New" Solid Lubricants, Mech.Engng., **95**, 27, (1973).
18. Vineall, G.J.C., A New Lubricant - Molybdenum Disulphide, The Motor, p.827, (June 29, 1955).

19. Spalvins, T. and Przybyszewski, J.S., Deposition of Sputtered Molybdenum Disulfide Films and Friction Characteristics of Such Films in Vacuum, NASA TN D-4269, (Dec. 1967)
20. Clauss, F.J., Solid Lubricants and Self-Lubricating Solids, Academic Press, New York, 1972.
21. Ducas, M., Les Lubrifiants Solides - Choix et Application, Journées internationales de tribologie appliquées à la micromechanique, Besançon, 14-15 June, 1973.
22. Wunsch, F., Solid Lubricants - Theory and Practice, Engineers Digest, **36**, 13, (1975) and **36**, 25, (1975).
23. Lipp, L.C., Solid Lubricants - Their Advantages and Limitations, Lubric. Eng., **32**, 574, (1976).
24. Lancaster, J.K., Solid Lubricants, RAE Tech.Memo MAT 325, (Dec.1979).
25. Vetter, H., Festschmierstoffe und ihre Anwendung in der Umformtechnik, Mineralöl-Anwendungstechnik, **3**, 1, (1980).
26. Campbell, M., Thompson, M.B. and Hopkins, V., Solid Lubricant Handbook for Use in the Space Industry, Midwest Research Institute, 1968.
27. Farr, J.P.G., Molybdenum Disulphide in Lubrication, Climax Molybdenum Company, Ltd. report, (1974).
28. Church, D., Molybdenum Disulphide in Lubrication, Molybdenum Mosaic, **5**, 8, (1982).
29. Winer, W.O., Molybdenum Disulphide as a Lubricant: a Review of the Fundamental Knowledge, Wear, **10**, 422, (1967).
30. Spengler, G., Molybdän-Disulfid als Schmiermittel, Erdöl und Kohle, **7**, 156, (1954).
31. Braithwaite, E.R., Occurrence, Extraction, Production and Uses of Molybdenum, Chapter 1 of Molybdenum: An Outline of its Chemistry and Uses, E.R. Braithwaite and J. Heber (eds.)., Elsevier, Amsterdam, 1994.
32. Materials and Technology, vol.3, Metals and Ores, T.J.W. van Thoor, ed., Longmans, London, 1970.
33. Grattan, P.A. and Lancaster, J.K., Abrasion by Lamellar Solid Lubricants, Wear, **10**, 453, (1967).
34. Giltrow, J.P. and Lancaster, J.K., The Role of Impurities in the Abrasiveness of MoS_2, Wear, **20**, 137, (1972).
35. Cyprus Amax Minerals Company Technical Information Sheet Molysulfide, Oct. 1995.
36. Ritsko, J.E., Laferty, J.M. and Hubbell, R.D., Evaluating of the Lubricating Properties of Chemically Upgraded MoS_2, Trans. ASME, **95**.F, 478, (1973).

37. Prokudina, V.K., Kalikhman, V.L., Golubnichaya, A.A., Borovinskaya, I.P. and Merzhanov, A.G., Synthetic Molybdenum and Tungsten Disulfides, Soviet Powder Metallurgy and Metal Ceramics, **17**, 450, (1978).
38. Djemal, G., Müller, N., Lachish, V. and Cahen, D., Photoelectrochemical Cells Using Polycrystalline and Thin Film MoS_2 Electrodes, Solar Energy Materials, **5**, 403, (1981).
39. Audas, R. and Irwin, J.C., Investigation of the Performance of an $MoS_2/I^-/I_2/C$ Electrochemical Solar Cell, J.Appl.Phys., **52**, 6954, (1981).
40. Kline, G., Kam, K.K., Ziegler, R. and Parkinson, B.A., Further Studies of the Photoelectrochemical Properties of the Group VI Transition Metal Dichalcogenides, Solar Energy Materials, **6**, 337, (1982).
41. Fornarini, L., Stirpe, F., Scrosati, B. and Razzini, G., Electrochemical Solar Cells with Layer-Type Semiconductor Anodes. Performance of n-MoS_2 Cells, Solar Energy Materials, **5**, 107, (1981).
42. Srivastava, S.K., Avasthi, B.N., Das, B. and Basu, S., High-Pressure Growth of Polycrystalline Molybdenum Disulphide, Materials Letters, **1**, 178, (1983).
43. Schneemeyer, L.F. and Cohen, U., Electrochemical Synthesis of Photoactive MoS_2, J. Electrochem. Soc., **130**, 1536, (1983).
44. Hjelm, P.J., Forsuk med Molybdaena och med Reduction af des Jord, K. Svenska Akad Nya Handl., **9**, 280, (1788).
45. McGraw-Hill Encyclopaedia of Science and Technology, vol. 8, p. 664, McGraw-Hill, New York, 1971.
46. Grattan, P.A. and Lancaster, J.K., Abrasion by Lamellar Solid Lubricants, Wear, **10**, 453, (1967).
47. Haltner, A.J. and Oliver, C.S., Effects of Chemical Properties of Molybdenum Disulfide on Frictional Behaviour, Symp. on Chem. of Friction and Wear, Divn. of Petroleum Chem., Am. Chem. Soc., (1958).
48. Killeffer, D.H. and Linz, A., Molybdenum Compounds: Their Chemistry and Technology, Interscience Publications, New York, 1952.
49. Guichard, M., Sur un Nouveau Sulfure de Molybdène Crystallisé, Compt. Rend., **130**, 137, (1900).
50. Dilworth, J.R., Molybdenum Sulphur Chemistry, Chapter 4 in Molybdenum: An Outline of its Chemistry and Uses, E.R. Braithwaite and J. Haber, (eds.), Elsevier, Amsterdam, 1994.
51. Clark, A.H., Evidence for the Geological Occurrence of Molybdenum Trisulphide, Nature Physical Science, **234**, 177, (1971).
52. Braithwaite, E.R. and Greene, A.B., A Critical Analysis of the Performance of Molybdenum Compounds in Motor Vehicles, Wear, **46**, 405, (1978).

53. Scott, D., A Study of Lubricant Soluble Molybdenum Sulphur Compounds as Lubricant Additives, Eurotrib 81, vol.III, Warsaw, (1981).
54. Lashkhi, V.L., Vipper, A.B., Mikutenok, Yu. A. and Kulagin, V.V., Effect of High-Temperature Antifriction Additives on Lubricating Properties of Motor Oil, Khim. i Tekhnol. Topliv i Masel, No. 10, 25, (Oct.1981). (In Russian).
55. Vipper, A.B., Lashkhi, V.L., Mikutenok, Yu. A., Durkin, V.A. and Kulagin, V.V., The Effect of Friction Modifiers on the Performance of Engine Oils, Treniye i Iznos, **2**, 934, (1981). (In Russian).
56. Vipper, A.B., Lashkhi, V.L., Matveevskii, R.M., Buyanovskii, I.A., Kulagin, V.V. and Natchuk, A.I., Friction Modifiers - Antifriction Engine Oil Additives, Khim.I Tekhnol. Topliv i Masel, No. 1, 56, (1981). (In Russian).
57. Lashkhi, V.L., Vipper, A.B. and Kulagin, V.V., Oil-Soluble Organic Compounds of Molybdenum as Lubricating Oil Additives, Khim. i Tekhnol. Topliv i Masel, No. 5, (1981). (In Russian).
58. Takeuchi, Y. and Nowacki, W., Detailed Crystal Structure of Rhombohedral MoS_2 and Systematic Deduction of Possible Polytypes of Molybdenite, Schweiz Mineral. Petrographische Mitteilungen, **44**, 105, (1964).
59. Bell, R.E. and Herfert, R.F., Preparation and Characterization of a New Crystalline Form of MoS_2, J.A.C.S, **79**, 3351, (1957).
60. Traill, R.J., A Rhombohedral Polytype of Molybdenite, Canadian Mineralogist, **7**, 524, (1963).
61. Graesar, S., Uber Funde der neuen rhomboedrischen MoS_2 - Modifikation (Molybdenit 3R) und von Tungstenit in den Alpen, Schweiz. Mineral. Petrographische Mitteilungen, **44**, 121, (1964).
62. Chianelli, R.R., Prestridge, E.B., Pecoraro, T.A. and DeNeufville, J.P., Molybdenum Disulfide in the Poorly Crystalline Rag Structure, Science, **203**, 1105, (16 March, 1979).
63. Climax Report No. L-260-1 and -2, Laboratory Investigation for Evaluation of MoS_2 from Various Sources, May, 1964.
64. Zelikman, A.N., Chistyakov, Yu. D., Indenbaum, G.V. and Krain, O.E., Crystal Structure Analysis of MoS_2 Synthesized by Various Methods, Soviet Physics - Crystallography, **6**, 308, (1961).
65. Woollam, J.A., Flood, D.J., Wagoner, D.E., Somoano, R. and Rembaum, A., Magnetic Field Effects on Superconductivity in Alkali Metal Intercalates of MoS_2, NASA TM-X-68109, (1972).
66. Bach, B. and Thomas, J.M., Novel Organometallic Sandwich Complexes Based on Transition-Metal Chalcogenides, J.Chem.Soc., Comm., p. 301, (1972).

67. Jamison, W.E., Electronic Effects on the Lubricating Properties of Molybdenum Disulfide and Related Materials, Proc. 2nd. Intl. Conf. on Solid Lub., (15-18 August, 1978), ASLE SP-6, p.1.
68. Selte, K. and Kjekshus, A., On the Structural Properties of the $Nb_{1+x}Se_2$ Phase, Acta Chem. Scand. **18**, 697, (1964).
69. Woollam, J.A. and Somoano, R.B., Physics and Chemistry of MoS_2 Intercalation Compounds, Materials Science and Engineering, **31**, 289, (1977).
70. Gamble, F.R., DiSalvio, F.J., Klemm, R.A. and Geballe, T.H., Superconductivity in Layered Structure Organometallic Crystals, Science, **168**, 568, (1970).
71. Marmaduke, B.A. and Donaghey, F.L., Engineering Evaluation of Solid Transition-Metal Sulfide Electrodes for Use in High Energy-Density Batteries, Inter-Society Energy Conversion Engineering Conference, p. 467, 11th Sept. 1976.
72. Woollam, J.A., Atomic Hydrogen Storage Method and Apparatus, US Patent No. 4,229,196, (21 Oct. 1980).
73. Magie, P.M., A Review of the Properties and Potentials of the New Heavy Metal Derivative Solid Lubricants, Lubric. Eng., **22**, 262, (1966).
74. Boes, D.J., New Solid Lubricants: Their Preparation Properties and Potential for Aerospace Applications, IEEE Trans., AS-2, no.2, 457, (1964).
75. Mikhailov, A.S., The Experimental Investigation of Electro-Chemical Oxidation and Dissolution of Molybdenite, Geokhimiya, **9**, 818, (1962). (RAE Library Translation No. 1030).
76. Elliott Brothers (London) Limited, Report on the Vacuum Evaluation of Slip Rings and Brushes for Space Applications, Contractor Report ESRO CR-1, (March, 1968).
77. Mellor, J.W., A Comprehensive Treatise on Inorganic and Theoretical Chemistry, vol. XI, Longmans Green, London, 1931.
78. Mashkovich, L.A., Khrisanova, L.B., Shmakova, E.S., Telegin, V.D., Kuteinikov, A.F., Novikov, I.I. and Fadeeva, T.M., Degradation of Molybdenum Disulfide During Heat Treatment of a Carbon-Containing Composition, Chemical and Petroleum Engineering, **11**, 1052, (1975).
79. Kalamazov, R.V., Ibragimov, A.T., Kal'kov, A.A. and Bogomolov, A.M., Reactions of Molybdenum Disulfide with Gases and Water Vapour, Soviet Powder Metallurgy and Metal Ceramics, **22**, 27, (1983).
80. Matsunaga, M., Homma, T. and Tanaka, A., Investigation of Vapour Adsorption on Molybdenum Disulphide Surfaces by Auger Electron Spectroscopy, ASLE Trans., **25**, 323, (1982).

81 Holinski, R., Lubricant Diffusion Embrittles Steel Threads, Engineers Digest, p. 27, (March, 1983).
82. Knappwost, A. and Thieme, F., Zur Kinetic des Reaktionseffektes bei der Feststoffschmierung von Eisen und MoS_2, Z. Phys. Chem., Neue Folge, **63**, 1, (1969).
83. Tsuya, Y., Shimura, H. and Matsunaga, M., A Study on Some Metal-Base Self-Lubricating Composites Containing Tungsten Disulfide, Lubric. Eng., **29**, 498, (1973).
84. Brainard, W.A., The Thermal Stability and Friction of the Disulfides, Diselenides and Ditellurides of Molybdenum and Tungsten in Vacuum (10^{-9} to 10^{-6} Torr), NASA TN D-5141, (April, 1969).
85. Cannon, P., Melting Point and Sublimation of Molybdenum Disulphide, Nature, **183**, (4675), 1612, (1959).
86. Scholz., W.G., Doane, D.V. and Timmons, G.A., Molybdenum by Direct Thermal Dissociation of Molybdenum Disulfide, Trans. Met. Soc. AIME., **221**, 356, (1961).
87. Zelikman, A.N. and Belyaevskaya, L.V., The Melting Point of Molybdenite, Zhurnal Neorganicheskoi Khimii, **1**, 2239, (1956).
88. Devine, M.J., Lamson, E.R. and Bowen, J.H., Inorganic Solid Film Lubricants, J. Chem. Eng. Data, **6**, 79, (1961).
89. McCabe, J.T., Molybdenum Disulphide - Its Role in Lubrication: A Brief Summary of the State of the Art, International Industrial Lubrication Exhibition Symposium, London, (8-11 March, 1965), p. 97.
90. Ross, S. and Sussman, A., Surface Oxidation of Molybdenum Disulphide, J. Phys. Chem., **59**, 889, (1955).
91. Buckley, D.H., Examination of Molybdenum Disulfide with LEED and Auger Emission Spectroscopy, NASA TN D-7010, (1970).
92. Godfrey, D. and Nelson, E.C., Oxidation Characteristics of Molybdenum Disulfide and Effect of Such Oxidation on its Role as a Solid-Film Lubricant, NACA TN No. 1882, (May, 1949).
93. Sliney, H.E., Decomposition Kinetics of Some Solid Lubricants Determined by Elevated-Temperature X-Ray Diffraction Techniques, Proc. USAF Aerospace Fluids and Lubricants Conference, San Antonio, (16-19 April, 1963).
94. Braithwaite, E.R., Solid Lubricants and Surfaces, Macmillan, London, 1964.
95. Braithwaite, E.R. and Greene, A.B., The Relevance of the Surface Chemistry of MoS_2, Particularly with Respect to its Wettability: Part I, Wear, **37**, 251, (1976).

96. Vineall, C.J.C. and Taylor, A., Molybdenum Disulphide Criticisms Answered, Sci. Lubric., **11**, 32, (1959).
97. Wyn-Roberts, D. New Frontiers for Space Tribology, Tribology Intl., **23**, 149, (1990).
98. Ballou, E.V. and Ross, S., The Adsorption of Benzene and Water Vapor by Molybdenum Disulfide, J. Phys. Chem., **57**, 653, (1953).
99. Cannon, P., The Primary Interactions Between Ammonia-Graphite and Ammonia-MoS_2, J. Chem. Phys., **58**, 130, (1961).
100. Johnston, R.R.M. and Moore, A.J.W., Water Adsorption on Molybdenum Disulphide Containing Surface Contaminants, J. Phys. Chem., **68**, 3399, (1964).
101. Cannon, P. and Norton, F.J., Reaction Between Molybdenum Disulphide and Water, Nature, **203**, (4946), 750, (1964).
102. Lancaster, J.K., A Review of the Influence of Environmental Humidity and Water on Friction, Lubrication and Wear, Tribology Intl., **23**, 380, (1990).
103. Bowden, F.P. and Tabor, D., The Friction and Lubrication of Solids, Part 1, Oxford University Press, London, 1954.
104. Haltner, A.J., Sliding Behaviour of Some Layer Lattice Compounds in Ultrahigh Vacuum, ASLE Trans., **9**, 136, (1966).
105. Jamison, W.E., Structure and Bonding Effects on the Lubricating Performance of Crystalline Solids, ASLE Trans., **15**, 296, (1972).
106. Matveevsky, R.M., The Experimental Study of the Shear Coefficient of Solid Lubricant Powders, Proc. ASLE Intl. Conf. on Solid Lubricants, Denver, Colorado, (24-27 Aug. 1971) ASLE SP-3, p. 198.
107. Puchkov, V.N. and Pashkov, A.I., Determination of the Shear Resistance of Solid Lubricant Coatings, Mechanical Sciences, **2**, **44**, (1974).
108. Bridgeman, P.W., Shearing Phenomena at High Pressure Particularly in Inorganic Compounds, Proc. Amer. Acad. of Arts and Sciences, **71**, 387, (1936).
109. Briscoe, B.J. and Smith, A.C., The Interfacial Shear Strength of Molybdenum Disulphide and Graphite Films, ASLE Trans. **25**, 349, (1982).
110. Akaoka, J. and Nitanai, A., Effects of Contact Pressure on Shear Strength and Durability of MoS_2 Films, JJSLE Intl. Edn., No. 2, P. 163, (1981).
111. Uemura, M., Okada, K., Mogami, A. and Okitsu, A., Effect of Friction Mechanisms on Friction Coefficient of MoS_2 in an Ultrahigh Vacuum, Lubric. Eng., **43**, 937, (1987).

112. Kanakia, M.D. and Peterson, M.B., Literature Review of Solid Lubrication Mechanisms, Interim Report BFLRF No. 213, US Army Belvoir Research Development and Engineering Center, November, 1986. (AD A185010).
113. Feng, I. Ming, Lubricating Properties of Molybdenum Disulfide, Lubric. Eng., **8**, 285, (1952).
114. Pritchard, C. and Midgley, J.W., The Effect of Humidity on the Friction and Life of Unbonded Molybdenum Disulphide Films, Wear, **13**, 39, (1969).
115. Gänsheimer, J., Influence de Certaines Vapeurs et Liquides sur les Propriétés de Frottement du Bisulfure de Molybdène, Energie Fluide et Lubrification, p. 85, (Oct. 1979).
116. Johnson, V.R. and Vaughn, G.W., Investigation of the Mechanism of MoS_2 Lubrication in Vacuum, J. Appl. Phys., **27**, 1173, (1956).
117. Matsunaga, M. and Nakagawa, T., Effect of Various Vapors on Coefficient of Friction of Clean Molybdenum Disulfide, ASLE Trans., **19**, 216, (1976).
118. Matsunaga, M. and Nakagawa, T., Frictional Behaviour of Various Kinds of Molybdenum Disulfide, Proc. 2nd ASLE Intl. Conf. on Solid Lubrication, Denver, Colorado, (15-18 Aug. 1978), ASLE SP-6, p. 45.
119. Matsunaga, M., Haguida, Y. and Nakagawa, T., The Effect of Organic Compounds on the Coefficient of Friction of Clean Molybdenum Disulfide, Proc. 2nd ASLE Conf. on Solid Lubrication, Denver, Colorado, (15-18 August, 1978), ASLE SP-6, p. 179.
120. Matsunaga, M. and Nakagawa, T., Frictional Behaviour of Clean Molybdenum Disulphide, Report. Inst. 2nd. Sci., Univ. of Tokyo, **27**, (Serial No. 177), 1, (March, 1979).
121. De Gee, A.W.J., Begelinger, A. and Salomon, G., Influence of the Atmosphere on the Endurance of Some Solid Lubricants, Proc. Instn. Mech. Engrs., **183**, Pt. 3I, p. 18, (1968-69).
122. Johnston, R.R.M. and Moore, A.J.W., The Burnishing of MoS_2 on the Metal Surface, Wear, **7**, 498, (1964).
123. Bartz, W.J. and Müller, K., Investigations on the Lubricating Effectiveness of Molybdenum Disulfide, Wear, **20**, 371, (1972).
124. Bisson, E.E. and Godfrey, D., Bonding of MoS_2 to Various Materials to Form a Solid Lubricating Film, NACA TN-2802, (1952).
125. Deacon, R.F. and Goodman, J.F., Orientation and Frictional Behaviour of Lamellar Solids on Metals, Proc. Instn. Mech. Engrs., **195C**, 344, (1957). (Proc. Conf. on Lubrication and Wear).
126. Brudnyi, A.I. and Karmodonov, A.F., Structure of Molybdenum Disulphide Lubricant Film, Wear, **33**, 243, (1975).

127. Takahashi, N. and Kashiwaya, S., High Resolution Transmission Electron Microscopy Observation of the MoS_2 - Au Interface Formed Due to Stick-Slip of MoS_2 Sliding on Au in Relation to the Friction Trace, Wear, **206**, 8, (1997).
128. Braithwaite, E.R., Friction and Wear of Graphite and Molybdenum Disulphide, Sci. Lubric., **18**, 13, (1966).
129. Lancaster, J.K., The Influence of Substrate Hardness on the Formation and Endurance of Molybdenum Disulphide Films, Wear, **10**, 103, (1967).
130. Lancaster, J.K., Anisotropy in the Mechanical Properties of Lamellar Solids, Wear, **9**, 169, (1966).
131. Silin A.A. and Aparin, V.I., The Role of Adhesion in the Orientation of Molybdenite During Friction, Mechanical Sciences, **1**, 92, (1979).
132. Kinner, G.H., Pippett, J.S. and Atkinson, I.B., A Laboratory Study of the Performance and Behaviour of Solid Film Lubricants Based on MoS_2 and Graphite Fluoride, RAE Tech Report 76026, (Feb. 1976).
133. Stupian, G.W. and Chase, A.B., Adhesion of Films of MoS_2 to Oxidised Metal Surfaces, Space and Missile Systems Organisation, USAF, Report No. SAMSO-TR-77-92, (AD AO 39775), (19 April, 1977).
134. Lancaster, J.K., Lubrication by Transferred Films of Solid Lubricants, Trans. ASLE, **8**, 146, (1965).
135. Finkin, E.F., Mechanical Aspects of Solid Lubricant and Elastomeric Coatings, Science and Technology of Surface Coatings, Academic Press, London, 1974.
136. Tsuya, Y., Microstructure of Wear, Friction and Solid Lubrication, Technical Report No. 81 of Mechanical Engineering Laboratory, Tokyo, 1975.
137. Gamulya, G.D., Dobroval'skaya, G.V., Lebedeva, I.L. and Yukhno, T.P., Orientation of MoS_2 Particles on the Frictional Surface of a Solid Lubricant Coating, Mechanical Sciences, No. 6, 67, (1978).
138. Holinski, R. and Gänsheimer, J., A Study of the Lubricating Mechanism of Molybdenum Disulfide, Wear, **19**, 329, (1972).
139. Lancaster, J.K., Dry Bearings: a Survey of Materials and Factors Affecting Their Performance, RAE Tech. Report 72101, (July, 1972).
140. Fleischauer, P.D., Fundamental Aspects of the Electronic Structure, Material Properties and Lubrication Performance of Sputtered MoS_2 Films, Thin Solid Films, **154**, 309, (1987).
141. Stupian, G.W., Feuerstein, S., Chase, A.B. and Slade, R., Adhesion of MoS_2 Powder Burnished on Metal Substrates, J. Vac. Sci. and Tech., **13**, 684, (1976).
142. Gänsheimer, J., A Review on Chemical Reactions of Solid Lubricants During Friction, ASLE Trans., **15**, 244, (1972).

143. Reid, J.V. and Schey, J.A., The Effect of Metallic Substrates on MoS_2 Film Formation, Proc. ASLE 2nd Intl. Conf. on Solid Lubrication, Denver, Colorado, (15-18 August, 1978), ASLE SP-6, p. 22.
144. Devine, M.J., Lamson, E.R. and Bowen, J.H., The Effect of the Chemical Composition of Metals in Solid Lubrication, Amer. Chem. Soc., Divn. of Pet. Chem., **8**, C-177, (1963).
145. Thelen, E. and Devine, M.J., Chemical Reactivity Factor for Metallic Sulfide Lubricants, ASLE Trans., **10**, 386, (1967).
146. Devine, M.J. and Sander, L.F., Interatomic Spacing Concept for Solid Lubricant - Metal Systems, Amer. Chem. Soc. Divn. of Pet. Chem., **13**, (1968).
147. Przybyszewski, J.S. and Spalvins, T., Friction and Contact Resistance During Sliding in Vacuum of Low Resistivity Metals Lubricated with Sputtered Molybdenum Disulfide Films, NASA TN D-5349, (July, 1969).
148. Barwell, F.T. and Milne, A.A., Use of Molybdenum Disulphide in Association with Phosphated Surfaces, Sci. Lubric., **3**, 10, (1951).
149. Lancaster, J.K., The Tribology Handbook, Section B5 on Solid Lubricants and Coatings, Butterworths, London, 1973.
150. Milne, A.A., Experiments on the Friction and Endurance of Various Surface Treatments Lubricated with Molybdenum Disulphide, Wear, **1**, 92, (1957 - 8).
151. McKannan, E.C. and Demorest, K.E., Dry Film Lubrication of Highly Loaded Bearings in Vacuum, Lubric. Eng., **20**, 134, (1964).
152. Peterson, M.B. and Johnson, R.L., Friction and Wear Investigation of Molybdenum Disulfide: 1 - Effect of Moisture, NACA TN-3055, (Dec. 1953).
153. Midgley, J.W., The Frictional Properties of Molybdenum Disulphide, J. Inst. Pet., **42**, (395), 316, (1956).
154. Barry, H.F. and Binkelman, J.P., MoS_2 Lubrication of Various Metals, Lubric. Eng., **22**, 139, (1966).
155. Peterson, M.B. and Johnson, R.L., Factors Influencing Friction and Wear with Solid Lubricants, Lubric. Eng., **11**, 325, (1955).
156. Pardee, R.P., The Effect of Humidity on Low-Load Frictional Properties of a Bonded Solid Film Lubricant, ASLE Trans., **15**, 130, (1972).
157. Gao, C., Bredell, L., Kuhlmann-Wilsdorf, D. and Makel, D.D., Micromechanics of MoS_2 Lubrication, Wear, **162-164**, 480, (1993).
158. Haltner, A.J., An Evaluation of the Role of Vapour Lubrication Mechanisms in MoS_2, Wear, **7**, 102, (1964).
159. Hopkins, V. and Campbell, M., Friction and Wear-Life of Selected Solid Lubricant Films at -100F, R.T. and 400F, Lubric. Eng., **25**, 430, (1969).

160. Gould, S.G. and Roberts, E.W., The In-Vacuo Torque Performance of Dry-Lubricated Ball Bearings at Cryogenic Temperatures, 23rd Aerospace Mechanisms Symp., Huntsville, Alabama, (3-5 May, 1989).
161. Rempe, W.H., Research and Development of Materials for Use as Lubricants in a Liquid Hydrogen Environment, ASLE Trans. **9**, 213, (1966).
162. McDaniel, R.H., The Effects of Reactor Radiation on Three High-Temperature Solid-Film Lubricants, Lubric. Eng., **21**, 463, (1965).
163. Jacobson, T.P., Dynamic Friction and Wear of a Solid Film Lubricant During Radiation Exposure in a Nuclear Reactor, NASA TN D-6940, (September, 1972).
164. Dukhovskoi, E.A., Ponomarev, A.N., Silin, A.A. and Tan'roze, V.L. The Occurrence of Anomalously Low Friction During Electron Bombardment of a Sliding Molybdenum Disulphide Surface, USSR J. Exp. and Theor. Phys., **4**, (20), 268, (1974). Royal Aircraft Establishment Library Translation 1829 (20 November, 1974).
165. Dukhovskoi, E.A., Onishchenko, V.S., Ponomarev, A.N., Silin, A.A. and Tal'roze, V.L., DAN SSSR, **217**, 599, (1974). (Quoted in Ref. 164).
166. Dukhovskoi, E.A., Ponomarev, A.N., Silin, A.A. and Tal'roze, V.L., Dan SSSR, **1**, 200, (1972). (Quoted in Ref. 164).
167. Aparin, V.I. and Dukhovskoi, E.A., Significant Enhancement of Orientation of Molybdenum Disulfide Surface Layer from Irradiation Resulting in Ultra-Low Friction when Paired with a Metal, DAN SSSR, **217**, 599, (1974).
168. Rittenhouse, J.B., Jaffe, L.D., Nagler, R.G. and Martens, H.E., Friction Measurements on a Low Earth Satellite, ASLE Trans., **6**, 161, (1963).
169. Langmuir, I., Vapour Pressure of Metallic Tungsten, Phys. Rev., **2**, 329, (1913).
170. Bowen, P.H. and Hickham, W.H., Outgassing Characteristics of Dry Lubricant Materials, Machine Design, **35**, 119, (1963).
171. Bartz, W.J., Some Investigation on the Influence of Particle Size on the Lubricating Effectiveness of Molybdenum Disulphide, ASLE Trans., **15**, 207, (1972).
172. Braithwaite, E.R., MoS_2 Second Thoughts, Ind. Lub. Trib., **21**, 241, (1969).
173. Devine, M.J., Lamson, E.R. and Stallings, L., Molybdenum Disulfide Diester Lubricating Greases, NLGI Spokesman, **27**, 320, (1964).
174. Durbridge, C.F.D., The Effect of Particle Size on the Lubricating Effect of Molybdenum Disulfide, Sci. Lubric., **17**, 3, (1965).
175. Giltrow, J.P. and Groszek, A.J., The Effect of Particle Shape on the Abrasiveness of Lamellar Solids, Wear, **13**, 317, (1969).

176. Sauer, H.J., Schowalter, R.E. and Mahale, S.V., Influence of Film Thickness on the Performance of Solid Lubricants, Lubric. Eng., **27**, 110, (1971).
177. McCain, J.W., A Theory and Tester Measurement Correlation About MoS_2 Dry Film Lubricant Wear, SAMPE Journal, p. 17, (1970).
178. Dayson, C., The Friction of Very Thin Solid Film Lubricants on Surfaces of Finite Roughness, ASLE Trans., **14**, 105, (1971).
179. Hopkins, V. and Campbell, M., Film Thickness Effect on the Wear Life of a Bonded Solid Lubricant Film, Lubric. Eng., **25**, 15, (1969).
180. Bahun, C.J. and Jones, J.R., Influence of Load, Speed and Coating Thickness on the Wear Life of a Bonded Solid Lubricant, Lubric. Eng., **25**, 351, (1969).
181. Hopkins, V., Discussion, p. 6, on Rabinowicz, E., Variation of Friction and Wear of Solid Lubricant Films with Film Thickness, ASLE Trans., **10**, 1, (1967).
182. Whitehouse, G.D., Nandan, D. and Whitehurst, C.A., The Effect of Film Thickness on Friction Coefficients for Solid Lubricants CaF_2, MoS_2 and Graphite, ASLE Trans., **13**, 159, (1970).
183. Bartz, W.J. and Xu, Jinfen, Wear Behaviour and Failure Mechanism of Bonded Solid Lubricants, Lubric. Eng., **43**, 514, (1987).
184. Salomon, G., De Gee, A.W.J. and Zaat, J.H., Mechano-Chemical Factors in MoS_2-Film Lubrication, Wear, **7**, 87, (1964).
185. De Gee, A.W.J., Salomon, G. and Zaat, J.H., On the Mechanisms of MoS_2 Film Failure in Sliding Friction, Trans. ASLE, **8**, 156, (1965).
186. Fusaro, R.L., A Comparison of the Lubricating Mechanisms of Graphite Fluoride and Molybdenum Disulfide Films, Proc. 2nd Intl. Conf. on Solid Lubrication, (15 -18 August 1978), ASLE SP-6, p. 59.
187. Fayeulle, S., Ehni, P.D. and Singer, I.L., Role of Transfer Films in Wear of MoS_2 Coatings, Paper V (ii), p. 129 in Mechanics of Coatings (Dowson, D., Taylor, C.M. and Godet, M., eds.) Elsevier, Amsterdam, (1990).
188. Haltner, A.J. and Oliver, C.S., Frictional Properties of Some Solid Lubricant Films Under High Load, J. Chem. Eng. Data, **6**, 128, (1961).
189. Calhoun, S.F., Meade, F.S., Murphy, G.P. and Young, R.L., Factors Affecting the Performance of Resin Bonded Solid Film Lubricants, Lubric. Eng., **21**, 97, (1965).
190. Lavik, M.T., Hubbell, R.D. and McConnell, B.D., Oxide Interaction - a Concept for Improved Performance with Molybdenum Disulfide, Lubric. Eng., **31**, 20, (1975).
191. Buckley, D.H., ibid, discussion of above paper.
192. Centers, P.W., The Role of Oxide and Sulfide Additions in Solid Lubricant Compacts, Trib. Trans., **31**, 149, (1988).

193. Gardos, M.N. and McConnell, B.D., Development of a High-Load, High-Temperature Self-Lubricating Composite - Parts I to IV, Proc. ASLE-ASME Lubrication Conference, New Orleans, LA, (5-7 October, 1981). ASLE SP-9.
194. Klenke, C.J., Tribological Performance of MoS_2 - B_2O_3 Compacts, Tribology Intl., **23**, 23, (1990).
195. Bartz, W.J., Tribological Behaviour of Three-Component Bonded Solid Lubricant Films, Proc. JSLE Intl. Trib. Conf., Tokyo, (8-10 July, 1985).
196. Rabinowicz, E., Friction and Wear of Materials, John Wiley & Sons, Inc., New York, (1965).
197. Bowden, F.P. and Tabor, D., The Friction and Lubrication of Solids, Part II, Clarendon Press, Oxford, (1964).
198. Barton, G.C. and Pepper, S.V., Transfer of Molybdenum Disulfide to Various Metals, NASA TP. 1019, (August, 1977).
199. Ali, D., Fitzer, E. and Ragoss, A., Preferential Crystallographic Orientation in Products of Technical Graphites, Conf. Ind. Carbon and Graphite, September 1957. Soc. Chem. Ind., London, (1958).
200. Fleischauer, P.D. and Bauer, R., The Influence of Surface Chemistry on MoS_2 Transfer Film Formation, ASLE Trans., **30**, 160, (1986).
201. Tyler, J.C. and Ku, P.M., Compressive and Tensile Properties of Molybdenum Disulfide Compacts, ASLE Trans., **10**, 28, (1967).
202. Janes, S., Neumann, K. and Sethna, D.N., Composite Materials Containing Particles and Fibres as Oil-Free Bearing Materials, Z. Werkstofftech., **9**, 204, (1978). (In German).
203. Connelly, M. and Rabinowicz, E., Detecting Wear and Migration of Solid-Film Lubricants Using Simultaneous Exoelectron Emission, ASLE Trans., **26**, 139, (1983).
204. Wahl, K.J. and Singer, I.L., Quantification of a Lubricant Transfer Process that Enhances the Sliding Life of a MoS_2 Coating, Trib. Letters, **1**, 59, (1995).
205. Vanshteyn, V.E. et al, A Method of Applying a Solid Lubricant Coating, Authors Cert. No. 150329, Bycull, Izobretonii, No. 18, p. 14, (1962). (Quoted in Reference 191).
206. Lancaster, J.K., Transfer Lubrication for High Temperatures: a Review, Trans. ASME, J. Tribol., **107**, 437, (Oct. 1985).
207. Drozdov, Yu.N. et al, Rotaprint Method of Lubricating Dry Reduction Gears, Russian Eng. J., **13**, 24, (1973).
208. Troyanovskaya, G.I., Use of Self-Lubricating Materials in Rotaprint Lubrication, Russian Eng. J., **14**, 50, (1974).

209. Devine, M.J., Lamson, E.R., Cerini, J.P. and Carroll, R.J., Solid Lubricants, Proc. Instn. Mech. Engrs., **182**, 311, (1967 - 68).
210. Van Wyk, J.W., Ceramic Airframe Bearings, Lubric. Eng. **31**, 558, (1975).
211. Lancaster, J.K., Etched-Pocket Dry-Bearing Materials, Tribology Intl., **18**, 139, (June, 1985).
212. Bowen, Paul H., Solid Lubrication of Gears and Bearings in a Space Environment, ASLE Trans., **7**, 227, (1964).
213. Drozdov, Yu. N. et al, Heat-Resistant Reduction Gear with Magnetic Supply of Solid Lubricant, Russian Eng. J., **56**, 13, (1976).
214. Drozdov, Yu. N. and Egorov, V.I., Wear Resistance of Dry Gear Transmissions in a Vacuum, Russian Eng. J., **57**, 20, (1977).
215. Pavlov, V.G., Assessment of Contact Stresses and of Film Thickness During Magnetic Powder Lubrication of Gears, Russian Eng. J., **59**, 3, (1979).
216. Vaisfeld, L.O., Work Performance of Gear Transmissions with Magnetic Powder Lubrication under Various External Conditions, Russian Eng. J., **59**, 8, (1979).
217. Macks, E.F., Nemeth, Z.N. and Anderson, W.J., Preliminary Investigation of Molybdenum Disulfide - Air Mist Lubrication for Roller Bearings Operating to DN Values of 1 x 10^6 and Ball Bearings Operating to Temperatures of 1,000°F, NACA/T1B/3109 RME 51G31, (October, 1951).
218. Nemeth, Z.N. and Anderson, W.J., Investigation of Temperature Limitation of Various Lubricants for High-Temperature 20-Millimetre-Bore Ball Bearings, NACA TN 3337, (January, 1955).
219. Nemeth, Z.N. and Anderson, W.J., Effect of Air & Nitrogen Atmospheres on the Temperature Limitations of Liquid and Solid Lubricants in Ball Bearings, Lubric. Eng., **11**, 267, (1955).
220. Cosgrove, S.L., Sibley, L.B. and Allen, C.M., Evaluation of Dry Powdered Lubricants at 1000F in a Modified Four-Ball Wear Machine, ASLE Trans., **2**, 217, (1960).
221. Wilson, D.S., Gray, S., Orcutt, F.K., Krause, H.H., Allen, C.M. and Macks, E.F., The Development of Lubricants for High-Speed Rolling Contact Bearings Operating Over the Range of Room Temperature to 1200 Degrees F, WADD TR 60-732, (1960).
222. Todd, M.J., Lead-Film Lubrication in Air and Vacuum, Proc. 1st. Europ. Space Trib. Symp., Frascati, Italy, ESA SP-111, p. 225, (Oct. 1975).
223. Müller, K., How to Reduce Fretting Corrosion - Effect of Lubricants, Tribology Intl., **8**, 57, (1975).
224. Groszek, A.J., Dispersions and Greases, US Patent No. 3,384,582, (1968).

225. Groszek, A.J. and Witheridge, R.E., Surface Properties and Lubricating Action of Graphite and MoS_2, ASLE Trans., **14**, 254, (1971).
226. Nittel, K.D., Seifen und MoS_2 - Schmierstoffe für die Kaltmassivumformung, Tribology 2000: 8th Intl. Colloquium, Esslingen, (14-16 January), 1992, Paper No. 18.14).
227. Kuznetsov, A.M. and Simonov, I.V., The Surface Microgeometry of Bores Broach Burnished Employing Solid Lubricants in Cold Metal Forming, Soviet Engng. Res., **2**, 45, (1982).
228. Goto, Y., Wakasugi, S. and Kozai, T., A Test for Investigating the Lubrication Properties of Solid Lubricants in Cold Metal Forming, J. Mech. Working Technology, **6**, 51, (1982).
229. Tanaka, K., Uchiyama, Y., Nakagawa, T. and Matsunaga, M., Wear and Frictional Behaviours of Molybdenum Disulfide Pellets, and Steel Lubricated with Molybdenum Disulfide Powder, Proc. 3rd Intl. Conf. on Wear of Materials, ASME, p. 637, (1981).
230. Matsunaga, M., Nakagawa, T. and Tennicki, M., The Wear and Frictional Behaviour of Molybdenum Disulfide Compacts, ASLE Trans., **26**, **64**, (1983).
231. Brendle, M. and Colin, G., The Frictional and Transfer Behaviour of Compacted Solid Lubricants on Smooth Metallic Surfaces, Proc. 7th Leeds/Lyon Symp. Friction and Traction, Leeds, 1980.
232. Brendle, M. and Colin, G., Friction-Induced Modifications at the Surface of Compacted Solid Lubricants, Proc. 8th Leeds/Lyon Symp. The Running-In Process in Tribology, Lyon, 1981.
233. Brendle, M. and Colin, G., Il Comportamento Tribologico di Lubrificanti Solidi Compattati su Superfici Liscie Metalliche, Tribologia i Lubrificazione, **17**, 25, (1982).
234. Bowden, F.P., Frictional Behaviour of Porous Metals Containing Molybdenum Disulphide, Res. Appl. in Industry, **3**, 383, (1950).
235. Bowden, F.P. and Rowe, G.W., Lubrication with Molybdenum Disulphide Formed from the Gas Phase, The Engineer, p. 667. (1957).
236. Tabor, D., private communication.
237. Brophy, J.E. and Ingraham, R.W., Application of Solid Lubricant Coatings to Surfaces, US Patent No. 2,902,417, 19 September, 1956.
238. Brophy, J.E., Ingraham, R.W. and Pettus, C., Low Friction Coatings with Good Bond Strength, Materials in Design Eng., (10-11 Sept. 1961).
239. Aladjem, A., letter to the editor, Materials in Design Eng., **55**, **23**, (1962).
240. Vest, C.E., Adaptation of an MoS_2 In Situ Process for Lubricating Spacecraft Mechanical Components, NASA TN D-2288, (May, 1964).

241. Bayer, R.G. and Trivedi, A.K., Molybdenum Disulfide Conversion Coating. Significance of Process Parameters on Performance, Metal Finishing, **75**, 47, (1977).

242. Nishimura, M., Nosaka, M., Miyakawa, Y. and Sakamoto, J., The Lubrication Characteristics of Molybdenum Disulfide Films Made by Chemical Process. Part 1. Optimum Conversion Conditions for Obtaining MoS_2 Films, JJSLE Intl. Edn., **1**, 119, (Nov. 1980).

243. Ibid, Part 2. Effect of Surrounding Atmosphere on the Lubrication Characteristics of a Film, p. 121.

244. Ibid, Part 3, Physical Analysis of a Film, p. 123.

245. Dukhovskoi, E.A., Lobashev, B.P., Makarov, Yu. V., Sarpov, N.I. and Silin, A.A., Lubricant Properties of Molybdenum Disulfide Diffusion Coatings with Various Crystalline Structures, Soviet Materials Science, p. **44**, (July, 1975).

246. Ermakov, A.T., et al, Lubricating Properties of Diffusion Type Coatings of Molybdenum Disulphide, Russian Engineering J., **14**, 36, (1974).

247. Aparin, V.I., Dukhovskoi, E.A., Makarov, Yu. V. and Silin, A.A., The Effect of the Type of Treatment of Molybdenum Before Sulfiding on the Wear Resistance of MoS_2 Coatings, Soviet Materials Science, **15**, 413, (1979).

248. Amsallem, C., Popakul, J. and Robert, Y., Traitement de Surface par Synthèse In Situ de Bisulfure de Molybdène pour Frotter à Sec Avec Un Très Bas Coéfficient de Frottement, I. Mech. E. Conf. Publ. 18, p. 241, (1973). (1st European Tribology Congress, London).

249. Atkins, D.R., Farr, J.P.G. and Rowe, G.W., Electrochemical Deposition of Solid Lubricant Films, ASLE Trans., **14**, 248, (1971).

250. Yajima, A., Matsuzaki, R., Eguchi, M. and Saeki, Y., The Formation of Molybdenum Disulfide by the Reaction Between Molybdenum Trioxide and Sulfur Dioxide in the Presence of Carbon, Bull. Chem. Soc. Japan, **54**, 3759, (1981).

251. Mizutani, Y., Shimura, Y. and Nagasawa, Y., Self-Lubricating Property of Fe-Mo-S Alloys in Vacuum, Proc. 3rd Intl. Conf. on Solid Lubrication, Denver, Colorado, (7-10 Aug. 1984). ASLE SP-14, p. 152.

252. Ponomarev, E.A., Neumann-Spallart, M., Hodes, G. and Lévy-Clément, C., Electrochemical Deposition of MoS_2 Thin Films by Reduction of Tetrathiomolybdate, Thin Solid Films, **280**, 86, (1996).

253. Sergeeva, L.M., Marchenko, E.A., Lobova, T.A. and Blagonravov, A.A., Effectivität der Anwendung Synthetischer Festkörperschmierstoffe mit Schichtstruktur, Schmierungstechnik, **12**, 337, (1981).

254. Braithwaite, E.R. and Heber, J.,(eds.), Molybdenum: An Outline of its Chemistry and Uses, p. 34, Elsevier, Oxford, (1994).
255. Feng, I. Ming, Perilstein, W.L. and Adams, M.R., Solid Film Deposition and Non-Sacrificial Boundary Lubrication, ASLE Trans, **6**, 60, (1963).
256. Isoyama,H. and Sakurai,T., Lubricating Mechanism of Di-u-thio-dithio-bis-(diethyldithiocarbamate)-dimolybdenum During Extreme Pressure Lubrication, Tribology Intl., **7**, 151, (1974).
257. Yamamoto, Y., Gondo, S., Kamakura, T. and Konishi, M., Organoamine and Organophosphate Molybdenum Complexes as Lubricant Additives, Wear, **120**, 51, (1987).
258. Farr, J.P.G., Molybdenum Disulphide in Lubrication. A Review, Wear, **35**, 1, (1975).
259. Kasrai, M., Cutler, J.N., Gore, K., Canning, G.W., Bancroft, G.M. and Tan, K.H., The Chemistry of Anti-Wear Films generated by the Combination of ZDDP and MODTC Examined by X-Ray Absorption Spectroscopy, World Tribology Congress, London, 8-12 September, 1997. (STLE Preprint No. 97-WTC-9).
260. Nagakari, M., Yamamoto, S., Sakurai, H. and Kubo, K., Study of the Influence of Organic Molybdenum Compound on Valve Train Wear Control Performance of Gasoline Engine Oil, World Tribology Congress, London, 8-12 September, 1997.
261. Mitchell, P.C.H., Oil-Soluble Mo-S Compounds as Lubricant Additives, Wear, **100**, 281, (1984).
262. Skeldon, P., Wang, H.W. and Thompson, G.E., Formation and Characterization of Self-Lubricating MoS_2 Precursor Films on Anodized Aluminium, Wear, **206**, 187, (1997).
263. Jin, Y.S. and Zhou, C.H., The Effectiveness of High Temperature Lubrication by In Situ Formation of Graphite/MoS_2 Films, Wear, **205**, 77, (1997).
264. Forbes, E.S., Upsdel, N.T. and Battesby, J., Current Thoughts on the Mechanism of Action of Tricresyl Phosphate as a Load-Carrying Additive, Proc. I. Mech. E. Tribology Convention, Paper C. 44/72, (1972).
265. Cann, P.M., Spikes, H.A. and Cameron, A., Thick Film Formation by Zinc Dialkyldithiophosphate, ASLE Trans., **26**, 48, (1983).
266. Morecroft, D.W., Reaction of n-Octadecane and Decoic Acid with Clean Iron Surfaces, Proc. Conf. on Chemical Effects at Bearing Surfaces, Swansea, Jan. 1971, p. 225.
267. Giltrow, J.P., Factors Affecting the Endurance of Rubbed Lubricant Films of Lamellar Solids, Brit. J. Appl. Phys., **18**, 831, (1967).

268. Brewe, E.D., Scibbe, H.W. and Anderson, W.J., Friction Torque of Ball Bearings with Burnished MoS_2 Films at 10^{-10} Torr at Several Loads Using Five Retainer Materials, NASA TN-D7023, (Dec. 1970).
269. Johnson, R.L., Buckley, D.H. and Swikert, M.A., Studies of Lubricating Materials in Vacuum, Proc. USAF-SwRI Aerospace Bearing Conf., San Antonio, p. 60, (March, 1964).
270. Jantzen, E., A Dry-Lubricated System for Space-Proofed Bearings, Proc. 1st Europ. Space Trib. Symp., Frascati, Italy, (9-11 April, 1975), ESA SP-111, p. 207.
271. Manon, J.C., Effects of Solid Lubrication on the Torque of a Ball Bearing, Proc. 1st. Europ. Space Trib. Sympo., Frascati, Italy, (9-11 April, 1975), ESA SP-111, p. 215.
272. New MoS_2 Coating Process Cuts Cold Extrusion Costs at Braun Engineering, AMAX Molysulfide Newsletter, **23**, 2, (1981).
273. Spalvins, T., Deposition of MoS_2 Films by Physical Sputtering and their Lubrication Properties in Vacuum, ASLE Trans., **12**, 36, (1969).
274. Spalvins, T., Sputtering Technology in Solid Film Lubrication, Proc. 2nd Intl. Conf. on Solid Lubrication, Denver, Colorado, (15-18 Aug. 1978), ASLE SP-6, p. 109.
275. Roberts, E.W., The Tribology of Sputtered Molybdenum Disulphide Films, Proc. I. Mech. E. Intl. Conf. Tribology - Friction, Lubrication and Wear Fifty Years On, London, (1-3 July, 1987), vol. 1, p. 503, Paper No. C172/87.
276. Spalvins,T., Lubrication with Sputtered MoS_2 Films: Principles, Operation, Limitations, NASA Tech. Memo. 105292, Cleveland, Ohio, (1991).
277. Lavik, M.T. and Campbell, M.E., Evidence of Crystal Structure in Some Sputtered MoS_2 Films, ASLE Trans., **15**, 233, (1972).
278. Roberts, E.W., Towards an Optimised Sputtered MoS_2 Lubricant Film, NASA 20th Aerospace Mechanisms Symposium, NASA Conf. Publ. 2423, (7-9 May 1986).
279. Rochat, G., Lammer, J., Hintermann, H.E., Niederhäuser, P. and Maillat, M., Lubrification Solide de Mobiles Micromechaniques, Paper No. F2.6, C.I.C. Conference, Geneva, 1979.
280. Spalvins, T., Lubrication with Sputtered MoS_2 Films, ASLE Trans., **14**, 267, (1971).
281. Spalvins, T., Structure of Sputtered Molybdenum Disulfide Films at Various Substrate Temperatures, ASLE Trans., **17**, 1, (1974).

282. Aubert, A., Nabot, J.Ph., Ernoult, J. and Renaux, Ph., Preparation and Properties of MoS_2 Films Grown by d.c. Magnetron Sputtering, Surf. Coat. Technol., **41**, 127, (1990).
283. Obeng, J.A. and Schrader, G.L., Reactive Sputtering of Molybdenum Disulphide Thin Films, Surf. Coat. Technol., **68/69**, 422, (1994).
284. Lince, J.R., Hilton, M.R. and Bommannavar, A.S., Oxygen Substitution in Sputter Deposited MoS_2 Films Studied by Extended X-Ray Absorption Fine-Structure, X-Ray Photoelectron Spectroscopy and X-Ray Diffraction, Surf. Coat. Technol., **43/44**, 640, (1990).
285. Dimigen, H., Hubsch, H. and Willich, P., Lubrication Properties of r.f. Sputtered MoS_2 Layers with Variable Stoichiometry, Thin Solid Films, **64**, 221, (1979).
286. Buck, V., The Performance of Unbonded MoS_2 for Space Applications, Proc. 2nd Europ. Space Mechanisms and Tribology Symp., Meersburg, Oct. 1985.
287. Stupp, B., Performance of Conventionally Sputtered MoS_2 Versus Co-Sputtered MoS_2 and Nickel, Proc. 3rd Intl. Conf. on Solid Lubrication, Denver, Colorado, (7-10 Aug. 1984). ASLE SP-14, p. 217.
288. Roberts, E.W., Williams, B.J. and Ogilvy, J.A., The Effect of Substrate Surface Roughness on the Friction and Wear of Sputtered MoS_2 Films, J. Phys. D., Appl. Phys., **25**, A65, (1992).
289. Gardos, M.N., Quality Control of Sputtered MoS_2 Films, Lubric. Eng., **32**, 463, (1976).
290. Niederhäuser, P., Maillat, M. and Hintermann, H.E., Moisture Resistant MoS_2-Based Composite Lubricant Films, Proc. 1st Europ. Symp. on Space Mechanisms and Tribology, Neuchâtel, Switzerland, (12-14 Oct. 1983). (ESA SP-196, Dec. 1983.
291. Fleischauer, P.D. and Tolentino, L.U., Structural Studies of Sputtered MoS_2 Films by Angle-Resolved Photoelectron Spectroscopy, Proc. 3rd Intl. Conf. on Solid Lubrication, Denver, Colorado, (7-10 Aug. 1984), ASLE SP-14, p. 223.
292. Roberts, E.W., The Lubricating Properties of Magnetron Sputtered MoS_2, European Space Agency Contract Report No. ESA (ESTL) 76, (Oct. 1987).
293. Spalvins, T., Characteristic Morphological and Frictional Changes in Sputtered MoS_2 Films, Proc. 3rd Intl. Conf. on Solid Lubrication, Denver, Colorado, (7-10 Aug. 1984), ASLE SP-14, p. 201.
294. Hirvonen, J.-P., Nastasi, M. and Jervis, T.R., MoS_2 Coatings - Lubrication for the Space Age, Tribologia, **10**, 47, (1991).
295. Buck, V., Morphological Properties of Sputtered MoS_2 Films, Wear, **91**, 281, (1983).

296. Buck, V., Preparation and Properties of Different Types of Sputtered MoS_2 Films, Wear, **114**, 263, (1987).
297. Fleischauer, P.D., Hilton, M.R. and Bauer, R., Effects of Microstructure and Adhesion on Performance of Sputter-Deposited MoS_2 Solid Lubricant Coatings, in Mechanics of Coatings, (Dowson, D., Taylor, C.M. and Godet, M., eds.), Elsevier Tribology Series 17, Amsterdam, 1990.
298. Fleischauer, P.D., Effects of Crystallite Orientation on Environmental Stability and Lubrication Properties of Sputtered MoS_2 Thin Films, ASLE Trans., **27**, 82, (1983).
299. Grosseau-Poussard, J.L., Moine, P. and Brendle, M., Shear Strength Measurements of Parallel MoS_2 Thin Films, Thin Solid Films, **307**, 163, (1997).
300. Lince, J.R., Hilton, M.R. and Bommannavar, A.S., EXAFS of Sputter-Deposited MoS_2 Films, Thin Solid Films, **264**, 120, (1995).
301. Martin, J.M., Pascal, H., Donnet, C., Le Mogne, Th., Loubet, J.L. and Epicier, Th., Superlubricity of MoS_2: Crystal Orientation Mechanisms, Surf. Coat. Technol., **68/69**, 427, (1994).
302. Donnet, C., Martin, J.M., Le Mogne, Th. and Belin, M., Super-Low Friction of MoS_2 Coatings in Various Environments, Trib. Intl., **29**, 123, (1996).
303. Spalvins, T., Bearing Endurance Tests in Vacuum for Sputtered Molybdenum Disulfide Films, NASA TM S-3193, (Jan. 1975).
304. Fleischauer, P.D. and Bauer, R., The Influence of Surface Chemistry on MoS_2 Transfer Film Formation, ASLE Trans., **30**, 160, (1986).
305. Anderson, M.J. and Roberts, E.W., An Assessment of Solid Lubricant Films for Use in High Temperature Space Applications, Proc. 5th European Space Mechanisms and Tribology Symp., ESTEC, Noordwijk, Holland, (28-30 Oct. 1992). (ESA SP-334, Apr. 1993, p. 379).
306. Stupp, B.C., Synergistic Effects of Metals Co-Sputtered with MoS_2, Thin Solid Films, **84**, 257, (1981).
307. Roberts, E.W. and Price, W.B., Advances in Molybdenum Disulphide Film Technology for Space Applications, Proc. 6th European Space Mechanisms and Tribology Symp., Zurich, Switzerland, (4-6 Oct. 1995). (ESA SP-374, Aug. 1995).
308. Zabinski, J.S., Donley, M.S., Walck, S.D., Schneider, T.R. and McDevitt, N.T., The Effects of Dopants on the Chemistry and Tribology of Sputter-Deposited MoS_2 Films, Trib. Trans., **38**, 894, (1995).
309. Kaye, S.P., Kheyrandish, H., Colligon, J.S. and Roberts, E.W., Ion Beam Modification of the Physical Properties of MoS_2 Films, Thin Solid Films, **228**, 252, (1993).

310. Fox, V.C., Hampshire, J. and Teer, D.G., The Tribological Properties of MoS_2/Metal Composite Coatings Deposited by Closed Field Magnetron Sputtering, World Tribology Congress, London, 8-12 Sept., 1997.
311. Huiwen Liu, Xushou Zhang, Deyang Yu and Xiaoping Wang, The Enhancement of Wear Life and Moisture Resistance of Sputtered MoS_2 Films by Metal Ion Implantation, Wear, **173**, 145, (1994).
312. Bhattacharya, R.S., Rai, A.K., McCormick, A.W. and Erdemir, A., High Energy (MeV) Ion Beam Modifications of Sputtered MoS_2 Coatings on Ceramics, Trib. Trans., **36**, 621, (1993).
313. Donley, M.S. and Zabinski, J.S., Tribological Coatings, Chapter 18 in Pulsed Laser Deposition of Thin Films, Chrisey, D.B. and Hubler, G.K. (eds.), John Wiley & Sons, Inc., New York, 1994.
314. Prasad, S.V., Zabinski, J.S. and McDevitt, N.T., Friction Behavior of Pulsed Laser Deposited Tungsten Disulfide Films, Trib. Trans., **38**, 57, (1995).
315. Zabinski, J.S., Donley, M.S., Dyhouse, V.J. and McDevitt, N.T., Chemical and Tribological Characterization of PbO-MoS_2 Films Grown by Pulsed Laser Deposition, Thin Solid Films, **214**, 156, (1992).
316. Gresham, R.M., Solid-Film Lubricants: Unique Products for Unique Lubrication, Lubric. Eng., **24**, 143, (1968).
317. Godfrey, D. and Bisson, E.E., Bonding of Molybdenum Disulfide to Various Materials to Form a Solid Lubricating Film; 1 - The Bonding Mechansim, NACA TN 2628, (Feb. 1952).
318. Benzing, R.J., McConnell, B.D. and Clow, W.L., A Methyl Phenyl Polysiloxane Bonded Lubricating Film, ASLE Intl. Conf. on Solid Lubricants, Denver, Colorado, (24-27 Aug. 1971), ASLE SP-3.
319. Hopkins, V. and Campbell, M., A Composite-Like Solid Film Lubricant, ASLE Preprint No. 76-AM-SC-2, (1976).
320. Hopkins, V., Gaddis, D.H. and Hubbell, R., MLF-5 An Inorganic Solid Lubricant Film, USAF - SwRI Aerospace Bearing Conference, San Antonio, Texas, (25-27 March, 1964).
321. Gaddis, D.H. and Hopkins, V., US Patent No. 3,361,666, Inorganic Solid Film Lubricants, (2 Jan. 1968).
322. Hartley, D. and Wainwright, P., Development of a Dry Film Tool Lubricant, Proc. 2nd ASLE Intl. Conf. on Solid Lubrication, Denver, Colorado, (15-18 Aug. 1978), ASLE SP-6, p. 281.
323. McConnell, B.D., Survey of Solid Film Lubricants, ASD Tech Report 61-22, (Sept. 1961).

324. Clow, W.L., Relation Between Inorganic Binder Wettability and Solid-Film Lubricant Wear-Life, ASML-TR-68-313, (Nov. 1968).
325. Lavik, M.T., High Temperature Dry Film Lubricant Research and Development, Proc. Air Force-Navy-Industry Lubricants Conf. Dayton, Ohio, (17-19 Feb. 1959). WADC Tech Report 59-244, p. 130, (Oct. 1959).
326. Clauss, F.J., Lubricants and Self-Lubricating Materials for Spacecraft Mechanisms, Lockheed Report LMSD-894812, (18 Apr. 1961).
327. Fitzsimmons, V.G. and Zisman, W.A., Dry-Film Lubricants from Molybdenum Disulfide Bonded with Microfibrous Boehmite, ASLE Trans., **12**, 117, (1969).
328. McConnell, B.D., Wieser, L.E. and Mecklenburg, K.R., Wear-Life Improvement of a Solid-Film Lubricant, Lubric. Eng., **24**, 81, (1968).
329. Olson, K.M. and Sliney, H.E., Additions to Fused-Fluoride Lubricant Coatings for Reduction of Low-Temperature Friction, NASA TN D-3793, (Jan. 1967).
330. Hopkins, V., Lavik, M.T., Hubbell, R.D. and Clow, W.L., Improved High-Temperature Solid Film Lubricants, AFML-TR-67-223, (July, 1967).
331. Vest, C.E. and Bazzarre, D.F., Co-Deposited Nickel-Molybdenum Disulphide, Metal Finishing, p. 52, (Nov. 1967).
332. Stupp, B.C. and Wright, J.W., Investigation of Binders for Solid Lubricants at Elevated Temperatures, Lubric. Eng., **19**, 463, (1963).
333. Stupp, B.C., Molybdenum Disulfide and Related Solid Lubricants, Lubric. Eng., **14**, 159, (1958).
334. Crump, R.E., Factors Influencing Wear and Friction of Solid Film Lubricants, Product Eng., **28**, p. C24, (Oct. 1957).
335. Gardos, M., The Synergistic Effects of Graphite on the Friction and Wear of MoS_2 Films in Air, Trib. Trans., **31**, 214, (1987).
336. Centers, P.W., discussion on 314. Trib. Trans., **31**, 224, (1987).
337. Gabel, M.K., Corrosion and Solid Film Lubricants, Lubric. Eng., **29**, 442, (1973).
338. DiSapio, A., Bonded Coatings Lubricate Metal Parts, Product Engineering, p. 48, (Sept. 1960).
339. Magie, P.M., Moly Bonded Films, Electromechanical Design, p. 50, (Apr. 1960).
340. Kawamura, M., Yoshida, K. and Aaki, L., Effects of Metal Surface Pretreatment on the Performance of Bonded Solid Film Lubricants, Proc., Interfinish 80, 10th World Congress on Metal Finishing, Japan, (Oct. 1980), p. 313.
341. Campbell, M.E., Loser, J.B. and Sneegas, E., Solid Lubricants, NASA SP-5059, (May, 1966).

342. DiSapio, A., Thickness and Coverage of Bonded Solid Film Lubricants: Measurement and Effect, Lubric. Eng., **34**, 94, (1978).
343. Peterson, M.B. and Finkin, E.F., Application of New and Improved Solid Lubricant Materials and Processes to Naval Aircraft, MTI Report No. 71 TR 48, (14 July, 1971).
344. Hopkins, V., Hubbell, R. and Kremith, R., Plasma Spraying, a New Method of Applying Solid Film Lubricants, Lubric. Eng., **24**, 72, (1968).
345. Jones, J.R. and Hoover, G.W., Abrasiveness of MoS_2 in Bonded Solid Lubricants, ASLE Trans., **14**, 55, (1971).
346. Gamulya, G.D., Lebedeva, I.L., Vvedensky, Yu. V. and Yukhno, T.P., Secondary Structure Formation and Wear Mechanisms for Solid Lubricant Coatings Under Friction in Vacuum, Wear, **171**, 143, (1994).
347. Hopkins, V. and Gaddis, D., Friction of Solid Film Lubricants Being Developed for Use in Space Environments, Lubric. Eng., **21**, 52, (1965).
348. Tsuya, Y. and Kitamura, M., Effect of Sliding Conditions on the Wear Rate of Organic and Inorganic Bonded Solid Lubricant Films, Proc.2nd. ASLE Intl. Conf. on Solid Lubrication, Denver, Colorado (15-17 Aug.1978) ASLE SP-6 p.85.
349. Haley, F.A. and McDaniel, R.H., Effects of Reactor Radiation on Ceramic-Bonded Solid-Film Lubricants, General Dynamics/Fort Worth Report MR-N-302 (HARF-63-6T), (Aug. 1963).
350. Taguchi, Y., Hirayama, S. and Hamano, M., Effects of Irradiation by Gamma Ray on the MoS_2 Dry Film Bonded with Phenolic Resin and Sodium Silicate, Lubric. Eng., **31**, 297, (1975).
351. Bertrand, R.G. and Vukasovich, M.S., Cleaning of MoS_2 Soils from Steel Surfaces, ASLE Preprint No. 76-AM-2A-2.
352. Davison, R.M. and Gilbert, T.I., Evaluation of MoS_2 Additions to a Bonded Solid Lubricant for Severe Ironing Applications, Lubric. Eng., **32**, 131, (1976).
353. Morgan, V.T., Porous Metal Bearings, Chapter 24 of "The Principles of Lubrication, A. Cameron (ed.), Longmans, London (1966).
354. Samuelson, F., Lubrication of Journal Bearings with Water-Base Lubricant, Proc. General Discussion on Lubrication & Lubricants, p. 269, I.Mech.E., London, 1937.
355. Rowell, H.W., The Lubrication of Laminated Synthetic Resinoid Bearing Materials, Proc. General Discussion on Lubrication & Lubricants, p. 266, I.Mech.E., London, 1937.
356. Bowden, F.P. and Shooter, K.V., Frictional Behaviour of Plastics Impregnated with Molybdenum Disulphide, Res. Appl. in Ind., **3**, 384, (1950).

357. Bowden, F.P., Frictional Behaviour of Porous Metals Containing Molybdenum Disulphide, Res. Appl. in Ind., **3**, 383, (1950).
358. Johnson, R.L., Swikert, M.A. and Bisson, E.E., Friction and Wear of Hot-Pressed Bearing Materials Containing Molybdenum Disulfide, NACA Tech. Note 2027, (Feb. 1950).
359. Pinchbeck, P.H., A Review of Plastic Bearings, Wear, **5**, 85, (1962).
360. Pratt, G.C., Plastic-Based Bearings, Chapter 8, Lubrication and Lubricants, E.R. Braithwaite (ed.), Elsevier, (1967).
361. Lancaster, J.K., Polymer-Based Bearing Materials: The Role of Fillers and Fibre Reinforcement, Tribology, **5**, 249, (1972).
362. Powell, P.C., The Selection and Use of Thermoplastics, Design Council Engineering Design Guide 19, Oxford University Press, (1977).
363. Chen, J.H. and Gage, P.E., Tribological Characteristics of MoS_2 and PTFE Filled Polymeric Bearings, Proc. 3rd Intl. Conf. on Solid Lubrication, Denver, Colorado, (7-10 Aug. 1984), ASLE SP-14, p. 320.
364. Modern Plastics Encyclopaedia, McGraw-Hill Inc., New York, (1976).
365. Design and Material Selection for Dry Rubbing Bearings, ESDU Data Item No. 87007, Engineering Sciences Data Unit, London, 1987.
366. Performance Guide Self-Lubricating Bearings, National Centre of Tribology, Risley, UK (1976).
367. Lancaster, J.K., Accelerated Wear Testing of PTFE Composite Bearing Materials, RAE Tech Report No. 78046, (1978).
368. Anderson, J.C., Wear of Commercially Available Plastic Materials, Tribology Intl., **15**, 255, (1982).
369. Fabre, G., Les Plastiques Comme Materiaux de Frottement, Ind. Plast. Mod., **11**, 40, (1959).
370. Sung N-H. and Suh, N.R., Effect of Fibre Orientation on Friction and Wear of Fibre-Reinforced Polymeric Composites, Wear, **53**, 129, (1979).
371. Dry Rubbing Bearings - a Guide to Design and Materials Selection, ESDU Data Item No. 68018, Engineering Sciences Data Unit, London, (1968).
372. Griffin, G.J.L., Orientation Effects in Filled Plastics Bearing Materials, ASLE Trans., **15**, 171, (1972).
374. Mitchell, D.C. and Pratt, G., Friction, Wear and Physical Properties of Some Filled PTFE Bearing Materials, Proc. Conf. on Lubrication and Wear, I.Mech.E., London, 1957.
375. Spengler, G., Hellwig, W. and Wunsch, F., Untersuchungen über das Reibungsverhalten von Festschmierstoffen in der Atmosphäre und in Hochvakuum, Deutsche Luft-und Raumfahrt, Forschungsberichte No 68-18,(Mar.1968).

376. Barry, R.C., Reinforced TFE and Moly Sulfide, Product Engineering, p. 76, (Mar. 1961).
377. Young, W.C., Clauss, F.J. and Drake, S.P., Lubrication of Ball Bearings for Space Applications, ASLE Trans., **6**, 178, (1963).
378. Scibbe, H.W., Brewe, D.E. and Coe, H.H., Lubrication and Wear of Ball Bearings in Cryogenic Hydrogen, NASA TM-X-61165, (1968).
379. Smith, G.R. and Vest, C.E., Lubrication of a Spacecraft Mechanism Using the Transfer Film Technique, Lubric. Eng., **27**, 422, (1971).
380. Mitchell, P.D., MoS_2 Enhances Wear Resistance of Nylon, Eng. Materials and Design, **4**, 502, (1961).
381. Steinbuch, R.T., Nylon 6 as a Bearing Material, Wear, **5**, 458, (1962).
382. Bowen, P.H., Friction and Wear Characteristics of Dry Lubricants, Machine Design, **35**, 195, (1963).
383. Mitchell, D.C. and Fulford, B.B., Wear of Selected Molybdenum Disulphide Lubricated Solids and Surface Films, Proc. Conf. on Lubrication and Wear, I.Mech.E., London, 1957, p. 376.
384. Nylon Bearings Roll Up Steel Mill Gains, Iron Age, (17 May, 1973).
385. Risdon, T.J. and Loban, R.D., Effect of Molybdenum Disulfide on the Wear Rates of Polymer Composites, Proc. 2nd ASLE Intl. Conf. on Solid Lubrication, Denver, Colorado, (15-18 Aug. 1978), ASLE SP-6, p. 230.
386. Risdon, T.J. and Loban, R.D., Adding Molysulfide to Polyamide-Imide Lowers Wear Significantly, AMAX Molysulfide Newsletter, **20**, (Oct. 1978).
387. Betts, S., Dackow, J. and Murray, S.F., Performance Characteristics of Self-Lubricating Polyimide Composites, Proc. 2nd ASLE Intl. Conf. on Solid Lubrication, Denver, Colorado, (15-18 Aug. 1978), ASLE SP-6, p. 195.
388. Sliney, H.E. and Jacobson, T.P., Graphite-Fiber-Reinforced Polyimide Liners of Various Compositions in Plain Spherical Bearings, Proc. 2nd ASLE Intl. Conf. on Solid Lubrication, Denver, Colorado, (15-18 Aug. 1978) ASLE SP-6, p. 258.
389. Vinogradova, O.V., Bazhenova, V.B., Krasnov, A.P., Gribova, I.A., Gulyaeva, T.A. and Nedel'kin, V.I., Tribochemical Processes in Polyphenylenesulphide Filled with Molybdenum Disulfide and Graphite, J. Friction and Wear, **17**, 110, (1996).
390. Lavik, M.T. and Hopkins, V., Development of Self-Lubricating Composites Utilising Carbonised Phenolic Matrix, AFML-TR-75-175-PT. (1975).
391. Serezevskii, V.N., Increased Wear and Scuffing Resistance of Slideways with the Use of Solid Lubricants, Russ. Eng. J., **58**, 16, (1978).

392. Schieffer, H.M. and Kubczak, G.V., Molysulfide Plays Key Role as Friction Modifier, AMAX Molysulfide Newsletter, **22**, (June, (1980).
393. Brendle, M.C.C., Frictional Properties of Solid Lubricants Modified By Polymer Grafting, Wear, **43**, 127, (1977).
394. Tremain, R.F.M., Overcoming Some Causes of Short Lives of Rings Fitted to Oil-Free Compressors, Proc. 2nd ASLE Intl. Conf. on Solid Lubrication, Denver, Colorado, (15-18 Aug. 1978), ASLE SP-6, p. 249.
395. Bessière, P. and Martre, M., L'Utilisation du Bisulfure de Molybdène dans la Metallurgie des Poudres, La Technique Moderne, **53**, 371, (1961).
396. Campbell, M.E. and Van Wyk, J.W., Development of Solid Lubricant Compacts for Use in Ball Bearing Separators, Lubric. Eng. **25**, 31, (1969).
397. Hubbell, R.D., McConnell, B.D. and Van Wyk, J.W., Development of Solid Lubricant Compacts for Use in Ball Bearing Separators, Lubric. Eng. **25**, 31, (1969).
398. Jones, J.R. and Gardos, M.N., Friction and Wear Characteristics of Lubricative Composites in Air and in Vacuum, Lubric. Eng., **27**, 47, (1971).
399. Lipp, L.C., Van Wyk, J.W. and Williams, F.J., Development of Solid Lubricant Compact Bearings for the Supersonic Transport, Lubric. Eng., **29**, 108, (1973).
400. Mecklenburg, K.R., Wear Rate Relationship for Three Lubricant Compact Materials, USAF Report No. AFML-TR-71-123, (July, 1971).
401. Mecklenburg, K.R., Establishing a Wear Rate Relationship for a Lubricant Compact Material, Lubric. Eng. **28**, 37, (1972).
402. Mecklenburg, K.R. and Benzing, R.J., Laboratory Experience with Long-Term Bearing Lubrication, ASLE Trans., **15**, 306, (1972).
403. Martin, P. and Murphy, G.P., High Speed Oscillating Tests of Lubricating Composites, Lubric. Eng., **29**, 484, (1973).
404. Suzuki, M., Moriyama, M., Mishimura, M. and Hasegawa, M., Friction and Wear of Self-Lubricating Composites at Temperatures to 450° in Vacuum, Wear, **162-164**, 471, (1993).
405. Tsuya, Y., Kazumori, V. and Kitamura, M., Optimum Concentration of Solid Lubricant in Compact, Lubric. Eng., **31**, 402, (1976).
406. Wu, Y., Wang, F., Cheng, Y. and Chen, N., A Study of the Optimization Mechanism of Solid Lubricant Concentration in Ni/MoS_2 Self-Lubricating Composite, Wear, **205**, 64, (1997).
407. Mizutani, Y., Imada, Y and Nakajima, K., Formation of Molybdenum Sulfide in Fe-Mo-S Alloys and its Effect on Frictional Property, Proc. JSLE-ASLE Intl. Lub. Conf., Tokyo, (9-11 June, 1975), p. 161.

408. Al'tman, V.A., Valakina, V.M., Gluskin, Ya.A., Memelov, V.L. and Moskvitchev, A.N., Effect of Alloying on Physicochemical Interaction of Components in Ni-MoS_2 System, Poroshkovaya Metallurgiya, No. 8, 59, (1978).
409. Koval'chenko, M.S. and Yulyugin, V.K., Densification Kinetics of Porous Metal Based on Molybdenum Disulfide in Hot Pressing, Soviet Powder Metallurgy and Metal Ceramics, **19**, 324, (1980).
410. Paderno, V.N., Yulyugin, V.K., Tkachenko, Yu. G., Martysenko, A.N. and Smirnov, V.P., Structure of Rubbing Surfaces of a Mo-MoS_2 Composite, Soviet Powder Metallurgy and Metal Ceramics, **19**, 501, (1980).
411. Tkachenko, Yu. G., Koval'chenko, M.S., Yulyugin, V.K., Bovkin, G.A. and Yurchenko, D.Z., Lubricating Action of MoS_2 on Rubbing Sintered Materials, Soviet Powder Metallurgy and Metal Ceramics, **17**, 552, (1978).
412. Ho, T.L., Role of Solid Lubrication in High-Energy Braking, I.Mech.E. Conf. on Oil-Immersed Brakes and Clutches, Paper No. C34/77, I.Mech.E., London, 1977.
413. Bhushan, B. and Gupta, B.K., Handbook of Tribology: Materials, Coatings and Surface Treatments, McGraw-Hill Inc., New York, 1991.
414. Paxton, R.R., Carbon, Graphite and Metal-Bonded Molybdenum Disulphide Solid Lubricant Bearings, Tribology Intl., **15**, 285, (1982).
415. Peterson, M.B., Advancements in Tribological Materials 1937-1987-2002, Proc. I.Mech.E. Intl. Conf. Tribology - Friction, Lubrication and Wear Fifty Years On, London, (1-3 July, 1987).
416. Sliney, H.E., A New Chromium Carbide-Based Tribological Coating for Use to 900°C with Particular Reference to the Stirling Engine, J. Vac. Sci. Technol. A., **4**, 2629, (1986).
417. DellaCorte, C. and Sliney, H.E., Tribological Properties of PM212: A High-Temperature Self-Lubricating, Powder Metallurgy Composite, NASA TM-102355, (1990).
418. Jones, J.R., Friction and Wear of Potentiometer Contacts in Vacuum, Wear, **11**, 355, (1968).
419. Gangopadhyay, A., Jahanmir, S. and Hegemann, B.E., Reduction in Friction Coefficient in Sliding Ceramic Surfaces by In-Situ Formation of Solid Lubricant Coatings, Paper III(i) in Mechanics of Coatings, Dowson, D. et al (eds.), Elsevier Tribology Series 17, Amsterdam, (1990).
420. Devine, M.J., Lamson, E.R. and Bowen, J.H., Antifriction Bearing Design Considerations for Solid Lubrication, ASME Preprint 63-MD-43, (1963).

421. Devine, M.J., Lamson, E.R. and Bowen, J.H., The Lubrication of Ball Bearings with Solid Films, ASME Preprint 61-LUBS-11, (1961).
422. Vest, C.E., Evaluation of Several Additional Dry Lubricants for Spacecraft Applications, Lubric. Eng., **30**, 246, (1974).
423. Devine, M.J. and Kroll, A.E., Aromatic Polyimide Compositions for Solid Lubrication, Lubric. Eng., **20**, 225, (1964).
424. Bowen, P.H., Boes, D.J., Grossett, K.W. and Bober, E.S., Lubricant Study in Ultrahigh Vacuum and in Various Gas Environments, NASA CR-65374, (1966).
425. Boes, D.J., Cunningham, J.S. and Chasman, M.R., The Solid Lubrication of Ball Bearings under High Speed - High Load Conditions from -225 to +1000F, Lubric. Eng., **27**, 150, (1971).
426. Ogawa, T., Konishi, K., Aihara, S. and Sawamoto, T., Development of Ball Bearings with Solid Film for High-Vacuum, High-Temperature, High-Speed Application, Lubric. Eng. **49**, 291, (1993).
427. Savage, R.H., Graphite Lubrication, J. Appl. Phys., **19**, 1, (1948).
428. Atkins, B.R. and Griffiths, D.P., Electrical Sliding Contacts and their Behaviour at High Altitudes, Proc. Conf. on Lubrication and Wear, I.Mech.E., London, 1957, p. 371.
429. Elsey, H.M., Treatment of High Altitude Brushes by Application of Metallic Halides, AIEE Trans., **64**, 577, (1945).
430. Johnson, J.L. and McKinney, J.L., Electrical-Power Brushes for Dry Inert-Gas Atmospheres, IEEE Trans. Part 5, Hybrids and Packaging, **7**, **62**, (1971).
431. Nelson, R.E., Improved Carbon Brushes for Aircraft Electrical Machines, WADC TR-57-170, (Jan. 1960).
432. O'Donnell, R.J., Harris, C.L. and Warwick, M.G., The Lubrication of Bearings and Slip Rings in Vacuum: Final Report, ESRO Contractor Report CR-91, (Sept. 1972).
433. Devine, M.J., Rolling Element Slip Rings for Vacuum Application, NASA TN D-2261, (Apr. 1964).
434. Vest, C.E. and Ward, B.W., Evaluation of Space Lubricants under Oscillatory and Slow Speed Rotary Motion, Lubric. Eng. **24**, 163, (1968).
435. Clauss, F.J., and Kingery, M.K., Sliding Electrical Contact Materials for Use in Ultrahigh Vacuum, Proc. 6th Struct. and Mat. Conf. AIAA, (Apr 1967), p.172.
436. Moberly, L.E. and Johnson, J.L., Electrical Sliding Contacts for Application in Space Environments, IEEE Trans. Supplement on Aerospace, p. 252, (1965).
437. Hawker Siddeley Dynamics Limited, Space Division, Final Report on De-Spin Mechanism, ESRO Contractor Report CR-30.

438. Lewis, E., Cole, S.R., Glosbrenner, E.W. and Vest, C.E., Friction, Wear and Noise of Slip-Ring and Brush Contacts for Synchronous Satellite Use, Proc. 1st. Europ. Space Tribology Symp., Frascati, Italy, (9-11 Apr. 1975), ESA SP-111, p. 367.
439. Maillat, M., Aubert, F., Hintermann, F.E. and Rocchi, R., Essais de Frottement de Contacts Electriques Glissants dans l'Ultravide, Laboratoire Suisse de Recherches Horlogères, Neuchâtel, Rapport No.980-1.1 ESRO(ESTL) 14 (20 Oct 1974).
440. Tanaka, A. and Kimura, Y., Development of Solid Lubricant Compacts for Use as Motor Brushes on a Scientific Satellite, Junkatsu, **23**, 231, (1978).
441. Tamaka, A. and Kimura, Y., Effects of Electric Current on Wear of a Brush Material Ag-Ni-MoS_2 Compact, JJSLE, (Oct. 1979).
442. Anderson, J.C. and Roberts, E.W., Satellite Slip Ring Performance - a Summary of ESTL Work 1974-81, ESA (ESTL) 53, (Aug. 1982).
443. Christy, R.L., Motor Brush Wear Test Results for Air and Vacuum Operation, Proc. 3rd. Intl. Conf. on Solid Lubrication, Denver, Colorado, (7-10 Aug. 1984), ASLE SP-14, P. 132.
444. Di Sapio, A. and Moloney, J., The Lubricating Performance of an 'in situ' Process MoS_2 Film in Air and in Liquids, ASLE Trans., **11**, 56, (1968).
445. Hopkins, V. and Campbell, M.E., Important Considerations in the Use of Solid Film Lubricants, Lubric. Eng., **27**, 396, (1971).
446. Novielli, V., A Solid Film Lubricant Applications Guide for the F-18 Finish Specification, Report NADC-80028-60, (21 May, 1980). (AD-A087955).
447. Black, A.L., Dunster, R.W. and Sanders, J.V., Molybdenum Disulphide Deposits - Their Formation and Characteristics on Steel Balls, Wear, **9**, 462, (1966).
448. Sastry, V.R.K., Phatak, S.O. and Sethuramiah, A., Experimental Study of the Influence of Liquid Lubricants on Burnished MoS_2 Films, Wear, **86**, 213, (1983).
449. Gänsheimer, J., Influence of Certain Vapors and Liquids on the Frictional Properties of Molybdenum Disulfide, ASLE Trans., **10**, 390, (1967).
450. Gänsheimer, J. and Holinski, R., Molybdenum Disulfide - Solid Lubricant and Chemical Additive, 1st. Europ. Trib. Congress, London, (25-27 Sept. 1973), I.Mech.E. Conf. Publication 18, p. 229, (1973).
451. Rolek, R.J., Cusano, C. and Sliney, H.E., The Influence of Temperature on the Lubricating Effectiveness of MoS_2 Dispersed in Mineral Oils, ASLE Trans., **28**, 493, (1985).

452. Crudu, I., Ibanescu, I., Neculaiasa, V., Babalau, D. and Rotaru, A., Influence of Domestic Molybdenum Disulphide Added to Lubricating Oils on the Coefficient of Friction, Constructia de Masini, **22**, 263, (1970).

453. Rylander, H.G., A Theory of Liquid-Solid Hydrodynamic Film Lubrication, ASLE Trans., **9**, 264, (1966).

454. Aoki, H. and Nakahara, T., Friction Characteristics of Fluid Film Bearings with MoS_2 Containing Lubricant, Proc. JSLE Intl. Trib. Conf., Tokyo, (8-10 July, 1985), p. 1127.

455. Herzig, H.L., Effect of Solid Additives on Fluid Lubricants, Lubric. Eng., **15**, 152, (1959).

456. Tsuya, Y., Saito, K. and Suzuki, H., Effect of Viscosity of Oil Containing Molybdenum Disulfide on its Load-Carrying Capacity, Lubric. Eng., **31**, 619, (1975).

457. Groszek, A.J., Preferential Adsorption of Compounds with Long Methylene Chains on Cast Iron, Graphite, Boron Nitride and Molybdenum Disulfide, ASLE Trans., **9**, 67, (1996).

458. Popinceanu, N.G., Gafitanu, M.D., Cretu, S.S., Diaconescu, E.N. and Hostuic, L.T., Rolling Bearing Fatigue Life and EHL Theory, Wear, **45**, 17, (1977).

459. Scott, D. and Blackwell, J., The Assessment of Recently Developed Lubricants for Rolling Elements, Lubric. Eng., **29**, 99, (1973).

460. Kuhnell, B.T. and Stecki, J.S., Effects of MoS_2 on the Life of Heavily Loaded Rolling Element Bearings, Trib. Trans., **31**, 324, (1988).

461. Ishibashi, A., Hoyashita, S. and Nakajima, A., Effects of MoS_2 Particles in Oil on the Pitting of Rolling/Sliding Contact Surfaces Under Nearly Full EHL Conditions, Proc. 5th Leeds-Lyon Symp. on Tribology, I.Mech.E., London, (1979), p. 282.

462. Soda, N. and Yamashita, M., Effect of Solid Lubricant in Oils on Fatigue under Rolling Contact, JJSLE, (Feb. 1976).

463. STLE Life Factors for Rolling Bearings, E.V. Zaretsky (ed.), STLE, Park Ridge, Illinois, (1992).

464. Hisakado, T., Tsukizoe, T. and Yoshikawa, H., Lubrication Mechanism of Solid Lubricants in Oils, J. Lubric. Tech., **105**, 245, (1983).

465. Cusano, C. and Sliney, H., Dynamics of Solid Dispersions in Oil During the Lubrication of Point Contacts. Part I - Graphite, ASLE Trans., **25**, 183, (1982).

466. Gänsheimer, J. and Holinski, R., MoS_2 in Oils and Greases under Boundary Conditions, ASME Trans., Ser. F., J. Lubric. Tech., **95**, 242, (1973).

467. Holinski, R., Lubrication Mechanism of Solid Lubricants in Oils, ASLE Trans., **18**, 263, (1975).

468. Risdon, T.J. and Gresty, D.A., An Historical Review of Reductions in Fuel Consumption of United States and European Engines with MoS_2, SAE Preprint No. 750674, (1975).
469. Müller, K. and Bartz, W.J., Motorsauberkeitserhöhung durch MoS_2 bzw. Graphit, Mineraloeltechnik, **24**, 1, (1979). (In German).
470. McDonnell, T.F. and Tempe, S.A., The Effects of Engine Oil Additives on Vehicle Fuel Economy, Emissions, Emission Control Components and Engine Wear, SAE Progress in Technology Series, No. 27, p. 245, (1982).
471. Shadow, G., Franke, W.-D. and Fehr, H., Nachweis der Wirkung Eines MoS_2-Konzentrates in Schmierölen von Verbrennungskraftmaschinen, Schmiertechnik u. Tribologie, **20**, 106, (1973).
472. Holinski, R. and Gänsheimer, J., Wirkungsweise von Festschmierstoffen in Olen und Fetten, VDI-Z, **114**, 131, (1972).
473. Rosenberg, R.C. and Campbell, W.E., The Effect of Mechanically Dispersed Solid Powders on Wear Prevention by White Oil at High Loads and Low Speeds, Lubric. Eng., **24**, 92, (1968).
474. Thorp, J.M., The Mechanism of Lubrication by Molybdenum Disulphide Dispersed in Oil and the Effect of a Zinc Dialkyldithiophosphate Additive, Wear, **23**, 63, (1973).
475. Bartz, W.J., Influence of Extreme Pressure and Detergent-Dispersant Additives on the Lubricating Effectiveness of Molybdenum Disulfide, Lubric. Eng. **33**, 139, (1977).
476. Bartz, W.J., Interrelations Between Molybdenum Disulphide and Oil Soluble Additives, NLGI Spokesman, **53**, 408, (1989).
477. Bartz, W.J. and Oppelt, J., Lubricating Effectiveness of Oil-Soluble Additives and Molybdenum Disulfide Dispersed in Mineral OIl, ASLE Preprint No. 79-LC-1C-3.
478. Crump, W.J.J., Some Experiments to Investigate the Effect of Molybdenum Disulphide on the Behaviour of Motor Vehicle Engines Particularly under Emergency Conditions, Proc. Konferenz über Tribologie vom Aspekt der Fahrzeuge, Budapest, (No. 1976).
479. Holinski, R., Dynamics of Boundary Layers, Lubric. Eng., **36**, 530, (1980).
480. Gresty, D.A., Kunz, E.J. and Risdon, T.J., The Effect of MoS_2 Based Lubricants on Automotive Gear Efficiency and Operating Temperatures, SAE Passenger Car Meeting, Detroit, (Sept. 1977). SAE Preprint No. 770834.
481. Anon, MoS_2 Additive Reduces Starting Current by 50%, AMAX Molysulfide Newsletter, **20**, (Oct. 1978).

482. Smith, R.K. and Marshek, K.M., Lubricants Containing Dispersed Molybdenum Disulfide Improve Worm Gear Efficiency, NLGI Spokesman, **48**, 47, (1984).
483. Anon, Molysulfide Increases Press Output by 50 per cent, AMAX Molysulfide Newsletter, **17**, (May, 1974).
484. Bagchi, H. and Basu, S.K., Selection of Cutting Fluid - Graphite or Molybdenum Base?, Ind. Lubric. & Tribol., **26**, 181, 1974.
485. Shekter, Yu. N., Machines and Tooling, **36**, 15, (1973).
486. Bannerjee, J.K., Some Effects of Molybdenum Disulphide (MoS_2) - Based Cutting Fluids on the Workpiece Surface Finish in Grinding and Conventional Machining Processes, Proc. Conf. Wear of Materials 1981, New York, (1981), p. 496.
487. Vukasovich, M.S. and Kelly, W.D., The Effect of MoS_2 on the Performance of Some Model Greases, Lubric. Eng., **28**, 125, (1972).
488. Barry, H.F. and Binkelman, J.P., Evaluation of Molybdenum Disulfide in Lubricating Greases, NLGI Spokesman, **30**, 45, (1966).
489. Risdon, T.J. and Sargent, D.J., Comparison of Commercially Available Greases With and Without Molybdenum Disulfide, Part I - Bench Scale Performance Tests, NLGI Spokesman, **33**, 82, (1969).
490. Risdon, T.J. and Binkelman, J.P., Oxidation Stability and Antifriction Bearing Performance of Lubricants Containing Molybdenum Disulfide, NLGI Spokesman, **32**, 115, (1968).
491. Barry, H.F., Risdon, T.J. and Sargent, D.J., Comparison of Commercially Available US Greases With and Without Molybdenum Disulfide, Lubric. Symp., Paisley Coll. of Technology, Paisley, Scotland, (23-24 Oct. 1969).
492. Risdon, T.J. and Sargent, D.J., Comparison of Commercially Available Greases With and Without Molybdenum Disulfide; Part II - Oxidation Stability and Ball Joint Tests, NLGI Spokesman, **34**, 358, (1971).
493. Risdon, T.J., Effect of MoS_2 Concentration on the Performance of Greases in the GMR Ball Joint Tester, NLGI Spokesman, **38**, 166, (1974).
494. Risdon, T.J., Some Energy Implications for the Use of MoS_2 in Greases, NLGI Spokesman, **41**, 150, (1977).
495. Risdon, T.J. and Gresty, D.A., The Effect of MoS_2 on Wear with Greases Containing an Abrasive Contaminant, NLGI Spokesman, **42**, 182, (1978).
496. Risdon, T.J., Evaluation of MoS_2 in Newer Grease Thickener Systems, NLGI Spokesman, **50**, 211, (1986).
497. Antony, J.P., Mittal, B.D., Naithani, K.P., Misra, A.K. and Bhatnagar, A.K., Antiwear/Extreme Pressure Performance of Graphite and Molybdenum Disulfide Combinations in Lubricating Greases, Wear, **174**, 33, (1994).

498. Kitchen, G.H., Lubrication of Small Motor Bearings for Unattended Service in Automotive Equipment, Lubric. Eng., **20**, 311, (1964).
499. Caterpillar Specifies Molybdenum Disulfide Greases, AMAX Molysulfide Newsletter, **17**, (1974).
500. Moly Greases Extend Bearing Life in Off-Highway Equipment, AMAX Molysulfide Newsletter, **25**, (No. 1), p. 1, (1983).
501. Jost, H.P. and Hicks, L.A., Molybdenum Disulphide in the Iron and Steel Industry, Iron and Steel, p. 3, (Jan. 1961).
502. Forsythe, R.E., Molybdenum Disulfide in the Steel Industry, Climax Molybdenum Company.
503. Fowle, T.I., Private communication.
504. Schneider, R., Anwendungstechnische Richtlinien zur MoS_2-Adhäsivpaste, Schmierungstechnik, **9**, 218, (1978).
505. Gänsheimer, J. and Wessely, J., Lubrication of Threads, Wear, **65**, 201, (1980).
506. Newnham, J., Hassell, B. and Behan, P., Evaluation of Titanium Bolts Coated with Ivadize Aluminium and Kalgard 2240c in Torque Tension Tests, ESA Contractor Report CR (P) 1304, (Nov. 1979).
507. Birger, I.A. and Iasilevitch, G.B., Threaded Connections, Mashinostroeniye, Moscow, (1973).
508. UK Civil Aviation Authority, Molybdenum Disulphide Lubricants - Effect on Turbine Engines, Airworthiness Notice No.12, Appendix No.38, (16 March, 1988).
509. Kawamura, M., Yoshida, K. and Aoki, I., Running-In Effect of Bonded Solid Film Lubricants on Conventional Oil Lubrication, Proc. 2nd ASLE Intl. Conf. on Solid Lubrication, Denver, Colorado, (15-18 Aug. 1978), ASLE SP-6, p. 101.
510. Vetter, H., Festschmierstoffe und ihre Anwendung in der Umformtechnik, Mineralöltechnik, **3**, 1, (1980).
511. Yunusov, M.G., Khromov, D.I., Karmadonov, A.F., Ryabchikova, O.A. and Shchurovskii, S.B., Improvement of the Life of Steel Cables Made from Wire Produced with a Lubricant Containing Molybdenum Disulphide, Stal, **7**, 636, (1975). (In Russian).
512. Bhattacharyya, D., Coad, J.N.O. and Dixon, M.J., Effects of MoS_2 Concentration on Friction, Wear, **86**, 57, (1983).
513. Nandi, U.K., Effects of Solid Lubricants in the Forward Extrusion Process, M. Tech. Thesis, University of Burdwan, (1977-78) (Reported in Ref.470).
514. Jamison, W.E. and Cosgrove, S.L., Friction Characteristics of Transition-Metal Disulfides and Diselenides, ASLE Trans., **14**, 62, (1971).

515. Jamison, W.E., Intercalated Dichalcogenide Solid Lubricants, Proc.3rd, Intl. Conf. on Solid Lubrication, Denver, Colorado, (7-10 Aug.1984) ASLE SP-14, p.73.
516. Mantell, C.L., Industrial Carbon, (2nd edn.) Van Nostrand, New York, (1946).
517. Modern Aspects of Graphite Technology, L.C.F. Blackman (ed.), Academic Press, New York, (1970).
518. Stock, A.J., Graphite, Molybdenum Disulfide and PTFE - a Comparison, Lubric. Eng., **19**, 333, (1963).
519. Bragg, W.L., Introduction to Crystal Analysis, Bell & Son, London, (1928).
520. Savage, R.H., Graphite Lubrication, J. Appl. Phys., **19**, 1, (1948).
521. Brennan, R.D., The Interlayer Binding in Graphite, J.Chem.Phys., **20**,40, (1952).
522. Bryant, P.J., Gutshall, P.L. and Taylor, L.H., A Study of Mechanisms of Graphite Friction and Wear, Wear, **7**, 118, (1964).
523. Peterson, M.B. and Johnson, R.L., Friction Studies of Graphite and Mixtures of Graphite with Several Metallic Oxides and Salts at Temperatures to 1,000°F, NACA Tech. Note 3657, (1956).
524. Kuriakose, A.K. and Margrave, J.L., Kinetics of the Reactions of Elemental Fluorine. IV. Fluorination of Graphite, J. Phys. Chem., **69**, 2772, (1965).
525. Fusaro, R.L. and Sliney, H.E., Graphite Fluoride $(CF_x)_n$ - a New Solid Lubricant, ASLE Trans., **13**, 56, (1970).
526. Martin, C., Sailleau, J. and Roussel, M., The Ultra-High Vacuum Behaviour of Graphite Fluoride Filled Self-Lubricating Materials, Wear, **34**, 215, (1975). (In French).
527. Fusaro, R.L., A Comparison of the Lubricating Mechanisms of Graphite Fluoride and Molybdenum Disulfide Films, Proc. 2nd Intl. Conf. on Solid Lubrication, (15 -18 Aug. 1978), ASLE SP-6, p. 59.
528. Atkinson, I.B. and Waghorne, R.M., Tribo-Chemistry of Graphite Fluoride Studied Using X-Ray Photoelectron Spectroscopy, Wear, **37**, 123, (1976).
529. Gisser, H., Petronio, M. and Shapiro, A., Graphite Fluoride as a Solid Lubricant, Proc. ASLE Intl. Conf. on Solid Lubricants, Denver, Colorado, (24-27 Aug. 1971), ASLE SP-3, p. 217.
530. Tsuya, Y., Uemura, H., Okamoto, Y. and Kurosaki, S., Co-Deposited Composite Metal-Graphite Fluoride Platings, ASLE Trans., **17**, 229, (1974).
531. Gänsheimer, J., Untersuchungen über die Schmierwirksamkeit verschiedener Sulfide and Selenide, Schmiertechnik, **12**, 278, (1965).
532. Flom, D.G., Haltner, A.J. and Gaulin, C.A., Friction and Cleavage of Lamellar Solids in Ultrahigh Vacuum, ASLE Preprint No. 64 LC-18.

533. Sliney, H.E., Solid Lubricant Materials for High Temperatures - a Review, Tribology Intl., **15**, 303, (1982).
534 Salomon, G., Begelinger, A., Van Bloois, F.I. and De Gee, A.W.J., Characterization and Tribological Properties of MoS_2 Powders and of Related Chalcogenides, ASLE Trans., **13**, 134, (1970).
535. Johnson, R.L. and Ludwig, L.P., Lubrication, Friction and Wear in Aircraft, NASA TM X-67872.
536. Radcliffe, S.J. and Parry, A.A., The Dispersion of Life of Bonded MoS_2 Solid Lubricant Coatings, Wear, **56**, 203, (1979).
537. Campbell, M.E., Solid Lubrication Technology: a Review, Mechanical Engineering, **90**, 28, 1968.
538. Spalvins, T., Coatings for Wear and Lubrication, Thin Solid Films, **53**, 285, (1978).
539. Bergmann, E., Malat, G., Müller, C. and Simon-Vermot, A., Friction Properties of Sputtered Dichalcogenide Layers, Tribology Intl., **14**, 329, (1981).
540. Kirner, K., Plasmagesprizte Feststoffschmiermittelschichten für öl-und fett-freie Lager, Bundesministerium für Forschung und Technologie, Forschungsbericht W76-16, (Nov. 1975).
541. Kalikhman, V.L., Golubnichaya, A.A., Gladchenko, E.P., Prokudina, V.K. and Shepinova, L.P., Crystalline Structure and Electrical Resistance of MoS_2-NbS, Alloys Produced by Self-Propagating High-Temperature Synthesis, Poroshkovaya Metallurgia, **10**, 57, (1982). (In Russian).
542. Marchenko, E.A. and Sergeeva, L.M., Certain Ways to Improve the Properties of Solid Lubricants Based on Refractory-Metal Dichalcogenides, Treniye i Iznos, **3**, 661, (1982). (In Russian).
543. Davis, B.T. and Presland, A.E. B., Tribolytic Deposition of Tungsten Disulphide, Nature Physical Science, **230**, 119, (1971).
544. Boes, D.J. and Bowen, P.H., Friction-Wear Characteristics of Self-Lubricating Composites Developed for Vacuum Service, ASLE Trans., **6**, 192, (1963).
545. Rempe, W.H., Research and Development of Materials for Use as Lubricants in a Liquid Hydrogen Environment, ASLE Trans., **9**, 213, (1966).
546. Boes, D.J., Unique Solid Lubricating Materials for High Temperature Air Applications, ASLE Trans., **10**, 19, (1967).
547. Kiparisov, S.S., Shvetsova, G.A., Lobova, T.A., Sergeeva, L.M., Pimenova, A.Z. and Volodina, G.A., Structure and Properties of a Self-Lubricating Material Based on Tungsten Diselenide, Soviet Powder Metallurgy and Metal Ceramics, **17**, 399, (1978).

548. Dayton, R.D. and Sheets, M.A., Evaluation of Grooved Solid Lubricated Bearings, AFAPL-TR-75-76, (Jul. 1975). AD-A023-473.
549. McConnell, B.D. and Mecklenburg, K.R., Solid Lubricant Compacts - an Approach to Long Term Lubrication Performance in Space, Lubric. Eng., **33**, 544, (1977).
550. Kay, E., The Corrosion of Steel in Contact with Molybdenum Disulphide, RAE Technical Report No. 65219, (Oct. 1965).
551. DeLaat, F.G.A., Testing Dry Film Lubricants for Space Use, Research/Development, p. 52, (Aug. 1968).
552. Gabel, M.K. and Peterson, M.B., A Study of Parameters Which Affect Corrosion Between Solid Film Lubricants and Aircraft Alloys, ASLE Preprint No. 76-AM-6C-2.
553. Heinicke, G., Tribochemistry, Akademie-Verlag, Berlin (1984).
554. Black, A.L., Properties of Bonded MoS_2 Coatings with Special Reference to Corrosion Resistance with Particular Reference to Fasteners, Seventeenth Annual Conference, Australasian Corrosion Association, Corrosion and the Mining Industry, Newcastle, Australia, (13-18 Nov. 1977).
555. Zhai, G.J., Liu, J-J., Zhu, B-L, Zhang, X-S. and Yang, S-R., The Role of Cerium in the Resistance of an MoS_2-Containing Composite Brush Plating Layer to Humid Atmosphere, Trib. Trans., **39**, 715, (1996).
556. Waterhouse, R.B., Fretting Corrosion, Pergamon Press, London, 1972.
557. Godfrey, D. and Bisson, E.E., Effectiveness of Molybdenum Disulfide as a Fretting-Corrosion Inhibitor, NASA TN-2180, (Sept. 1950).
558. Weismantel, E.E., Friction and Fretting with Solid Film Lubricants, Lubric. Eng., **11**, 97, (1955).
559. Spalvins, T., Energetics in Vacuum Deposition Methods for Depositing Solid Film Lubricants, Lubric. Eng., **25**, 436, (1969).
560. Campbell, M., Solid Lubricants - Where They Stand Today, Chemical Eng., **80**, 56, (1st October,1973).
561. Powers, T.E., Molybdenum Disulphide in Nylon for Wear Resistance, Modern Plastics, (June. 1970).

~~~~~~~~~~~~~~~~
~~~~~~~~~~~~~~~~

SUBJECT INDEX

~~~~~~~~~~~~~~~~~~~~~
~~~~~~~~~~~~~~~~~~~~~

www.ingramcontent.com/pod-product-compliance
Ingram Content Group UK Ltd.
Pitfield, Milton Keynes, MK11 3LW, UK
UKHW020231290926
15470UKWH00002B/472